Scheduling and Operation of Virtual Power Plants

Scheduling and Operation of Virtual Power Plants

Technical Challenges and Electricity Markets

Edited by

Ali Zangeneh
Shahid Rajaee Teacher Training University
Tehran, Iran

Moein Moeini-Aghtaie
Sharif University of Technology
Tehran, Iran

Elsevier
Radarweg 29, PO Box 211, 1000 AE Amsterdam, Netherlands
The Boulevard, Langford Lane, Kidlington, Oxford OX5 1GB, United Kingdom
50 Hampshire Street, 5th Floor, Cambridge, MA 02139, United States

Copyright © 2022 Elsevier Inc. All rights reserved.

MATLAB® is a trademark of The MathWorks, Inc. and is used with permission.
The MathWorks does not warrant the accuracy of the text or exercises in this book.
This book's use or discussion of MATLAB® software or related products does not constitute endorsement or sponsorship by The MathWorks of a particular pedagogical approach or particular use of the MATLAB® software.

No part of this publication may be reproduced or transmitted in any form or by any means, electronic or mechanical, including photocopying, recording, or any information storage and retrieval system, without permission in writing from the publisher. Details on how to seek permission, further information about the Publisher's permissions policies and our arrangements with organizations such as the Copyright Clearance Center and the Copyright Licensing Agency, can be found at our website: www.elsevier.com/permissions.

This book and the individual contributions contained in it are protected under copyright by the Publisher (other than as may be noted herein).

Notices

Knowledge and best practice in this field are constantly changing. As new research and experience broaden our understanding, changes in research methods, professional practices, or medical treatment may become necessary.

Practitioners and researchers must always rely on their own experience and knowledge in evaluating and using any information, methods, compounds, or experiments described herein. In using such information or methods they should be mindful of their own safety and the safety of others, including parties for whom they have a professional responsibility.

To the fullest extent of the law, neither the Publisher nor the authors, contributors, or editors, assume any liability for any injury and/or damage to persons or property as a matter of products liability, negligence or otherwise, or from any use or operation of any methods, products, instructions, or ideas contained in the material herein.

Library of Congress Cataloging-in-Publication Data
A catalog record for this book is available from the Library of Congress

British Library Cataloguing-in-Publication Data
A catalogue record for this book is available from the British Library

ISBN: 978-0-323-85267-8

For information on all Elsevier publications
visit our website at https://www.elsevier.com/books-and-journals

Publisher: Joseph P. Hayton
Acquisitions Editor: Graham Nisbet
Editorial Project Manager: Aleksandra Packowska
Production Project Manager: Debasish Ghosh
Designer: Victoria Pearson

Typeset by VTeX

*Dedicated to our families
and all our teachers who educate and support us*

Contents

List of contributors		xiii
Preface		xvii
Abbreviations and acronyms		xxi

1 Introduction and history of virtual power plants with experimental examples — 1
Altaf Q.H. Badar, Piyush Patil, and M.J. Sanjari
- 1.1 Introduction — 1
- 1.2 Distributed generation — 2
- 1.3 Virtual power plant (VPP) — 5
- 1.4 Research on VPP — 21
- 1.5 Summary — 21
- References — 22

2 VPP and hierarchical control methods — 27
Jinho Kim
- 2.1 Introduction — 27
- 2.2 Hierarchical control methods — 27
- 2.3 Conclusion — 33
- References — 34

3 Bidding strategy in the electricity market — 37
Morteza Shafiekhani, Ali Zangeneh, and Farshad Khavari
- 3.1 Introduction — 37
- 3.2 Virtual power plant — 37
- 3.3 Optimal bidding of VPP in electricity market — 39
- 3.4 Hypotheses and problem objectives — 42
- 3.5 Case study — 49
- 3.6 Conclusion — 53
- Appendix 3.A — 53
- Nomenclature — 54
- References — 55

4 Optimization model of a VPP to provide energy and reserve — 59
Niloofar Pourghaderi, Mahmud Fotuhi-Firuzabad, Moein Moeini-Aghtaie, and Milad Kabirifar
- 4.1 Introduction — 59

	4.2 Optimization model of VPP to provide energy	60
	4.3 Optimization model of VPP to provide reserve	76
	4.4 Examples for VPP optimization model providing energy and reserve	84
	4.5 Conclusion	104
	Nomenclature	104
	References	107
5	**Provision of ancillary services in the electricity markets**	**111**
	Mehrdad Setayesh Nazar and Kiumars Rahmani	
	5.1 Introduction	111
	5.2 Problem modelling and formulation	113
	5.3 Solution algorithm	116
	5.4 Numerical results	116
	5.5 Conclusions	127
	Nomenclature	128
	References	129
6	**Frequency control and regulating reserves by VPPs**	**131**
	Taulant Kërçi, Weilin Zhong, Ali Moghassemi, Federico Milano, and Panayiotis Moutis	
	6.1 Introduction	131
	6.2 Taxonomy	138
	6.3 Examples	145
	6.4 Conclusions	157
	References	158
7	**VPP's participation in demand response aggregation market**	**163**
	Ali Shayegan-Rad and Ali Zangeneh	
	7.1 Introduction	163
	7.2 Single-level model of DSO without DR programs	164
	7.3 Single-level scheduling model of DSO with DR programs	165
	7.4 Bi-level scheduling model between DSO and VPP-DRA	167
	7.5 Numerical studies and discussions	170
	7.6 Conclusion	176
	Nomenclature	176
	References	177
8	**VPP's participation in demand response exchange market**	**179**
	Ali Shayegan-Rad and Ali Zangeneh	
	8.1 Introduction	179
	8.2 VPP scheduling framework	180
	8.3 VPP scheduling model	180
	8.4 Uncertainties arising from VPP scheduling	183
	8.5 Numerical studies and discussions	184

	8.6 Conclusion	190
	Nomenclature	190
	References	191
9	**Uncertainty modeling of renewable energy sources**	193
	Davood Fateh, Mojtaba Eldoromi, and Ali Akbar Moti Birjandi	
	9.1 Introduction	193
	9.2 Modeling of RESs	195
	9.3 Modeling of VPP	196
	9.4 Classification and description of uncertainties in VPP	196
	9.5 Optimization approaches of VPP with uncertainties	198
	9.6 Problem formulation	200
	9.7 Tools used to solve optimization problems of VPP with uncertainties	202
	9.8 Case study	203
	9.9 Conclusion	206
	References	206
10	**Frameworks of considering RESs and loads uncertainties in VPP decision-making**	209
	Zeal Shah, Ali Moghassemi, and Panayiotis Moutis	
	10.1 Introduction	209
	10.2 Proposals for handling uncertainty within a VPP	214
	10.3 Taxonomy	218
	10.4 Conclusions and path forward	222
	References	223
11	**Risk-averse scheduling of virtual power plants considering electric vehicles and demand response**	227
	Omid Sadeghian, Amin Mohammadpour Shotorbani, and Behnam Mohammadi-Ivatloo	
	Nomenclature	227
	11.1 Introduction	229
	11.2 Problem formulation	232
	11.3 Case study	239
	11.4 Conclusions	250
	References	254
12	**Optimal operation strategy of virtual power plant considering EVs and ESSs**	257
	Milad Kabirifar, Niloofar Pourghaderi, and Moein Moeini-Aghtaie	
	12.1 Introduction	257
	12.2 Modeling of EVs	259
	12.3 Modeling of ESSs	267
	12.4 VPP operation strategy modeling in the presence of EVs and ESSs	274

	12.5 Examples for VFP optimal operation strategy considering EVs and ESSs	278
	12.6 Conclusion	290
	Nomenclature	291
	References	293
13	**EVs vehicle-to-grid implementation through virtual power plants**	**299**
	Moein Aldin Parazdeh, Navid Zare Kashani, Davood Fateh, Mojtaba Eldoromi, and Ali Akbar Moti Birjandi	
	Nomenclature	299
	13.1 Introduction	300
	13.2 Vehicle-to-grid (V2G)	301
	13.3 Bidirectional converters for V2G systems	302
	13.4 Bidirectional AC–DC converter (BADC)	302
	13.5 Bidirectional DC–DC converter (BDC)	303
	13.6 Modeling the problem	308
	13.7 Case study	318
	13.8 Conclusion	321
	References	322
14	**Short- and long-term forecasting**	**325**
	Hüseyin Akçay and Tansu Filik	
	14.1 Introduction	325
	14.2 Wind speed forecasting from long-term observations	326
	14.3 Conclusion	338
	References	339
15	**Forecasting of energy demand in virtual power plants**	**343**
	Farshad Khavari, Jamal Esmaily, and Morteza Shafiekhani	
	Introduction	343
	Load behavior	344
	Different weather parameters	347
	Different methods for clustering analysis	348
	Different methods for STLF	350
	Fitness criteria	351
	Case study	353
	Conclusion	356
	References	357
16	**Emission impacts on virtual power plant scheduling programs**	**359**
	Nazgol Khodadadi, Amin Mansour-Saatloo, Mohammad Amin Mirzaei, Behnam Mohammadi-Ivatloo, Kazem Zare, and Mousa Marzband	
	Nomenclature	359
	16.1 Introduction	360

	16.2 Problem formulation	**364**
	16.3 Simulation and numerical results	**367**
	16.4 Conclusions	**373**
	References	**374**
17	**Multi-objective scheduling of a virtual power plant considering emissions**	**377**
	Morteza Shafiekhani and Ali Hashemizadeh	
	17.1 Introduction	**377**
	17.2 Problem formulation	**379**
	17.3 Case studies	**385**
	17.4 Conclusion	**394**
	Nomenclature	**395**
	References	**396**

Author index   **399**
Subject index   **415**

List of contributors

Hüseyin Akçay
Department of Electrical and Electronics Engineering, Eskişehir Technical University, Eskisehir, Turkey

Altaf Q.H. Badar
Electrical Engineering Department, National Institute of Technology Warangal, Warangal, India

Mojtaba Eldoromi
Electrical Engineering Department, Shahid Rajaee Teacher Training University, Tehran, Iran

Jamal Esmaily
Electrical Engineering Department, Shahid Rajaee Teacher Training University, Tehran, Iran

Davood Fateh
Electrical Engineering Department, Shahid Rajaee Teacher Training University, Tehran, Iran

Tansu Filik
Department of Electrical and Electronics Engineering, Eskişehir Technical University, Eskisehir, Turkey

Mahmud Fotuhi-Firuzabad
Faculty of Electrical Engineering Department, Sharif University of Technology, Tehran, Iran

Ali Hashemizadeh
School of Management and Economics, Beijing Institute of Technology, Beijing, China

Milad Kabirifar
Sharif University of Technology, Tehran, Iran

Navid Zare Kashani
Electrical Engineering Department, Shahid Rajaee Teacher Training University, Tehran, Iran

Taulant Kërçi
School of Electrical & Electronic Engineering, University College Dublin, Dublin, Ireland

Farshad Khavari
Electrical Engineering Department, Shahid Rajaee Teacher Training University, Tehran, Iran

Nazgol Khodadadi
Faculty of Electrical and Computer Engineering, University of Tabriz, Tabriz, Iran

Jinho Kim
Electrical & Computer Engineering Department, Auburn University, Auburn, AL, United States

Amin Mansour-Saatloo
Faculty of Electrical and Computer Engineering, University of Tabriz, Tabriz, Iran

Mousa Marzband
Department of Mathematics, Physics and Electrical Engineering, Northumbria University, Newcastle, England, United Kingdom

Federico Milano
School of Electrical & Electronic Engineering, University College Dublin, Dublin, Ireland

Mohammad Amin Mirzaei
Faculty of Electrical and Computer Engineering, University of Tabriz, Tabriz, Iran

Moein Moeini-Aghtaie
Faculty of Energy Engineering Department, Sharif University of Technology, Tehran, Iran

Ali Moghassemi
Department of Electrical Engineering, South Tehran Branch, Islamic Azad University, Tehran, Iran

List of contributors

Behnam Mohammadi-Ivatloo
Faculty of Electrical and Computer Engineering, University of Tabriz, Tabriz, Iran
Department of Electrical and Electronics Engineering, Mugla Sitki Kocman University, Mugla, Turkey

Amin Mohammadpour Shotorbani
Faculty of Electrical and Computer Engineering, University of Tabriz, Tabriz, Iran

Ali Akbar Moti Birjandi
Electrical Engineering Department, Shahid Rajaee Teacher Training University, Tehran, Iran

Panayiotis Moutis
Wilton E. Scott Institute for Energy Innovation, Carnegie Mellon University, Pittsburgh, PA, United States

Moein Aldin Parazdeh
Electrical Engineering Department, Shahid Rajaee Teacher Training University, Tehran, Iran

Piyush Patil
Electrical Engineering Department, Yeshwantrao Chavan College of Engineering, Nagpur, India

Niloofar Pourghaderi
Sharif University of Technology, Tehran, Iran

Kiumars Rahmani
Faculty of Electrical Engineering, Shahid Beheshti University, Tehran, Iran

Omid Sadeghian
Faculty of Electrical and Computer Engineering, University of Tabriz, Tabriz, Iran

M.J. Sanjari
School of Engineering and Built Environment, Griffith University, Gold Coast, QLD, Australia

Mehrdad Setayesh Nazar
Faculty of Electrical Engineering, Shahid Beheshti University, Tehran, Iran

Morteza Shafiekhani
Department of Electrical Engineering, Faculty of Engineering, Pardis Branch, Islamic Azad University, Pardis, Tehran, Iran

Zeal Shah
Electrical & Computer Engineering Department, University of Massachusetts Amherst, Amherst, MA, United States

Ali Shayegan-Rad
MAPNA Electric and Control, Engineering & Manufacturing Co. (MECO), MAPNA Group, Karaj, Iran

Ali Zangeneh
Electrical Engineering Department, Shahid Rajaee Teacher Training University, Tehran, Iran

Kazem Zare
Faculty of Electrical and Computer Engineering, University of Tabriz, Tabriz, Iran

Weilin Zhong
School of Electrical & Electronic Engineering, University College Dublin, Dublin, Ireland

Preface

Researches on virtual power plants (VPPs) have extensively increased since the last decade. The increasing penetration of distributed energy resources (DERs) is a global trend in power systems leading to some changes and new challenges in the power system infrastructure, operation, and control. The common characteristics of DERs are generally private ownership, unscheduled generation, as well as small and distributed capacity, that result in their uncoordinated operation in power systems. The uncoordinated operation of DERs may either cause some problems or at least reduce the generation benefits of DERs. Besides, they do not participate in the electricity markets due to their small and nondispachable power generation. Thus, a new entity called virtual power plant (VPP) has emerged to collect the dispersed generation of small resources and coordinate their operation. It acts in the same way as the traditional power plants. VPPs will provide not only a coordinated energy management for generation and storage resources, but also a demand side management for consumers. They will be a new concept of power grids in the near future. Although there is a growing literature about various aspects of VPPs, there is also a great need for a comprehensive explanation that spans different subjects of the VPPs.

This book tries to cover the wide area of VPPs, ranging from required infrastructures and planning to operation and control. The extensive amount of knowledge and experience presented in this book can be a useful source for engineers, researchers, managers, and policy makers to learn about the different aspects of virtual power plants, including modeling and analysis. The book generally explains the fundamentals and contemporary subjects in operation, scheduling, and electricity markets issues of virtual power plants. The editors designated the topics of the book and invited some expert contributors to write the chapters.

In the following chapters, readers will be introduced to different issues regarding virtual power plants.

Chapter 1 introduces the concept of VPPs, their role, duties, authorities, and regulations in power systems.

Chapter 2 reviews the concept and principle of hierarchical control methods for a VPP consisting of heterogeneous distributed energy resources. Hierarchical control methods enable the VPP to utilize the various capabilities of the DERs by sharing data between higher- and lower-level controls, thereby increasing the flexibility of the VPP in providing electricity services in a power system. A conceptual hierarchical control structure, functions, and applications of the network assets within a VPP, along with an example, are presented in this chapter.

Chapter 3 is about the optimal bidding strategy of a VPP in the electricity market. The problem of the optimal bidding strategy can be single- or multi-level. This chapter is based on a bi-level model in which the upper level represents VPP profit maximization while the lower level deals with the day-ahead market clearing problem.

Chapter 4 addresses an optimization model of VPP to provide energy and reserve. The VPP can schedule the DERs to provide energy and different types of reserve services in order to trade in wholesale electricity markets or provide network services. In this regard, diverse DERs' energy and reserve models are mentioned and VPP optimization models to aggregate DERs in order to provide energy and reserve and make profit in wholesale markets are regarded. Furthermore, two examples for the VPP optimization problem are provided to show how the model can be utilized by a VPP.

Chapter 5 presents a two-stage optimization process for optimal day-ahead and real-time scheduling of a VPP that participates in the energy and ancillary service markets. The first-stage problem optimizes the bidding strategy of a VPP that comprises active energy, reactive energy, and reserve markets' bidding strategies, and a risk-averse formulation is presented. The second-stage problem optimally schedules the VPP facilities in the active and reactive energy markets for a real-time horizon.

Chapter 6 discusses frequency control, one of the most typical and important system services that can be provided by VPPs. As VPPs comprise multiple different resources, which are dispersed over potentially vast areas, procuring regulating reserves and realizing frequency control is a challenging task. This chapter defines frequency control as a service offered by VPPs, and illustrates the ways this service may be planned and realized.

Chapter 7 illustrates the ability of VPPs to aggregate demand response programs as an effective electricity source along with their other energy sources. Since scheduling demand response (DR) provided by individual customers will not be an easy task, individual responsive customers are planned by a subentity in VPP, called VPP demand response aggregator (VPP-DRA). This chapter develops two different programming approaches to model DR applications in the VPP, namely single-level scheduling approach (SLSA) and bi-level scheduling approach (BLSA).

Chapter 8 develops an economic model of a VPP to take part in a demand response exchange market. The VPP persuades its customers to allocate part of their demand for participating in the energy and RR demand response (DR) programs. To this end, it signs an incentive contract with responsive loads based on price–quantity curves. Four different scenarios are applied to evaluate the capability and effectiveness of the presented programming model.

Chapter 9 investigates the state-of-the-art approaches, techniques, and challenges within the uncertainty modeling of renewable energy resources (RESs) in a practical power grid. Probabilistic methods also emerge as a serious research direction for these studies. Accurate and effective modeling of RES uncertainty involves holistic improvement.

Chapter 10 examines how and to what extent uncertainty affects VPPs and how they can handle it. In this chapter, the sources of uncertainty that are more substantial to the operation of a VPP are presented, and it is discussed that how researchers have addressed, quantified, and controlled it, and what the path forward is.

Chapter 11 deals with a scheduling framework for VPPs by considering the related operational and security constraints. Furthermore, the uncertainty in the electricity market is taken into account and the resulted risk due to the uncertainty is controlled by the conditional value-at-risk (CVaR) measure. CVaR is used to improve the cost of the worst-case scenarios and solve the problem for different risk levels.

Chapter 12 presents an optimal operation strategy for a VPP by considering electric vehicles (EVs) and other energy storage systems (ESSs). However, the aggregation and coordination of EVs (especially EVs with vehicle-to-grid (V2G) capability) offer new opportunities to grid operators by providing the possibility of charging/discharging cycle control of their batteries. The optimal coordination and scheduling of these technologies along with other DERs can accommodate the uncertain and intermittent renewable energy sources (RESs) in the network, as well as improve service reliability and power quality.

Chapter 13 studies the vehicle-to-grid (V2G) concept, a promising technology that allows fixed or parked voltage batteries to act as distributed sources, to store or release energy at appropriate times. This bidirectional power exchange is achieved using bidirectional electronic power converters that connect the network to the EV battery. In this chapter, an energy management model for VPPs is developed, and the cost and diffusion effects of VPP formation and EV penetration are analyzed. Different types of AC–DC and DC–DC bidirectional converter topologies that facilitate active V2G currents are reviewed and compared. In addition, this chapter discusses the different classes of reported charger/discharge systems for V2G applications.

Chapter 14 presents a practical study on the wind speed measurements collected from a meteorological station in the Marmara region of Turkey. Experimental results will show that the trimming of diurnal, weekly, monthly, and annual trend components in the data significantly enhances estimation accuracy. This scheme builds on data detrending, covariance factorization via a recent subspace method, and one-step-ahead and/or multi-step-ahead Kalman filter predictors.

Chapter 15 investigates an hourly prediction of the load for a time ranging from one hour to several days. In this chapter, at first, historical weather and load data have been classified, which consists of modeling the data for each class by intervention analysis based on statistical methods and knowledge about electrical demand curves, then a proper tool for load forecasting has been chosen. In addition, a real case study based on a modern method has been simulated and discussed.

Chapter 16 explains that emission limitations play a significant role in the flexibility of VPPs to participate in the energy markets. To this end, in this chapter, an emission-constrained optimal self-scheduling model is proposed for the participation of a CHP-based VPP in the electricity wholesale market.

Chapter 17 points out that VPPs reduce the need to build very large units by collecting the capacity of smaller units and therefore play an important role in reducing emissions from large units. Paying attention to the issue of emission in a virtual power plant can affect its profit. This chapter deals with the economic scheduling of the virtual power plant by considering emission in the form of a multiobjective problem with two different methods and tries to select the best strategy for the VPP power plant, so that it leads to the highest profit and the lowest emission.

Abbreviations and acronyms

ADMM Alternating direction method of multipliers
AGC Automatic generation control
AND Active distribution network
ANN Artificial neural network
ARIMA Autoregressive integrated moving-average
ARMA Autoregressive moving-average
BADC Bidirectional AC–DC converter
BC Bilateral contract
BDC Bidirectional DC–DC converter
BESS Battery energy storage system
BLSA Bi-level scheduling approach
BM Balanced market
BPNN Backpropagation neural network
CCFC Capability-coordinated frequency control
CHP Combined heat and power
CLC Cluster-level control
CPP Conventional power plant
CVPP Commercial VPP
DAM Day-ahead market
DG Distributed generation
DER Distributed energy resource
DLC Device-level control
DMS Distribution management system
DoD Depth of discharge
DR Demand response
DSO Distribution system operator
ELM Extreme learning machine
EMM Energy management model
EMS Energy management system
ESS Energy storage system
EV Electric vehicle
EVPL Electric vehicle parking lot
FCCP Fuzzy chance constraint programming
FOR Feasible operation region
FRR Frequency restoration reserve
GA Genetic algorithm
GAMS General algebraic modeling system
GenCo Generation companies
GHG Greenhouse gas
HCS Hierarchical control strategy
HVAC Heating, ventilation and air conditioning
IA Interval analysis
IBR Inverter-based resource
ICT Information and communication technology
IGTD Information gap decision theory

ISO Independent system operator
KKT Karush–Kuhn–Tucker
LC Load curtailment
LMP Locational marginal price
LPF Low-pass filter
LR Load recovery
LS Load shifting
LTLF Long-term load forecasting
MAE Mean absolute error
MAPE Mean absolute percentage error
MATLAB Matrix laboratory
MCS Monte Carlo simulation
MILP Mixed-integer linear programming
MG Microgrid
MHP Micro hydro power
MPEC Mathematical problem with equilibrium constraints
MPPT Maximum power point tracking
MTLF Mid-term load forecasting
PBUC Probabilistic price-based unit commitment
PCC Point of common coupling
PDF Probability density function
PEM Point estimation method
PFC Primary frequency control
PLC Plant-level control
PSO Particle swarm optimization
PV Photovoltaic
PWM Pulse width modulation
RBFNN Radial basis function neural network
RES Renewable energy sources
RHA Rolling horizon approach
RMSE Root mean square error
RoCoF Rate of change of frequency
RR Regulation reserve
SDN Smart distribution network
SHP Small hydro plant
SFC Secondary frequency control
SLSA Single-level scheduling approach
SOC State of charge
STLF Short-term load forecasting
SVM Support vector machine
THD Total harmonic distortion
TVPP Technical VPP
TSO Transmission system operator
V2G Vehicle-to-grid
V2H Vehicle-to-home
VPP Virtual power plant
VSM Virtual synchronous machine
WT Wind turbine

Introduction and history of virtual power plants with experimental examples

Altaf Q.H. Badar[a], Piyush Patil[b], and M.J. Sanjari[c]
[a]Electrical Engineering Department, National Institute of Technology Warangal, Warangal, India, [b]Electrical Engineering Department, Yeshwantrao Chavan College of Engineering, Nagpur, India, [c]School of Engineering and Built Environment, Griffith University, Gold Coast, QLD, Australia

1.1 Introduction

The depletion of conventional energy sources and increase in energy demand has caused renewable energy sources (RES) and their applications to gain more and more attention from researchers and industries around the globe. The various factors for the surge in RES are:

- High resources' potential,
- Eco-friendliness,
- Popular application,
- Government policies for using RES, which are an important part of energy strategies throughout the world.

The various types of RES are solar, wind, fuel cell, tidal, biomass, etc. The rise in the use of renewable energy gave birth to a new technology, which has been termed as distributed generation (DG). The technologies used in DG and RES have merits like reliability, economy, and flexibility. DG technologies are being developed at a fast pace with advanced technology entering the market with immediate implementation. As with every technology, DG has its own advantages and limitations. The wider use of RES as DG has given birth to new challenges for the network operators. However, challenges can also be seen as opportunities for reengineering the electrical power system and integration of new components in it.

One of the basic disadvantages of RES/DG generating units is their inability to operate in unison, which brings down the collective importance of these units. Thus, a virtual utility termed as virtual power plant (VPP) is introduced to realize the aggregated benefits of DG and other entities within the grid.

The VPPs can be cloud-based and in such cases they may be not physically present. They aggregate distributed power plants in their capacities, controllable loads, and storage units to operate as a single entity [1] as shown in Fig. 1.1. Various distributed energy resources (DER) are included for enhancing power generation and energy trading. A VPP manages the energy production and consumption of its aggregated

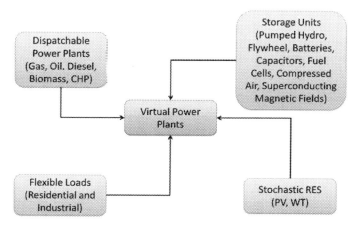

Figure 1.1 Aggregated units in a VPP.

constituents to optimize their performance. The generating units can be conventional power plants (CPP) or RES. Fig. 1.1 gives in detail the different units that can be included in a VPP.

A VPP creates a system that integrates various power source types to improve the reliability of the system. The sources covered under VPP can have micro-CHP, small wind turbine (WT), solar PV, small hydro plant (SHP), run of the river, biomass plant, diesel generator, or battery energy storage system (BESS) [1]. These plants together form a cluster, which may be dispatchable or non-dispatchable DG, but are controlled through a central controller. A VPP even has controllable or flexible loads under its control.

VPPs have the capacity to supply power in a short time and can be used especially during peak power consumption. VPP brings in better flexibility and efficiency into the operation and deliverability of the power system. This helps to handle the fluctuations in the system in a better way. The control of a VPP can sometimes be complicated, and optimality of operation, scheduling, or energy trading may not be achieved. Communication plays a very important role in maintaining the cluster of entities under a VPP. The future is promising for VPP, as it will bring in multiple features into the current power system.

1.2 Distributed generation

Distributed generation (DG) refers to the production of electricity near the consumption place or near the energy source. Another way to imagine DG is the spreading of generating units throughout the grid. DG allows smaller energy producers to become a part of the grid. RES plays an important part in the emergence of DG technologies and may be obtained from resources like wind, solar, tidal, waves, biomass, geothermal, etc. [2].

The regularly used RES/DG technologies and their ratings are listed below in Table 1.1, and Table 1.2 lists the differences between DG and a traditional system.

Table 1.1 Types of DG technology and their ranges [3].

Technology	Ratings
WT	200 W – 3 MW
PV	20 W – 100 kW
SHP	1 MW – 100 MW
Micro Hydro Power (MHP)	25 kW – 1 MW
Tidal Energy	100 kW – 5 MW
Geothermal Energy	5 MW – 100 MW
BESS	0.5 MW – 5 MW
Biomass	100 kW – 20 MW

Table 1.2 DG vs traditional system.

DG	Traditional system
DG can be installed at the load/resource end	Plants are installed near the source
Primary energy source may vary depending on availability	Primary energy source remains almost constant as the raw material supply remains constant
Generated power is relatively small	Generated power is relatively high/bulk
Power flow due to introduction of DG becomes bidirectional, as generating sources are spread out	Power flow is unidirectional

The major advantages of using DGs are:
- Lesser T&D losses,
- Power quality is improved,
- Higher reliability of the grid,
- Better voltage profile,
- Greenhouse gas (GHG) emissions are less.

In recent years, with the enhancements in research and wide applications of DG, along with the introduction of competitive markets, it is anticipated that DG technology shall capture a major share of the traditional generation system. For example, by 2020, the EU countries aim to capture approximately 20% of power production through RES [4]. In the future, more policies shall be introduced by governments in favor of DG technologies combined with their flexibility, which will attract more and more interest towards the installation of DG. The increase in the number of users will lead to major problems such as governing the power flow and stability of the power system. Hence an adaptive technology has to be implemented, so as to integrate the penetration of DGs in power systems [5].

The role of DG in the aggregation of VPP is immense. A VPP has different components like controllable loads, generation, etc. The generation requirement of a VPP can be easily satisfied through the inclusion of DG. The DG with its inherent characteristics and nearness to the load center is highly preferred to be a part of VPP, especially to enhance its bidding capacity. The VPP can also participate in ancillary markets and demand response programs through the active participation from DGs.

1.2.1 Impact of DG on power system

Some of the major disruptions that shall be introduced by DGs will be in the following fields [2]:

- Overcurrent protection,
- Islanding [6],
- Voltage dips,
- Stability,
- Optimal placement [7],
- Line loss reduction.

1.2.2 Impediment with DG technology

It has been a trend in the power system to move towards clean and decentralized energy transformation [8,9]. DGs are, however, plagued with problems like fluctuation, randomness, and intermittence. These characteristics of DG further give rise to challenges of reliability and stability in the grid. The introduction of DG into the power system leads to a change in power flow, congestion, flicker, harmonics, etc. DG is also paired with immature technology combined with high costs, which implies economic problems for current grid-connected DG projects [10,11].

A major hurdle for the implementation of DG is the high cost. However, over time the costs have decreased significantly. DG is also associated with the concept of islanded operation. DGs are also designed to operate in the standalone mode, making them useful for the local community. The DGs can therefore operate in connected or islanded mode. New technologies are required to implement these operating modes of DGs.

1.2.3 Enhancement of DGs

Nowadays DG technologies have enhanced a lot [5]. This propels the user to adopt DGs more easily. Some of the major enhancements are:

- Increase in efficiency of solar cells up to 24%,
- Generation of wind energy up to several MW,
- Ability to replace conventional CHPs by micro-, bio-, and multi-fuel-CHPs,
- Advancement in fuel cell technologies,
- Larger and more efficient storage devices,
- Introduction of new RES like tidal generators, SHP, etc.

1.3 Virtual power plant (VPP)

1.3.1 Introduction

A VPP has a positive impact on the grid to complement the existing classic CPPs by adding newer suppliers in the system. These suppliers may have small and distributed power systems which are linked to set up virtual pools which can be operated through a centralized control station. A VPP contains multiple DG plants (such as WT, MHP, etc.) and any other power source that is able to cooperate within a local area and be controlled through a centralized station. Fig. 1.2 illustrates a model of the VPP.

A literature review shows that there are few definitions of VPP, and they are mentioned below:

- *"The concept of VPP derives from the definition of the virtual utility"* [12].
- *"VPP is defined as a group of interconnected decentralized residential micro combined heat and power (CHP), using fuel cell technology, installed in multi-family houses, small enterprise, and public facilities, for individual heating, cooling and electricity production"* [13].
- *"VPP is defined as an aggregation of different type generation units (such as electrical and thermal units, CHP units)"* [14].
- *"VPP is defined as a cluster of dispersed generator units, controllable loads, and storages systems, which are aggregated to operate as a unique power plant"* [15].
- *"VPP is defined to aggregate a few DGs to the distribution grid, which has the capability of selling both thermal and electrical energy to neighboring customers"* [16].
- *"VPP is defined as a group to aggregate the capacity of many diverse DERs. This group can create a single operating profile from a composite of the parameters characterizing each DERs and can incorporate the impact of the network on aggregate DERs output"* [17].
- *"VPP is composed of several various technologies with various operating patterns, with which they can connect to different node of distribution network"* [18].
- *"VPP is defined as multi-technology and multi-site heterogeneous entities, which can also operate in isolated networks"* [19].
- "A flexible representation of a portfolio of DERs, not only aggregating the capacity of many diverse DERs, but also creating a single operating profile from a composite of the parameters characterizing each DER and incorporating spatial constraints" [18].

Additionally, some definitions focus on use of software as a vital element of a VPP: *"Virtual Power Plants rely upon software systems to remotely and automatically dispatch and optimize generation or demand side or storage resources in a single, secure web-connected system"* [20].

A new concept of dynamic VPP has recently found its place in the research community. Dynamic VPPs are also termed as clusters. The dynamic VPPs cooperate in a temporary manner while referring to the real time situation in the electricity markets and to the forecast of expected feed-in from various aggregated units. The dynamic VPP will be reformed once the product has been delivered [21].

Figure 1.2 Basic structure of VPP.

Definition of VPP also depends on sales of electricity capacity of VPP as, "sales of electricity capacity which, rather than being "physical" divestitures, are "virtual" and held by a single or by multiple firms which dominate the electricity market. The firms retain for themselves the control and management of the plant. However, the market offers contracts which tend to replicate an output which is similar to that of a physical power plant" [22]. Taking into consideration the above discussions and definitions, a VPP can be defined as: "A portfolio of DERs, including generating units and controllable loads (1), which are connected by a control system (2) based on information and communication technology (ICT). The VPP acts as a single visible entity in the power system (3), is always grid-tied and can be either static or dynamic" [23].

The requirements identified for management of a VPP within a grid will require [1]:

- Improved monitoring and control of the available distribution network for guaranteed performance and better reliability, as well as security of power supply.
- Modeling, designing, and testing of advanced components based on the changed requirements due to addition of VPP in the system.
- The grid should be analyzed for identifying any shortcomings in its operations, and guidelines should be proposed for the reinforcement and development of power network with a VPP.

A tabulated review of VPP parameters is presented in [24] considering different members in a VPP (like solar plants, conventional power plants, etc.), model type (like price-based unit commitment, energy management, etc.), target period, trading methods, etc. A similar comparison is also present in [25] and [26].

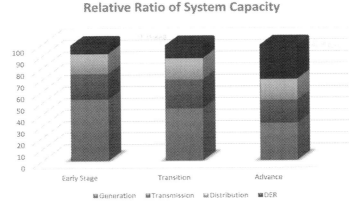

Figure 1.3 Relative ratio of system capacity.

1.3.2 Energy storage systems in VPP

To overcome the issues mentioned earlier, systems having greater reserve capacity are required. This may be achieved through enhancing energy storage technologies. Some of the energy storage technologies which are presently in use are:

- hydraulic,
- compressed air,
- superconducting magnetic fields,
- fluid battery pack,
- flywheel, etc.

The mechanical storage technologies when compared with the BESS lack better and quicker response time. There are various strategies proposed and developed for efficient utilization of BESS [27,28]. To have the interconnection of DGs and achieve a stable grid with the help of new and advanced controllers, the output power of a BESS should be adjusted and monitored continuously for charging and discharging of power [28]. However, currently BESSs are more expensive and require maintenance. There are other storage technologies also available, such as pumped storage power plants, but this technology mostly depends on construction sites and environment.

The presence of conventional generation, transmission, distribution, and DER as a ratio of total capacity of the power system is shown in Fig. 1.3 [29]. Conventional generation shared a considerable portion of the power system in the early stages. However, now the conventional generations are gradually being replaced by RES. Optimal sizing of these RES resources [30] is of prime importance in the formation of VPPs. In the future, DER will be an important part of the grid. It will have the responsibility of supporting the grid and take part in the energy markets. In case a large number of DGs along with dispatchable BESS and distributed large-capacity load can be inte-

Table 1.3 Different control approaches to VPP [3].

Method	Key feature	Advantages
Centralized Controlled VPP	• The logic applied for controlling the system is a duty of the VPP • DER and market, as well as production planning, are separate from each other	Market demand can be met by using DER
Distributed Controlled VPP	• It has a different hierarchical model • Local VPP maintains the coordination and supervision of a specific/limited number of DERs • Higher level of VPP makes necessary decisions	It simplifies the communication and the responsibilities of each VPP.
Fully Distributed Controlled VPP	In this structure, every DER works independent of each other. It participates and reacts to different states arising in the power system and electricity markets	Independent of each other

grated to the grid, the DER will be controllable and its characteristics will be similar to CPP. Micro-grid (MG) also integrates DGs, BESS, and load in itself. However, MG has limitations of spreading geographically. Thus, the new concept of a VPP was envisioned and introduced. VPP must gather information from many application domains, e.g., demand response schemes, RES, BESS, energy market operation, etc.

1.3.3 Types of VPP

VPP has been undergoing many changes recently, specifically because of the economic value that is attached to it. Along with the economic benefits, the VPP also adds security and reliability in power system operations. The different methods to classify VPPs are presented below:

1) Control

Based on the control of VPPs, they can be classified as presented in Table 1.3.

VPPs exist with different topology and control structures as presented in Table 1.3 and Figs. 1.4–1.6.

2) Objective

Another classification of VPPs brings two categories of the VPP to the fore: the technical VPP (TVPP) and the commercial VPP (CVPP) [11].

TVPP has DERs which are present within a specified geographic region. TVPP cooperates for managing local power systems for the distribution system operator (DSO). It also provides system balancing, as well as ancillary services, for the transmission

Figure 1.4 Centralized controlled VPP.

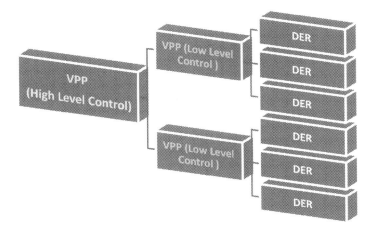

Figure 1.5 Distributed controlled VPP.

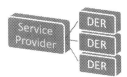

Figure 1.6 Fully distributed controlled VPP.

system operator (TSO). The TVPP is responsible for facilitating the management of restrictions of local networks [31]. DSO, combined with TVPP, presents an active distribution network (ADN). The TVPP improves reliability by allowing the DERs to operate in varied conditions. Some of the defined functions of TVPP are [11]:

- Continuous condition monitoring, fault identification, and maintenance;
- Asset management;
- Analysis and optimization of the system.

CVPP mainly deals with the cost and operating characteristics of its DER units. CVPP takes active part in energy markets either through trading or provision of services. The CVPP allows market access to smaller DER units. CVPP's main objective is aggregation of entities for commercial benefits and not for system stability. Unlike TVPP, CVPP can have participants from a wide geographical area. CVPP helps in

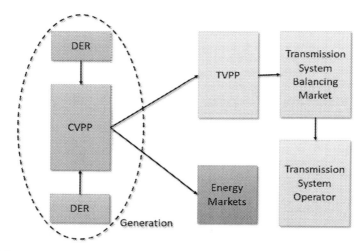

Figure 1.7 TVPP and CVPP in a power system [18].

optimized scheduling based on forecasted demand and available generation capacity. Basic CVPP functions are:

- Managing DER characteristics;
- Production and consumption forecasting along with optimized generation scheduling;
- Helping individual DERs participate in energy markets through bidding and selling;
- Forecasting of production and demand [31].

A representation of the relation between the TVPP and CVPP can be ascertained from Fig. 1.7.

CVPP aggregates DERs to create a portfolio so as to participate in the energy markets. It can represent DER from any geographic region in the power system [18]. On the other hand, TVPP helps DERs to be represented with the system operator. The DERs can participate in system management through the TVPP.

3) Market Segment

In this classification, a unique methodology has been utilized to differentiate the types of VPP currently existing in the world, based on findings by Pike Research [20]. The classifications are: DR-based VPP, supply side VPP, mixed asset VPP, and wholesale auction VPP. DR-based VPP are visible in the US with a long existing demand response market. The supply side VPPs are native to Germany with a lot of them being R&D projects. Mixed asset VPP brings out the basic objective of VPP, utilizing demand response and DG, and sharing the grid resources optimally. Lastly, the wholesale auction VPPs are currently present in Europe, which include the large power plants to supply base loads.

1.3.4 VPP in collation with other technologies

1.3.4.1 Micro-grid (MG)

MG comprises of a collection of loads, along with small energy generation units, combined together as a single entity which is controllable and responsible to meet the thermal and electric demands of its local population [32]. Akin to an MG, VPP is a collection of DGs along with BESS and loads which can be controlled. Also, VPP and MG both are able to integrate demand response programs [31]. However, VPP is a wider concept as compared to MG. The main differences between MG and VPP are:

A. Design

MG focuses more towards self-management through the use of DGs and controllable loads. On the other hand, VPP is aggregated on the basis of participation, which strives to implement collective control of DGs, controllable loads, and BESS devices. The major shift in focus is inherent as the VPP is not geographically bounded like the MG. This helps VPP to participate in energy markets.

B. Interface

MG utilizes a coupling switch for connecting to the grid. VPP is a virtual collection and uses an open protocol to get connected to the grid. VPP is virtually existing and therefore does not require specific hardware for its connection to the grid. VPP, being a collection of sources, can participate in operations of power systems.

C. Area

MG's physical presence depends on DGs, transmission lines, load, etc., which are its parts. This indicates that the physical spread of an MG is dependent on the geographical placement of its components [29]. However, in the case of VPP, which is cloud based, the spread depends on the utilization of communication networks and technology, making it have a large presence geographically.

D. Functions

The MG is supposed to enhance the power quality, reliability, security, etc., of the system. VPP acts in a different way and is used to improve controllability of the power system through frequency regulation, demand, or emergency response and ancillary services, like the energy market [29].

E. Other

VPPs and MGs are usually used as substitutes for integrating DERs in the power system. Still there exist major differences between the two [20,33]. Based on the connectivity with the grid, the VPP is always connected to the grid and is a continuous participant in its operation. On the other hand, MG can operate in grid connected mode, as well as islanded mode, and can thus be out of the grid at times. Another difference appears on the basis of the capacity. MG has a lower capacity as it is bounded by physical boundaries, whereas VPP commands a higher capacity.

1.3.4.2 Conventional power plant (CPP)

Stability and controllability of output are the main features of CPPs like thermal or hydro generation which is helpful in frequency regulation as well as peak shaving.

VPP, on the other hand, acts as a single dispatchable and trading unit in the power system and energy markets, respectively. Their main differences are listed below:

A. Environment

CPPs are usually based on fossil fuels and have bulk generation, which is harmful for the environment. They may occupy and affect a large geographical region like a hydro plant which destroys the local topology. CPPs are located near the source of raw material and thus energy may be lost in transmission, which results in increased generation, further polluting the environment. VPPs are usually composed of RES and implement clean technology along with low carbon power production.

B. Power Capacity

CPPs consist of a fixed number of generating units and have bulk production. VPP is quite flexible and its capacity can vary. VPPs can be expanded by allowing more DER units to be aggregated into the system.

C. Control System

CPP has a well-established control system operation and delivers satisfactory performance with a slow regulation rate. VPP is flexible and has a better response and regulation rate compared to CPP. VPP is therefore better equipped to efficiently participate in the energy markets as compared to CPP.

It should be noted here that in some cases, the CPP may itself be a part of a large VPP.

1.3.4.3 Virtual synchronous machines (VSM)

VSMs are similar to traditional synchronous machines, in which they can mimic their dynamic response. VSMs are used in control strategies of power system operation. VSMs can also be implemented for auxiliary services, like reactive power control, oscillation damping, etc. [34].

The comparison between VPP and VSM is presented based on the following points:

A. Response Time

VSM responds to transient dynamics and tries to maintain power inertia of power electronic devices. On the other hand, VPP is responsible for steady state behavior of power systems and responds to the economic operation of energy markets. Thus, the response time of VSM is quite less than VPP.

B. Control Scope

VSM is more concerned with the control strategies of a single DG and its power electronic interface, while VPP should control and coordinate multiple DGs, loads, and BESS. VSM has a more restricted scope of control whereas VPP has a more spread out control area.

1.3.5 Obstacles for implementing VPP

VPP has a role in enabling individual DERs to participate in the working of power systems and energy markets. But while doing so, there arise challenges like allocation of resources, operation and control system, communication, power transactions, delivery of services, etc.

A. Resource Allocation

VPP is an aggregation of different entities within the power system. These entities may be DGs, BESS, load, etc. In [35], the authors propose a purely distributed control strategy such that multiple DGs can easily compose a VPP. The impacts of incorporating a BESS in a VPP are investigated through a generic methodology in [36]. Also, an augmented model is implemented for performing adequacy studies related to generation. The authors of [37] present a system having 1000 heating, ventilating, and air-conditioning (HVAC) units which are modeled to provide continuous regulation reserves (CRR). In [38], the aggregation and control of multiple responsive load populations as a single VPP is studied. The combinations of these loads with and without weight allocation are studied for composing an optimal VPP model. These research publications usually focus on providing various options in the development and operation of a VPP [39]. Some publications, however, have presented a plan for optimal allocation of VPP. Resource allocation is an important factor in the development of a VPP, especially considering that individual DERs are not static. The dynamic, as well as practical, characteristics of different entities within the VPP should be considered for resource allocation. There should be some well-established evaluation method and index system.

B. Operation and Control System

In [40], an agent-based approach involving EVs and WT is used. EVs are used as a storage medium. Optimal energy scheduling and maximization of profit is achieved through this combination while considering the uncertainty associated with WT. The optimal operation of energy markets in the grids is proposed through multi-level negotiation mechanism in [41]. A large scale VPP (LSVPP) operation is considered in [42]. The operation of LSVPP is optimized through a new algorithm for a day-ahead scheduling of thermal and electrical loads. Internal dispatch of entities in a VPP for maximized profits is studied in [43]. The state problem is solved through a mixed-integer linear programming (MILP) model. The role of small generation units and loads especially for frequency stabilization is presented in [44]. A novel distributed coordination algorithm is utilized for direct control of these smaller entities in a VPP. These generation units and loads are also proposed to be used as secondary control reserves. The operation and control of a VPP is affected due to regulation rate, sustainable time, and other factors related to DER. The control of VPP cannot be perfected, and thus there are certain variations in its performance. VPPs are also required to provide for different ancillary services across multiple time scales, which require various control requirements. The VPP controls have to utilize the DER characteristics. Interaction with the grid is also important for optimal control and operation of a VPP.

A VPP has to manage and communicate with each DER in real time through its control system, which is a complex process because of the large number of units involved. The control algorithm therefore plays a vital role in efficient operation of a VPP. Advanced metering infrastructure (AMI) should be utilized to apply the latest technologies for real-time metering that would help VPPs utilize the DER flexibility [45] to its maximum potential.

C. Information and Communication Technology (ICT)

The concept of IoT for communication is used in [46], for a small VPP. Arduino is a low cost and low voltage platform, implemented for controlling high-voltage power lines through MQTT, an IoT protocol. Data interoperability is also very important for secure operation of VPPs, as it comprises various entities like DERs, load, BESS, etc., which may be produced by different manufacturers. The properties of data generated in a VPP, like data rate, scale, variety, etc., should be integrated through various dispersed sources for transfer of information. It is urgently needed to develop standardized protocols and utilize latest communication technologies for interconnection and control of VPP components through rapid transmission and aggregation of data.

An interoperable solution is not present, which acts as a barrier for the scalability of VPPs [47]. Interoperability is related to all conceptual and ICT related challenges, like common standards for communication, implied on various participants and their complex interactions which are needed for a proper realization of the proposed solution in [48]. A variety of custom made solutions, like Kraftwerke's "Next Box", are being evolved for various projects to overcome these problems [49].

D. Power Transactions

A weekly self-scheduling of VPP is considered in [50], whose components include RES, BESS, and a CPP. An MILP model is used to maximize weekly profits while considering its commitment in the energy markets, technical constraints, and intermittency of RES. The uncertainties associated with WT/RES and energy markets are dealt with in [51], while applying a robust optimization method. A two-stage stochastic MILP maximizes the profits of VPP in day-ahead and balancing markets in [52]. The power transactions or energy markets have in themselves an uncertain output which varies with time, which combined with varying output of DER makes it very complicated for a VPP to participate in such transactions in an optimal way. VPP has to also take into consideration the benefit/profit of individual players within the aggregation. VPP is also a candidate to participate in energy markets for providing ancillary services to the grid due to its flexibility, but has its own complications involved.

E. Economic

During the aggregation of DERs into a VPP, it should be economically beneficial for its operators, over the stand-alone operation. The price mechanisms in force shall play a pivotal role in the outcome for taking decisions, along with existing RES support schemes. A flat or block rate tariff structure may not be very attractive for the aggregation of a VPP. The energy pricing structure may be used as an incentive for DER units to actively participate in a VPP [53]. An incentive-based approach can be implemented as a motivation to participate in the formation of a VPP. DERs are better placed for providing services efficiently, when aggregated. Markets in a regulatory environment will not be a welcome signal for VPP formation; however, incentive-based schemes can reduce the effect of regulatory framework. VPP also has to face the market barriers for entering energy markets. In the current markets, VPP participation

Introduction and history of virtual power plants with experimental examples

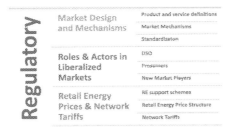

Figure 1.8 Classification of regulatory barriers.

is not facilitated due to factors like high entry barriers, reserve capacity requirements, etc., in the existing markets. Lastly, the markets may also not be able to fully utilize the distinct features of VPP like flexibility, etc. A VPP may therefore not be compensated enough for such services in the existing markets.

F. Regulatory

There are different electric grid setups in different parts of the world. Some of the electric grids are regulated while others are deregulated. In deregulated electric grids, the markets may function in a variety of ways. The rules and regulations in these regions do vary by a huge amount and, considering the structure of a VPP, it may have problems in operations due to regulatory reasons. The regulatory barriers can be classified into three categories, as shown in Fig. 1.8.

The pricing strategies in a VPP are again a major issue, since the total output of a VPP is an aggregation of different DERs and controllable loads along with BESS. The pricing problem is further extrapolated when the VPP participates in the energy market. The provision of ancillary services by VPP can be fulfilled through the combined efforts of its participants which may require certain binding regulations. The order in which it participates in the markets should also be formally regularized. Regulations associated with energy prices in the retail market and other network tariffs have a large influence on how the VPPs will be deployed in the distribution grid. Schemes related to RES implementation also affect the formation of VPP.

VPP is a spread through a wide geographical region and hence its categorization within a grid also requires proper regulation for efficient operation.

G. Cyber-Attack

VPP is a cloud-based virtual entity and thus is dependent on the communication technologies being used in the day. A grid or VPP is not secure in its current form and is prone to getting attacked virtually. A lethal cyber-attack on the VPP will render it vulnerable to being left impractical. The cyber-attacks could be colluding or non-colluding depending on the participants involved in the attack. Cyber-attacks can even be used to disrupt the economic operation of a VPP [54]. VPP, being a distributed utility and only centrally controlled, also provides inroads for cyber-attackers. VPP therefore requires some dependable cyber-physical security system for its secure implementation and satisfied operation.

H. Uncertainties involved with the operation of VPP

VPP is responsible for aggregating a large number of entities, operating under a single central control [55]. VPP can have under itself large scale power generation and it even participates in the energy trading markets at different levels. There are different uncertainties that will stop the VPP from precisely estimating important parameters related to its optimal operations. These uncertainties are categorized as (i) renewable energy source uncertainty, (ii) market price uncertainty, and (iii) load demand uncertainty.

1.3.6 VPPs around the world

According to a report made by Pike Research in 2012, global VPP capacity was estimated to increase by almost 65%, from 55.6 GW to 91.7 GW. However, in terms of revenue, it were to increase to $6.5 billion in 2017 from $5.3 billion in 2011 [56]. According to the clean-tech market intelligence firm predictions, the global VPP revenues could reach up to $12.7 billion. As per the investigation done by Polls and Surveys (P&S), the global growth of the VPP market can increase from $191.5 million in 2016 to $1.1875 billion in 2023.

The increase in demand of energy industries has made VPPs an important player, and they are increasingly becoming a vital element, which will be growing rigorously in times to come. Table 1.4 mentions some of the VPP projects all over the globe.

A. European Virtual Fuel Cell Power Plant (VFCPP)

The European Commission funded a project involving 31 decentralized stand-alone residential fuel cell systems [57]. The technology used was high temperature (HT) PEM fuel cell, based on PBI (polybenzimidazole) high temperature PEM membrane technology. The project faced a number of challenges, like

- Cost,
- Simplicity,
- Compatibility of output temperature with existing heating systems, for improving chances of tri-generation.

B. Power Matcher VPP

Power Matcher (PM), as the name itself indicates, performs the work of a plant by matching the power generation and demand, i.e., supply and demand matching (SDM). The concept is market-based, with a high DG penetration. It was proposed by the Energy Research Center of the Netherlands (ECN) [65]. PM also has an objective to provide a mechanism for trading energy through itself. The architecture is shown in Fig. 1.9 [58].

The project was executed in a power matching city of Hoogkerk, in the Netherlands [66], where 25 homes were integrated with the grid through smart grid/energy technology. The energy was captured through hybrid heat pumps, PV panels, EVs, along with wind and gas turbines.

Table 1.4 VPP projects around the globe.

Name	Period	Country	Type of DER	Functions
VFCPP	2001–2005	Germany, the Netherlands, Spain, etc.	Fuel cell	Use fuel cells as VPP at household locations [57]
PM VPP	2005–2007	The Netherlands	μ-CHP	Provide the market mechanism [58]
FENIX	2005–2009	UK, Spain, France, etc.	μ-CHP, PV,	Use wind power enabled DER-based systems [45]
EDISON	2009–2012	Denmark	EV	Increase usage of wind power, balancing of power [59]
FLEX POWER	2010–2013	Denmark	Wind energy	Balancing the active power from demand and small-scale generation [60]
WEB2EN-ERGY	2010–2015	Germany, Poland, etc.	CHP, PV wind power, biogas power, hydropower	Implement and approve "smart distribution" [61]
TWENTIES	2012–2015	Belgium, Germany, France, etc.	Wind power	Innovative alliance of wind power plant [62]
CON EDISON VPP	2016–2018	The USA	PV, battery storage	Integrate the grid with combined power from solar and storage systems of the residents [63]
SA VPP	2016–2018	Australia	PV, batteries	Reduce customers' energy bill, provide grid support services and security of supply [64]

C. FENIX VPP

FENIX is a European collaborative project between 20 partners and partially funded by the European Commission with a budget of around 14.7 million euro [67].

Figure 1.9 Architecture of a power matcher.

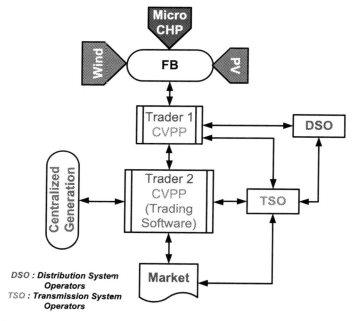

Figure 1.10 Architecture of FENIX.

The basic characteristics of FENIX have three interdependent subjects of research whose outcome forms the foundation for proper operation of futuristic highly decentralized power systems. These include:

- Distributed system control architecture;
- Information and communication architecture;
- Supporting market and commercial structure.

The aim of FENIX was to integrate DER in an economic way for operation and development of electricity networks. Its architecture is given in Fig. 1.10. This architecture was designed in such a way that it could be implemented at other places in Europe. FENIX has been designed such that it can:

- Utilize the uniqueness of any European country;
- Introduce new technical VPP (TVPP) functionalities which will optimize the changes in the energy markets, including energy producers and distributors.

The objectives were to limit:
- Impact on existing information systems;
- Number of interactions between actors.

Three new elements were introduced into the FENIX architecture:
- The FENIX box interfaces with DER of the local system, thus allowing remote access for control and monitoring purposes;
- Commercial VPP (CVPP) plans and optimizes the use of DER units;
- TVPP application at the distribution management system (DMS) level performs functions, like validation of generation schedules, dispatching of DER to monitor the electrical parameters, etc.

The architecture is customized to be compliant with specifications of the northern and southern scenarios while considering specific regulatory framework and association between the participating actors [68]. The northern scenario case integrates about 4.3 MW of DG and about 2.5 MW of flexible DG, notably CHPs with gas engines in Woking Borough (UK).

The southern scenario, on the other hand, of the Alava distribution network (Spain) includes different technologies (CHP, wind, etc.). The overall installed capacity is approximately 150 MW, of which about 50% is flexible, for the southern applications.

D. EDISON

The aim of EDISON project is to utilize integrated EVs, including plug-in hybrid EVs, in the Danish electricity grid which will provide the additional power needed as complementary power for swelling usage of wind power in the grid [69]. Due to the intermittency and randomness of wind-based DER, EVs prove to be advantageous for providing power to the grid.

EDISON motivated the owners of EV to participate in the project. It offered a time based metering and billing system, which indirectly allows the EV owners to benefit from electricity rates. The conceptual architecture and grid stakeholder models are shown in Figs. 1.11 and 1.12, respectively.

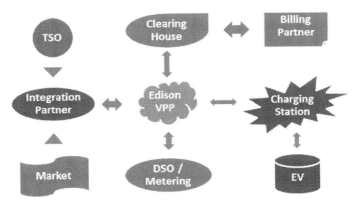

Figure 1.11 Architecture of EDISON.

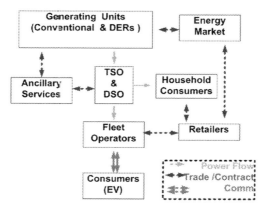

Figure 1.12 EDISON grid stakeholder model.

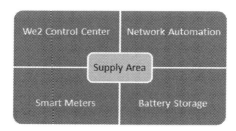

Figure 1.13 Architecture of Web2Energy.

The Danish EDISON project proved the non-rigid charging/discharging of EV. The intelligent handling of the VPP system brought benefits not only to EV owners, but also to the electricity grid and society. The initial focus was on the concept of smart charging of EV. The VPP is dependent on charging of EV batteries which is further dependent on cost of energy, constraints related to grid, renewable energy availability, etc. The VPP has to also rely on a proper and standardized ICT architecture.

E. Web2Energy

Web2Energy tries to implement smart metering, smart energy management, and smart distribution automation, which are all very important pillars of a smart distribution system [70].

The architecture of Web2Energy is shown in Fig. 1.13, while its details are as follows [71]:

- 20/0.4 kV network of the HSE AG around Darmstadt,
- 9×20/0.4 kV transformer (remote supervision in selected network part),
- 5 CHP plants,
- 12 storage batteries,
- 12 photovoltaic plants,
- 3 wind parks,
- 2 hydropower plants,

- 3 large controllable loads,
- 200 household consumers.

1.3.7 VPPs under construction

A. ConEdison VPP

The ConEdison VPP plans to integrate hundreds of PV arrays placed in residential regions and a BESS into the grid. A new business model will be tested and verified which shall implement combined integration. SunPower and Sunverge are designing the software which will autonomously control the system. The total estimated cost of the project is around $15 million.

B. South Australian VPP (SA VPP)

A huge VPP is being established at South Australian VPP (SA VPP). It will help the customers reduce the electricity bills. On the other hand, it shall contribute to maintaining stability of the grid and managing peak load demands from the grid. Upon completion, it shall be the largest existing VPP with around (i) 1000 residential customers having 5 MW of installed PV panels and (ii) a storage capacity of 7 MWh. SA VPP is a joint collaboration of Australia energy retailer AGL and the federal government's Australian Renewable Energy Agency (ARENA), etc.

1.4 Research on VPP

The VPPs are a matter of research for multiple fields, regions, etc. This section introduces a brief summary.

The total number of publications related to VPP currently stands as 1255 [72]. The research publications in recent years are tabularized in Table 1.5.

Table 1.5 Some of the recent publications on virtual power plant (VPP).

No.	Topic	Reference
1	Energy Trading	[73], [74], [75], [76]
2	Load Scheduling	[74], [77]
3	Techno Economic Analysis	[78]
4	Renewables Integration	[25], [79], [80]

There are multiple research publications; however, the above Table 1.5 would help the readers understand the current applications of VPP.

1.5 Summary

In today's world, the conventional energy sources and power demand have an inverse relation: on the one hand, day by day the conventional energy sources are depleting and, on the other hand, the energy demand is increasing. The time has come to find new energy sources to meet the future energy crisis. There is no doubt that in upcoming years the government policies and revolution in the renewable energy industry will lead to an increase in the number of isolated DERs.

In future VPP can play an important role in the power sector. VPP has the capability to integrate the DERs and, at the same time, optimize their working. Proper coordination of VPP can fully optimize the generating stations and DERs on the consumer side which can eventually benefit the investors and society in its social and economic reforms.

References

[1] J.M. Morales, A.J. Conejo, H. Madsen, P. Pinson, M. Zugno, Integrating Renewables in Electricity Markets: Operational Problems (Vol. 205), Springer Science & Business Media, 2013.
[2] P. Patil, M.R. Ramteke, Impact of distributed generation on power distribution system: over-current protection by phase angle estimation, in: 2015 IEEE Power, Communication and Information Technology Conference (PCITC), IEEE, October 2015, pp. 844–849.
[3] L.I. Dulău, M. Abrudean, D. Bică, Distributed generation and virtual power plants, in: 2014 49th International Universities Power Engineering Conference (UPEC), IEEE, September 2014, pp. 1–5.
[4] European Commission's Directorates-General for Research and Innovation, Energy and the Joint Research Centre, A European strategic energy technology plan (SET-Plan): towards a low carbon future, 2007.
[5] K. El Bakari, W.L. Kling, Virtual power plants: an answer to increasing distributed generation, in: 2010 IEEE PES Innovative Smart Grid Technologies Conference Europe (ISGT Europe), IEEE, October 2010, pp. 1–6.
[6] M. Vatani, M.J. Sanjari, G.B. Gharehpetian, Islanding detection in multiple-DG microgrid by utility side current measurement, International Transactions on Electrical Energy Systems 25 (9) (2015) 1905–1922.
[7] M. Vatani, D.S. Alkaran, M.J Sanjari, G.B. Gharehpetian, Multiple distributed generation units' allocation in distribution network for loss reduction based on a combination of analytical and genetic algorithm methods, IET Generation, Transmission & Distribution 10 (1) (2016) 66–72.
[8] Z. Yang, J. Liu, S. Baskaran, C.H. Imhoff, J.D. Holladay, Enabling renewable energy—and the future grid—with advanced electricity storage, JOM 62 (9) (2010) 14–23.
[9] E. Romero-Cadaval, B. Francois, M. Malinowski, Q.C. Zhong, Grid-connected photovoltaic plants: an alternative energy source, replacing conventional sources, IEEE Industrial Electronics Magazine 9 (1) (2015) 18–32.
[10] Ł.B. Nikonowicz, J. Milewski Virtual power plants-general review: structure, application and optimization, Journal of Power Technologies 92 (3) (2012) 135–149.
[11] H. Saboori, M. Mohammadi, R. Taghe, Virtual power plant (VPP), definition, concept, components and types, in: 2011 Asia–Pacific Power and Energy Engineering Conference, IEEE, March 2011, pp. 1–4.
[12] S. Awerbuch, A. Preston (Eds.), The Virtual Utility: Accounting, Technology & Competitive Aspects of the Emerging Industry (Vol. 26), Springer Science & Business Media, 2012.

[13] E.A. Setiawan, Concept and Controllability of Virtual Power Plant, Kassel University Press GmbH, 2007.
[14] F. Bignucolo, R. Caldon, V. Prandoni, S. Spelta, M. Vezzola, The voltage control on MV distribution networks with aggregated DG units (VPP), in: Proceedings of the 41st International Universities Power Engineering Conference (Vol. 1), IEEE, September 2006, pp. 187–192.
[15] P. Lombardi, M. Powalko, K. Rudion, Optimal operation of a virtual power plant, in: 2009 IEEE Power & Energy Society General Meeting, IEEE, July 2009, pp. 1–6.
[16] R. Caldon, A.R. Patria, R. Turri, Optimal control of a distribution system with a virtual power plant, in: Bulk Power System Dynamics and Control, Cortina d'Ampezzo, Italy, 2004, p. 18.
[17] C. Kieny, B. Berseneff, N. Hadjsaid, Y. Besanger, J. Maire, On the concept and the interest of virtual power plant: some results from the European project Fenix, in: 2009 IEEE Power & Energy Society General Meeting, IEEE, July 2009, pp. 1–6.
[18] D. Pudjianto, C. Ramsay, G. Strbac, Virtual power plant and system integration of distributed energy resources, IET Renewable Power Generation 1 (1) (2007) 10–16.
[19] H. Morais, P. Kádár, M. Cardoso, Z.A. Vale, H. Khodr, VPP operating in the isolated grid, in: 2008 IEEE Power and Energy Society General Meeting-Conversion and Delivery of Electrical Energy in the 21st Century, IEEE, July 2008, pp. 1–6.
[20] P. Asmus, Microgrids, virtual power plants and our distributed energy future, The Electricity Journal 23 (10) (2010) 72–82.
[21] M. Sonnenschein, O. Lünsdorf, J. Bremer, M. Tröschel, Decentralized control of units in smart grids for the support of renewable energy supply, Environmental Impact Assessment Review 52 (2015) 40–52.
[22] L.M. Ausubel, P. Cramton, Virtual power plant auctions, Utilities Policy 18 (4) (2010) 201–208.
[23] G. Plancke, K. De Vos, R. Belmans, A. Delnooz, Virtual power plants: definition, applications and barriers to the implementation in the distribution system, in: 2015 12th International Conference on the European Energy Market (EEM), IEEE, May 2015, pp. 1–5.
[24] M. Shabanzadeh, M.K. Sheikh-El-Eslami, M.R. Haghifam, An interactive cooperation model for neighboring virtual power plants, Applied Energy 200 (2017) 273–289.
[25] N. Naval, R. Sánchez, J.M. Yusta, A virtual power plant optimal dispatch model with large and small-scale distributed renewable generation, Renewable Energy 151 (2020) 57–69.
[26] S.M. Nosratabadi, R.A. Hooshmand, E. Gholipour, A comprehensive review on microgrid and virtual power plant concepts employed for distributed energy resources scheduling in power systems, Renewable & Sustainable Energy Reviews 67 (2017) 341–363.
[27] M.J. Sanjari, H. Karami, Optimal control strategy of battery-integrated energy system considering load demand uncertainty, Energy 210 (2020) 118525.
[28] A.Q. Badar, M.J. Sanjari, Economic analysis and control strategy of residential prosumer, International Transactions on Electrical Energy Systems 30 (9) (2020) e12520.
[29] X. Wang, Z. Liu, H. Zhang, Y. Zhao, J. Shi, H. Ding, A review on virtual power plant concept, application and challenges, in: 2019 IEEE Innovative Smart Grid Technologies-Asia (ISGT Asia), IEEE, May 2019, pp. 4328–4333.
[30] S. Saranchimeg, M.J. Sanjari, N.K. Nair, Toward a resilient network by optimal sizing of large scale PV power plant in microgrid-case study of Mongolia, in: 2018 Australasian Universities Power Engineering Conference (AUPEC), IEEE, November 2018, pp. 1–6.
[31] L.F. Ramos, L.N. Canha, Virtual power plants and their prospects, Available: www.preprints.org, 2019.

[32] E. Mashhour, S.M. Moghaddas-Tafreshi, A review on operation of micro grids and virtual power plants in the power markets, in: 2009 2nd International Conference on Adaptive Science & Technology (ICAST), IEEE, January 2009, pp. 273–277.
[33] D. Pudjianto, C. Ramsay, G. Strbac, Microgrids and virtual power plants: concepts to support the integration of distributed energy resources, Proceedings of the Institution of Mechanical Engineers. Part A, Journal of Power and Energy 222 (7) (2008) 731–741.
[34] O. Mo, S. D'Arco, J.A. Suul, Evaluation of virtual synchronous machines with dynamic or quasi-stationary machine models, IEEE Transactions on Industrial Electronics 64 (7) (2016) 5952–5962.
[35] H. Xin, D. Gan, N. Li, H. Li, C. Dai, Virtual power plant-based distributed control strategy for multiple distributed generators, IET Control Theory & Applications 7 (1) (2013) 90–98.
[36] A. Bagchi, L. Goel, P. Wang, Adequacy assessment of generating systems incorporating storage integrated virtual power plants, IEEE Transactions on Smart Grid 10 (3) (2018) 3440–3451.
[37] N. Lu, Y. Zhang, Design considerations of a centralized load controller using thermostatically controlled appliances for continuous regulation reserves, IEEE Transactions on Smart Grid 4 (2) (2012) 914–921.
[38] D. Wang, H. Jia, C. Wang, N. Lu, M. Fan, W. Miao, Z. Liu, Performance evaluation of controlling thermostatically controlled appliances as virtual generators using comfort-constrained state-queueing models, IET Generation, Transmission & Distribution 8 (4) (2014) 591–599.
[39] K. El Bakari, W.L. Kling, Development and operation of virtual power plant system, in: 2011 2nd IEEE PES International Conference and Exhibition on Innovative Smart Grid Technologies, IEEE, December 2011, pp. 1–5.
[40] M. Vasirani, R. Kota, R.L. Cavalcante, S. Ossowski, N.R. Jennings, An agent-based approach to virtual power plants of wind power generators and electric vehicles, IEEE Transactions on Smart Grid 4 (3) (2013) 1314–1322.
[41] Z. Vale, T. Pinto, H. Morais, I. Praça, P. Faria, VPP's multi-level negotiation in smart grids and competitive electricity markets, in: 2011 IEEE Power and Energy Society General Meeting, IEEE, July 2011, pp. 1–8.
[42] M. Giuntoli, D. Poli, Optimized thermal and electrical scheduling of a large scale virtual power plant in the presence of energy storages, IEEE Transactions on Smart Grid 4 (2) (2013) 942–955.
[43] M. Zdrilić, H. Pandžić, I. Kuzle, The mixed-integer linear optimization model of virtual power plant operation, in: 2011 8th International Conference on the European Energy Market (EEM), IEEE, May 2011, pp. 467–471.
[44] S. Ruthe, C. Rehtanz, S. Lehnhoff, Towards frequency control with large scale virtual power plants, in: 2012 3rd IEEE PES Innovative Smart Grid Technologies Europe (ISGT Europe), IEEE, October 2012, pp. 1–6.
[45] FENIX Objectives, [Online]. Available: http://www.fenix-project.org/.
[46] D. Zubov, An IoT concept of the small virtual power plant based on Arduino platform and MQTT protocol, in: Proc. 2016 International Conference on Applied Internet and Information Technologies, 2016, pp. 95–103.
[47] S. Jaffe, M. Torchia, J. Feblowitz, R. Nicholson, The VPP: Integrating Operations Technology and IT Transformation of the Power Industry, IDC Energy Insights, MA, USA, 2011.
[48] EnergyVille, Research into sustainable energy and smart energy systems, [Online]. Available: http://homes.esat.kuleuven.be/~energyville/sites/default/files/EnergyVille%20industriedag%20Interoperability.pdf, 2014.

[49] Next Kraftwerke GmbH, [Online]. Available: https://www.nextkraftwerke.com/.
[50] H. Pandžić, I. Kuzle, T. Capuder, Virtual power plant mid-term dispatch optimization, Applied Energy 101 (2013) 134–141.
[51] M. Rahimiyan, L. Baringo, Strategic bidding for a virtual power plant in the day-ahead and real-time markets: a price-taker robust optimization approach, IEEE Transactions on Power Systems 31 (4) (2015) 2676–2687.
[52] M.A. Tajeddini, A. Rahimi-Kian, A. Soroudi, Risk averse optimal operation of a virtual power plant using two stage stochastic programming, Energy 73 (2014) 958–967.
[53] A. Delnooz, D. Six, K. Kessels, M.P.F. Hommelberg, e-Harbours: identification and analysis of barriers for virtual power plants in harbour regions, in: 2012 9th International Conference on the European Energy Market, IEEE, May 2012, pp. 1–6.
[54] P. Li, Y. Liu, H. Xin, X. Jiang, A robust distributed economic dispatch strategy of virtual power plant under cyber-attacks, IEEE Transactions on Industrial Informatics 14 (10) (2018) 4343–4352.
[55] S. Yu, F. Fang, Y. Liu, J. Liu, Uncertainties of virtual power plant: problems and countermeasures, Applied Energy 239 (2019) 454–470.
[56] Richard Martin, Revenue from Virtual Power Plants Will Reach $5.3 Billion by 2017, Forecasts Pike Research, (April 18, 2012) [Online]. Available: https://www.businesswire.com/news/home/20120418005337/en/Revenue-from-Virtual-Power-Plants-Will-Reach-5.3-Billion-by-2017-Forecasts-Pike-Research.
[57] European Virtual Fuel Cell Power Plant, System Development, Build, Field Installation and European Demonstration of a Virtual Fuel Cell Power Plant Consisting of Residential Micro-CHP's, [Online].
[58] Power Matcher Introduction, [Online]. Available: http://www.powermatcher.net/.
[59] EDISON, [Online]. Available: http://www.edison-net.dk/.
[60] Flexpower project description, [Online]. Available: http://www.eaenergianalyse.dk/reports/1027_flexpower_project_description.pdf.
[61] Smart Grids – the vision of the electricity networks of the future, [Online]. Available: https://www.web2energy.com/en/about-theproject/.
[62] Twenties project, [Online]. Available: http://www.twenties-project.eu.
[63] ConEdison to pilot $15 million virtual power plant project, [Online]. Available: http://analysis.newenergyupdate.com/energy-storage.
[64] Sunverge Customer AGL Brings Online World's Largest Residential Virtual Power Plant, [Online]. Available: http://www.sunverge.com/.
[65] F.W. Bliek, A. van den Noort, B. Roossien, R. Kamphuis, J. de Wit, J. van der Velde, M. Eijgelaar, The role of natural gas in smart grids, Journal of Natural Gas Science and Engineering 3 (5) (2011) 608–616.
[66] J.K. Kok, C.J. Warmer, I.G. Kamphuis, PowerMatcher: multiagent control in the electricity infrastructure, in: Proceedings of the Fourth International Joint Conference on Autonomous Agents and Multiagent Systems, July 2005, pp. 75–82.
[67] J. Martí, Overview FENIX findings and recommendations on LSVPP, [Online]. Available: http://www.fenix-project.org/.
[68] J. Jansen, Financial and Socio-economic Impacts of Embracing the FENIX Concept, [Online]. Available: http://www.fenix-project.org/.
[69] D. Gantenbein, B. Jansen, D. Dykeman, P.B. Andersen, E.B. Hauksson, F. Marra, et al., WP3-Distributed Integration Technology Development, [Online]. Available: http://www.edison-net.dk/.
[70] Management summary of the project 'Web2Energy', [Online]. Available: https://www.web2energy.com/en/results/managementsummary/.

[71] S. Horchler, W2e Final Presentation, [Online]. Available: http://www.web2energy.com/.
[72] [Online]. Available: http://app.dimensions.ai.
[73] A. Baringo, L. Baringo, J.M. Arroyo, Day-ahead self-scheduling of a virtual power plant in energy and reserve electricity markets under uncertainty, IEEE Transactions on Power Systems 34 (3) (2018) 1881–1894.
[74] X. Kong, J. Xiao, C. Wang, K. Cui, Q. Jin, D. Kong, Bi-level multi-time scale scheduling method based on bidding for multi-operator virtual power plant, Applied Energy 249 (2019) 178–189.
[75] M. Shafiekhani, A. Badri, M. Shafie-Khah, J.P. Catalão, Strategic bidding of virtual power plant in energy markets: a bi-level multi-objective approach, International Journal of Electrical Power & Energy Systems 113 (2019) 208–219.
[76] W. Guo, P. Liu, X. Shu, Optimal dispatching of electric-thermal interconnected virtual power plant considering market trading mechanism, Journal of Cleaner Production 279 (2021) 123446.
[77] Y. Liu, M. Li, H. Lian, X. Tang, C. Liu, C. Jiang, Optimal dispatch of virtual power plant using interval and deterministic combined optimization, International Journal of Electrical Power & Energy Systems 102 (2018) 235–244.
[78] H. Wang, S. Riaz, P. Mancarella, Integrated techno-economic modeling, flexibility analysis, and business case assessment of an urban virtual power plant with multi-market co-optimization, Applied Energy 259 (2020) 114142.
[79] M. Rahimi, F.J. Ardakani, A.J. Ardakani, Optimal stochastic scheduling of electrical and thermal renewable and non-renewable resources in virtual power plant, International Journal of Electrical Power & Energy Systems 127 (2021) 106658.
[80] T. Sikorski, M. Jasiński, E. Ropuszyńska-Surma, M. Węglarz, D. Kaczorowska, P. Kostyla, et al., A case study on distributed energy resources and energy-storage systems in a virtual power plant concept: technical aspects, Energies 13 (12) (2020) 3086.

VPP and hierarchical control methods

Jinho Kim
Electrical & Computer Engineering Department, Auburn University, Auburn, AL, United States

2.1 Introduction

A virtual power plant (VPP) is a coalition of heterogeneous distributed energy resources (DERs) including energy storage systems (ESSs) that can provide a wide range of ancillary services to power systems. Information and communication technologies are needed for VPP applications to deal with the complexity of increasing DERs at distribution levels and relieve geographical limits of forming a controllable entity as a VPP. With proper communication protocols, a VPP is able to manage DERs physically apart by having a hierarchical control strategy (HCS) that can virtually aggregate the capacities and flexibilities of various DERs within a VPP for the purposes of enhancing power system operation. Thereby, a group of DERs can be a VPP that behaves like a conventional-centralized power plant through an HCS.

An HCS is a concept that can complete the capabilities of a VPP to provide the ancillary services as a centralized dispatchable power plant. The HCS can provide opportunities for VPPs to contribute to primary and secondary controls dealing with other stability and operation aspects of a power system, such as voltage control, frequency control, frequency restoration, power exchange, and resource allocation. These services primarily aim to maintain the instantaneous balance of electrical supply and demand. Essential ancillary services required in power systems are shown in Fig. 2.1. Within a control area, individual distributed power sources—capable of providing some of the services mentioned in Fig. 2.1—can be coordinated to provide a wide range of ancillary services as a single power plant via an HCS. To organize the complexity of having various DERs within a VPP, an HCS may consist of multi-level controllers based on different timescales of ancillary services and applications of distributed power sources with communication protocols.

2.2 Hierarchical control methods

Fig. 2.2 shows an example of an HCS for network assets—DERs—in a control area. The assets can be controlled and coordinated with multi-level controls: device-level control (DLC), plant-level control (PLC), and cluster-level control (CLC). The individual control layers are distinguished by the different timescale of services provided.

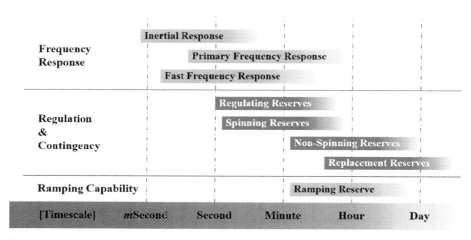

Figure 2.1 Essential ancillary services required in electric power systems [1].

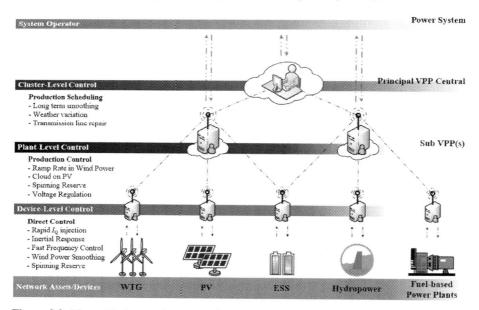

Figure 2.2 Hierarchical control structure for network assets in a control area.

At the DLC, individual network assets, such as DERs and ESSs, are likely to have their own control schemes to directly respond to the sudden and large variation of electrical variables locally measured because increasing penetration of renewable DERs will require DERs to contribute to the primary services for the system stability. In general, these services are required to be automated or even immediate responsive: rapid reactive current (I_Q) injection, emulated inertial response, fast frequency control, wind power smoothing, spinning reserve, etc.

At the PLC, a group of network assets can form a VPP by being coupled via a communication protocol. Then, the network assets can be coordinated to control the production of the VPP and provide ancillary services at a point of common coupling (PCC) or a target point, which is remote from the corresponding assets. Thus, the services provided by the PLC can extend the spectrum of the ancillary services that probably the DERs within a VPP are subject to provide. A PLC will be subjected to provide the services that have relatively slower and longer responses than the DLC provides. The services can be realized by a PLC communicating with the DLCs within the same VPP and assigning them control signals to achieve a certain ancillary service. The PLC can provide ramp rate control under wind power or PV variations, spinning reserve, voltage regulation, etc. At this level, a VPP can directly communicate with the power system to comply with any requirements from the system operator. Or, the VPP can be treated as a sub-VPP—as a unit below a VPP at a higher tier—that is ordered by a principal VPP central at the cluster-level control (CLC).

At the CLC, a group of sub-VPPs can form a principal VPP, central to manage or schedule the collective production from all sub-VPPs for the long-term responses: long-term production smoothing, weather variation, transmission line repair, etc. In order to realize an HCS in a certain control area consisting of various network assets, proper communication protocols should be studied and secured within the control area. Also, the characteristics and applications of the network assets in the area should be analyzed so that they can be properly utilized.

2.2.1 Applications of network assets

Various network assets such as inverter-based resources (IBRs), ESSs, and conventional generators found in distribution systems can be coordinated by forming a VPP to comply with any services required by a system operator. However, the services that can be provided from the VPP depend on the characteristics and applications of the network assets within a VPP. Table 2.1 shows applications of IBRs and ESSs in an electric power system.

Due to the virtue of power electronics, photovoltaic (PV) power plant and type-3 and type-4 wind turbine generators (WTGs) can rapidly respond to the commands from their DLCs or PLC. However, these IBRs are non-dispatchable assets because solar and wind resources are not controllable. Thus, these IBRs are suitable for the provision of short-term ancillary services such as rapid I_Q injection or controlled inertial response.

There are various energy storage technologies from supercapacitor to pumped storage hydropower. Basically, ESSs are dispatchable—depending on the state of charge—and capable of providing primary frequency control by utilizing the energy secured. Even the inverter-interfaced ESSs are capable of responding to system disturbances or command signals from an upper-level control. Also, a large power-to-gas (P2G) infra-system can provide huge-flexible storage within a control area and a wide range of ancillary services in collaboration with inverter-interfaced fuel cell systems and electric vehicles.

Table 2.1 Applications of IBRs and ESSs.

Applications / IBRs & ESSs	PV [2,3]	Type 3&4 WTG [4-6]	Capacitor or Inductor	Super-capacitor ESS [7]	Fly-wheel ESS [8]	Pumped Hydro	Advanced Pumped Hydro [9]	Large Hydro	Large P2G [10]
Annual smoothing of loads, PV, wind, and small hydro									○
Smoothing weather effects: load, PV, wind, and small hydro									○
Weekly smoothing of loads, weather variations									
Daily load cycle, transmission line repair						○	○	○	○
Peak load lopping, standing reserve						○	○	○	○
Spinning reserve, ramp rate control, clouds on PV			○	○	○	○	○	○	○
Wind power smoothing, spinning reserve		○	○	○	○		○		○
Inertial response, fast frequency control		○	○		○		○		
Rapid I_Q injection	○								

← Service Duration →

Thus, the flexibility of a VPP in terms of ancillary services depends on the selection of DERs into a VPP based on applications of them and hierarchical control methods.

2.2.2 Implementation of hierarchical control method

Hierarchical control can be referred to as a structural control that enables the network assets to have a harmonious-collective response at a PCC complying with requests from system operators or optimizing the operation of the network assets as a single VPP or a group of VPPs. Fig. 2.3 shows an example of a power system with a hierarchical control structure. In general, the hierarchical control structure can be divided into three levels: device level, plant level, and cluster level.

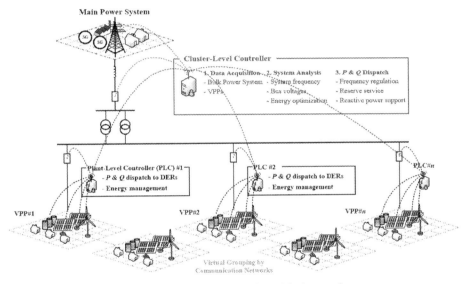

Figure 2.3 An example of a power system with a hierarchical control structure.

The inverter-based DERs can employ an independent control scheme or default control scheme that allows directly responding to the sudden and large variations in the system frequency and local voltages. Typically, these functions can be realized by benchmarking the governor and exciter of conventional synchronous power plants. Also, the DERs at the device level can be grouped—by available communication networks—to be assigned with active and reactive powers from a PLC to provide ancillary services as a single power plant.

At the PLC, the DERs within a VPP can be coordinated via a PLC to provide ancillary services and manage resources within a corresponding VPP. For this, a PLC will need access to the operational data of the DERs and control commands—probably from an external controller at a higher tier or determined by the status of electrical variables measured at a target point. The spectrum of the services to be provided from a PLC depend on the combination of DERs within the VPP.

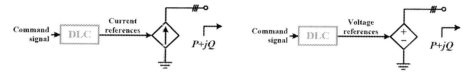

Figure 2.4 Equivalent circuits for inverter-based DERs.

At the cluster level, a group of VPPs can form a unified VPP to be centralized to optimize resources included within the WPP and provide long-term ancillary services: long-term production smoothing, weather variation, transmission line repair, etc. Thus, the DERs, which are forming a VPP, may have opportunities to participate in power markets.

The inverter-based DERs are able to provide the functions that have to be immediately responsive or automated to changes in the system frequency and the voltage at a PCC regarding the power system stability [5–9]. In general, the control functions such as emulated inertial response, fast frequency control, primary frequency control (limited to the assets able to comply with an ordered active power command from an upper-level control system), I_Q injection, power factor correction, and voltage regulation are suitable for being employed in the device level controllers. The inverter-based DERs can be represented by grid-following current or voltage source with the corresponding DLC as shown in Fig. 2.4. For system stability, the DLC can employ frequency and voltage-droop characteristics (only for dispatchable DERs) in Fig. 2.5 to provide automated controls according to the status of the system frequency and local voltage. The droop characteristics form a gain and width of a linear band and its limits. For the frequency-droop characteristic in Fig. 2.4(a), the gain (or slope) can be set as a constant or adaptive value which is above the slope used to be assigned to conventional power plants; obviously, the more DERs are inside a VPP, the greater the slope. Also, the available maximum and minimum powers can be determined by accounting for the headroom of each DER. For the voltage droop characteristic in Fig. 2.4(b), the gain and the limits can be determined in a manner similar to those of the frequency-droop characteristic. The gain can be adaptive according to the request from the system operator or the proportion of the DERs within a VPP. Then, the available reactive powers can be determined by the total headroom for reactive power in the VPP. In addition, the DERs can be centralized by complying with command signals from a PLC.

PLC is a conceptual system that enables the DERs within a VPP to be coordinated and provide the collective output at a PCC or target point through the exchange of data between the PLC and the DLCs. Through a PLC, a group of DERs may largely provide the two services: active power and reactive power controls. Regarding these services, the PLC may enable the group to provide slower but more sophisticated services than individual DLCs do. As a single power plant, the DERs within a VPP can provide active power and reactive power controls through a PLC as shown in Figs. 2.6(a) and 2.6(b). The typical droop characteristics shown in Fig. 2.5 can be employed in the

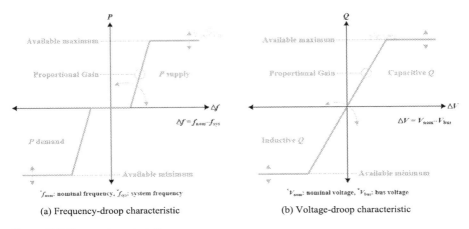

Figure 2.5 Droop characteristics.

PLC for a VPP to provide a collective harmonious response at a PCC or target point. In the PLC, the resultant active command can be distributed to DERs considering various factors: available headroom of each IBR, SOCs of ESSs, and plant-level energy optimization. On the other hand, for the reactive power distribution, an identical-partial voltage set-point can be added to the initial voltage set-point of the IBRs to adjust their reactive power.

The PLC can be ordered by command powers from an upper-level controller for secondary actions or economic scheduling responding to the power market. In addition, the ramp rate of active power at a VPP can be managed by simply adopting a ramp-rate limiter.

2.3 Conclusion

The concept of a VPP has become prominent technology to provide the flexibility required in future power systems with a high penetration level of renewables. The flexibility that a VPP can provide will mainly depend on the development of hierarchical control methods for the VPP.

The hierarchical control method for a VPP is able to simplify the complexity of heterogeneous network assets in terms of active and reactive power controls at the PCC. The virtue of hierarchical control lays in the division of control works at different control layers according to the timescale of target services; thus this conceptual control scheme enables a group of sources to provide a wide range of ancillary services (depending on control strategies at each control layer and the combination of the network assets) systematically. The flexibilities that a VPP can provide will depend on the functions developed at each control layer in an HCS.

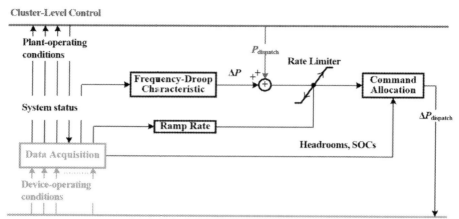

*ΔP: additional power reference, *$P_{dispatch}$: dispatched power from CLC, *$\Delta P_{dispatch}$: distributed power to DLC

(a) Active power control

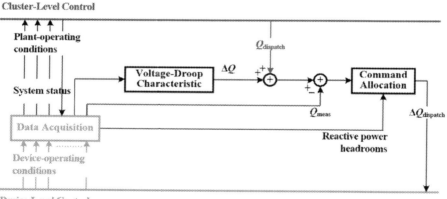

*ΔQ: additional reactive power reference, *$Q_{dispatch}$: dispatched power from CLC, *$\Delta Q_{dispatch}$: distributed power to DLC

(b) Reactive power control

Figure 2.6 An example of a hierarchical control scheme.

References

[1] Paul Denholm, Yinong Sun, Trieu Mai, An Introduction to Grid Services: Concepts, Technical Requirements, and Provision from Wind, Technical Report, NREL, January 2019.
[2] Mohamed Shawky El Moursi, Weidong Xiao, Jim L. Kirtley, Fault ride through the strategy on dynamic characteristics of photovoltaic power plant, IET Generation, Transmission & Distribution 7 (9) (Sept. 2013) 1027–1036.

[3] Mayur Basu, Vincentius Raki Mahindara, Jinho Kim, Robert M. Nelms, Eduard Muljadi, Comparison of active and reactive power oscillation damping with PV plants, IEEE Transactions on Industry Applications (Feb. 2021).
[4] E. Muljadi, V. Gevorgian, M. Singh, S. Santoso, Understanding inertial and frequency response of wind power plants, in: 2012 PEMWA, Denver, CO, USA, Jul. 2012.
[5] D. Yang, J. Kim, Y.C. Kang, E. Muljadi, N. Zhang, J. Hong, S.H. Song, T. Zheng, Temporary frequency support of a DFIG for high wind power penetration, IEEE Transactions on Power Systems 33 (3) (May 2018) 3428–3437.
[6] J. Kim, E. Muljadi, J.W. Park, Y.C. Kang, Flexible IQ–V scheme of a DFIG for rapid voltage regulation of a wind power plant, IEEE Transactions on Industrial Electronics 64 (11) (Nov. 2017) 8832–8842.
[7] J. Kim, V. Gevorgian, Y. Luo, M. Mohanpurkar, V. Koritarov, R. Hovsapian, E. Muljadi, Supercapacitor to provide ancillary services with control coordination, IEEE Transactions on Industry Applications 55 (5) (Sept. 2019) 5119–5127.
[8] Gastón Orlando Suvire, Marcelo Gustavo Molina, Pedro Enrique Mercado, Improving the integration of wind power generation into AC microgrids using flywheel energy storage, IEEE Transactions on Smart Grid 3 (4) (Dec. 2012) 1945–1954.
[9] J. Kim, E. Muljadi, V. Gevorgian, M. Mohanpurkar, Y. Luo, R. Hovsapian, V. Koritarov, Capacity-coordinated frequency control scheme of a virtual power plant with renewable energy sources, IET Generation, Transmission & Distribution 13 (16) (Jul. 2019) 3642–3648.
[10] R. Hovsapian, Role of Electrolyzers in Grid Services, Washington, DC. Available at https://www.hydrogen.energy.gov/pdfs/htac_may17_08_hovsapian.pdf, May 2017.

Bidding strategy in the electricity market

Morteza Shafiekhani[a], Ali Zangeneh[b], and Farshad Khavari[b]
[a]Department of Electrical Engineering, Faculty of Engineering, Pardis Branch, Islamic Azad University, Pardis, Tehran, Iran, [b]Electrical Engineering Department, Shahid Rajaee Teacher Training University, Tehran, Iran

3.1 Introduction

In the past, the power system had two main features, one being the centralized control of production units and the other being the inactive or passive distribution network. With advances in renewable resource technologies and the growing need to utilize such resources, distributed generation resources have replaced a significant portion of the energy produced by conventional power plants. It should be noted that the provision of these services by conventional production units requires large investments in both the transmission and distribution sectors. Therefore, in order to make the most of distributed energy sources, a new model has been proposed for the operation of power devices based on the joint operation of distributed generation units and variable loads that can replace conventional power plants. The new operation model for distributed energy resources enables interaction with the transmission grid and participation in energy markets and network management. On the other hand, it is predicted that future distribution networks will receive thousands of distributed generation units and variable loads, so there is no possibility of communication between these units and the transmission system operator, which is necessary, because the cost of establishing these connections is very high. Instead, decentralized management based on the aggregation of a number of distributed energy sources can reduce the potential problems and operating complexities of the system, so the concept is called virtual power plant (VPP). In general, a virtual power plant is defined as a set of controllable production resources, uncontrollable production resources, variable loads, and storage resources that act together as a unit [1].

3.2 Virtual power plant

The main purpose of a virtual power plant is to provide access to distributed energy sources (renewable and non-renewable) with low power range to the energy market. Due to random changes in the energy produced by renewable sources such as wind and solar, the risk of participation of a single unit in the energy market is very high;

indeed, if the power is not provided, this unit has to buy power at a high price from the balancing markets.

Another problem with distributed energy sources is their low production capacity to participate in the energy market because participating in energy or ancillary services markets requires having a minimum exchange capacity. To solve these problems, the virtual power plant combines several distributed energy units into one cluster. Therefore, the existence of several production units together causes the production to be balanced, or in other words to reduce the existing uncertainties related to each distributed production unit. Now a group of distributed energy sources together have access to different energy markets such as the day-ahead energy market, balancing market, and ancillary services market. There are two kinds of VPPs, technical and commercial.

3.2.1 Commercial VPP

A commercial virtual power plant generally has a characteristic and aggregate output that represents the characteristics and costs of its constituent units. In this model, the effect of distribution networks on the characteristics of the aggregated virtual power plant is not considered. Services and activities of this model include exchange with wholesale market, balancing, signing bilateral contracts [2].

3.2.2 Technical VPP

A technical virtual power plant includes a number of distributed energy sources located in a geographical area. This model, like the commercial model, has a feature and aggregate output that represents the characteristics and costs of its constituent units, but this model includes the real-time effect of distribution network on the aggregate characteristic of the units. The scope of activity of this model includes local system management for the distribution system operator by providing ancillary and balancing services in the distribution system [3].

3.2.3 Modeling of VPP

In terms of modeling, the components of a virtual power plant can include four general types: controllable or conventional units, variable loads, storage equipment, and stochastic or uncontrollable production units consisting of renewable sources. Virtual power plant production units can use renewable and non-renewable energy sources. The primary goal of a virtual power plant is to coordinate the production and consumption of its units in order to increase efficiency and optimal performance. Controllable units in a virtual power plant are generally small production units such that their fuel is mostly from fossilized resources. Storage equipment can store energy in one hour and drain it in other hours for better use. Uncontrollable generating units, also known as stochastic generation sources, use renewable sources such as wind and solar to generate electricity. Therefore, the output of these resources has many uncertainties, so, due to the uncontrollable nature of these resources, units or support capacity must be available to compensate for any deviation of these resources from the expected production. Among the renewable sources, wind turbine power devices have special importance and the widespread use of wind generators indicates the economic viability of this technology in windy areas. But participation of a wind farm in the electricity market has various risks and uncertainties, including costs due to imbalances in production and failure to meet the obligations of these resources. Therefore, to reduce the mismatch between the amount of committed energy and the amount of energy produced by the wind farm or a virtual power plant in general can be combined with a controllable unit. One of the leading ways to jointly use renewable production resources with conventional power plants with high response speed is to cover the output fluctuations of renewable production resources. Another way is to use one or more storage units. This method, while making renewable production resources flexible, leads to the transfer of energy exchange from low- to high-price hours. In general, the simultaneous use of different resources to participate in the electricity market leads to better performance than in the case of separate producers.

3.3 Optimal bidding of VPP in electricity market

The optimal bidding strategy problem of a virtual power plant is a little more complicated than in the case of GenCos. Because the virtual power plant is made up of different components and hence it is a bit difficult to plan.

In this section, the studies conducted on optimal bidding strategy of virtual power plants in the market will be reviewed. The optimal bidding problem of VPP in the market can be examined from many aspects. In this chapter, the focus will be on single-level, bi-level and tri-level models.

3.3.1 Optimal bidding strategy of VPP based on single level model

In this model, the virtual power plant has an objective function and seeks to optimize it. Generally, this function represents the profit of the virtual power plant. As it is known, one of the important features of the virtual power plant is the ability to participate in the energy market, so the owners of these power plants can move in order to maximize their profits by determining the optimal bidding strategy to the network. More details about single level model of VPP are presented in Chapter 17 of this book. References [4] and [5] have examined the optimal bidding problem by a virtual power plant in a joint energy and spinning reserve markets. The proposed non-equilibrium model is based on the price-based unit commitment. This model considers the demand–supply equilibrium constraint and the constraints of the virtual power plant itself. The optimal bidding problem of a virtual power plant with a wind farm, a photovoltaic unit and a gas unit is presented in [6]. The problem is expressed as an integer complex linear programming model. VPP participates in day-ahead market and has long-term bilateral contracts. A new algorithm that optimizes the scheduling of thermal and electrical units of a large virtual power plant is presented in [7]. This large virtual power plant has small production units, small consumption units, storage units, and combined heat and power units. This model does not take into account any uncertainties and is solved by a mixed-integer linear programming model. Reference [8] has used optimal bidding strategy of VPP in day-ahead market and studied uncertainties by use of point estimate method. In this reference, purchasing and selling is carried out in the day-ahead market and the main goal is to maximize the expected profit of the virtual power plant by participating in the day-ahead market and meeting the demand of controllable loads. Network constraints are also included in the model. The programming problem is a mixed-integer nonlinear programming model. In [9], a stochastic scheduling problem for a virtual power plant is modeled to meet the thermal and electrical load considering the network security constraints and uncertainties of electrical and thermal loads, wind speed, solar radiation, and market price. Reference [10] presents a new virtual power plant (VPP) model that integrates all available full-scale distributed renewable generation technologies. The proposed VPP operates as a single plant in the wholesale electricity market and aims to maximize profit from its operation to meet demand. In [11], an optimal bidding strategy of a VPP participating in the day-ahead frequency regulation market (FRM) and the energy market (EM) is proposed.

3.3.2 Optimal bidding strategy of VPP based on bi-level model

In all single-level research in the field of virtual power plants, this set of units participates as a price-taker unit in the electricity market. This means that the previous virtual power plants do not have the ability to change the market clearing price to their advantage because their output power is small compared to the capacity of other conventional power plants. The bidding model for a price-taker producer in the electricity market is different from the model of a price-maker because the price-maker producer has the ability to influence market prices, or in other words has market power. On the other hand, the market clearing price can affect the amount of cleared power for the price-maker producer. To solve this dependence, the problem can be formulated as a bi-level model. As this chapter is based on the bi-level model, more details about this model and its formulations are represented in Section 3.4.

At the upper level of the model, profit maximization of the price-maker producer is expressed, and at the lower level the market clearing process is stated. The lower-level problem can be replaced by first-order optimality conditions such as Karush–Kuhn–Tucker conditions [12]. Therefore, the problem of elementary bi-level optimization becomes a one-level problem and it is called a mathematical program with equilibrium constraints [13]. The mathematical program with equilibrium constraints to obtain an optimal bidding strategy for conventional units has been highly regarded. The model used in this chapter is based on references [14] and [15] and the studied virtual power plant is of commercial type, so the network structure is not modeled. Contrary to this chapter, in references such as [3], [16], and [17] the virtual power plant is technical and is connected to the upstream network through several busbars. The model used in these references is different from the model used in this chapter.

VPP in this chapter is commercial. The only similarity between these references is that they have a bi-level model in which the upper-level problem refers to the profit of the virtual power plant and the lower-level problem refers to the social welfare.

Reference [14] presents a bi-level tri-stage optimization model to improve the bidding strategy of a commercial virtual power plant in the day-ahead market, which includes wind farms, storage devices, and residential or commercial demands, and its goal is profit maximization. The uncertainties are caused by wind farm production, consumer demand, competitors' cost curves, and balancing market prices. The upper level of the proposed model includes the internal scheduling of virtual power plants in order to maximize profits, and the lower level represents day-ahead market clearing process by the independent system operator. In other words, this level is the social welfare problem. This level is modeled linearly, so it can be said that it is convex, so it is possible to replace it with first order KKT conditions. The resulting model is a mathematical model with equilibrium constraints that can be transformed into an integer complex linear programming model form by the KKT optimization conditions and strong duality theory. In [15], a novel two-stage robust Stackelberg game is proposed to solve the problem of day-ahead energy management for aggregate prosumers considering the uncertainty of intermittent renewable energy output and market price. Reference [16], with the aim of promoting DERs role in energy markets, proposes a new framework for the optimal bidding strategy of a technical virtual power plant (TVPP) in different markets. Reference [17] presents a two-layer energy management model (EMM) in the smart distribution network (SDN) considering flexi-renewable virtual power plants (FRVPPs) that participate in the day-ahead energy and reserve markets. In [18], considering the VPP trading environment, a bi-level optimization model is established to determine ESS's optimal location and capacity.

3.3.3 Optimal bidding strategy of VPP based on tri-level model

This model has been implemented for wind units in several ways, but the problem of finding an optimal bidding strategy of a virtual power plant based on the tri-level model has been little investigated. In this model, the virtual power plant, in addition to having strategic behavior in the day-ahead market, also has strategic behavior in the balancing market. In other words, the market clearing process takes place once in the day-ahead market and once in the balancing market. Reference [19] proposes a model to find the equilibria in the short-term electricity market with large-scale wind power penetration. The behavior of each strategic player is modeled through a three-stage mathematical problem with equilibrium constraints (MPEC) where the upper-level problem maximizes the profit of the strategic player, the first lower-level problem describes the clearing process of the day-ahead market, and the second lower-level problem describes the clearing process of the real-time market while considering the network constraints. Reference [20] proposes a stochastic optimization model to obtain the optimal bidding strategy for a strategic wind power producer in the short-term electricity market. The upper level problem of the model maximizes the profit of the wind power producer, while the first lower level problem represents the market clearing processes of the day-ahead and the second one for real-time market.

3.4 Hypotheses and problem objectives

The studies for this chapter are based on the assumption that the virtual power plant is able to offer power in the day-ahead and balancing markets. Therefore, the main purpose of this chapter is to obtain an optimal bidding model for the virtual power plant with the aim of maximizing the profit from the sale of power to the day-ahead and balancing markets. The proposed model is explained by a bi-level model. The first level is related to maximizing the profit of the virtual power plant and the lower-level is a market-clearing problem with the aim of maximizing the level of social welfare.

3.4.1 Problem formulation

The day-ahead electricity market is one of the most common models of daily trade in competitive electricity markets. In such markets, producers and consumers of power are faced with the fundamental problem of bidding and offering to sell or buy from the market on a daily basis in order to increase their profits or reduce costs. The pool market is cleared once a day for the day ahead on an hourly basis. So a producer or consumer of power to obtain his optimal bidding strategy must use an appropriate optimization model taking into account the technical and economic constraints of its units. These constraints for a virtual power plant include forecasting the amount of power generation by renewable units, constraints for start-up and shut-down cost of units, minimum up- and down-time of conventional generation units, and existing relationships for planning storage units according to the constraints on these units. In order to participate in the day-ahead market, the virtual power plant must decide on both the amount of production capacity and the bid price because these two quantities must be presented to the market operator a day before. However, the virtual power plant does not have accurate information about the actual amount of production available at the time of delivery. The main purpose of this chapter is to provide a suitable method for deciding on the amount of power and price offered to the network in order to maximize profits. By participating in the balancing market, the virtual power plant can buy or sell the high deviations of its production from the set amount in the day-ahead market. It should also be noted that most of the exchange power with the network is done in the day-ahead market, so the main focus of the virtual power plant will be on price changes in the day-ahead market.

3.4.2 The main objectives

A bi-level problem for optimizing bidding strategy by a virtual power plant over a 24-hour period is presented. In this chapter, the virtual power plant is assumed to be a price maker. Therefore, this possibility is provided to the virtual power plant so that it can increase its profit by using the market conditions by changing the clearing price. Balancing market prices are forecasted for each hour. It should be noted that the main purpose of the virtual power plant is to increase profits in the day-ahead market and at the same time reduce the costs associated with purchasing from the balancing market.

Bidding strategy in the electricity market

It should be noted that the virtual power plant operates only strategically in the day-ahead market and participating in the balancing market is only for compensating for its deviations by purchasing or selling in this market because the largest volume of transactions takes place in the day-ahead market.

3.4.3 Bi-level problem

The problem of an optimal bidding strategy of VPP is expressed in the following bi-level model. In the first level, the profit maximization model of the virtual power plant is formulated and stated as follows:

Maximize

$$\sum_{t=1}^{T}(\sum_{b_v=1}^{B_v} \Omega_{tb_v}^{DA}.P_{tb_v}^{DA} + \Omega_t^{Bal}\left(dn_{reg}.P_t^{dn} - up_{reg}.P_t^{up}\right) - Cost_t^{cu}) \qquad (3.1)$$

subject to:

$$\sum_{b_v=1}^{B_v} P_{tb_v}^{DA} + P_t^{Ch} + P_t^{dn} = P_t^{wind} + P_t^{solar} + P_t^{up} + P_t^{VPPu} + P_t^{Dch}. \qquad (3.2)$$

The upper-level problem is represented by Eqs. (3.1)–(3.14) and the lower-level problem is represented by Eqs. (3.15)–(3.19). The dual variables related to the lower level constraints are shown after the colon opposite to each constraint. The objective function (3.1), which should maximize the expected profit, includes the following parts:

- The expression $\Omega_{tb_v}^{DA}.P_{tb_v}^{DA}$ represents the revenue of the virtual power plant resulting from the sale of power to the day-ahead market, which is equal to the product of the power sold at the marginal price of the node that the virtual power plant in it is located and which is obtained at the lower level after clearing the market. Nodal marginal prices are obtained from Eq. (3.19).
- The term $\Omega_t^{Bal}\left(dn_{reg}.P_t^{dn} - up_{reg}.P_t^{up}\right)$ represents the revenue and cost of a virtual power plant resulting from the sale and purchase of energy in the balancing market. This exchange is done due to the deviation of the amount of production from the predicted amount by the renewable units and the uncertainties related to them to compensate for the lack of production or to sell excess energy to the grid.

The term $Cost_t^{cu}$ is used to describe the start-up cost and operating cost of a conventional unit of VPP.

Eq. (3.2) indicates the power balance inside the virtual power plant. We also have the following constraints:

$$s_t^{VPPu} - s_{t-1}^{VPPu} \le v_t^{VPPu}, \qquad (3.3)$$

$$P_t^{VPPu} = \sum_{j=1}^{J} P_{tj}^{VPPu}, \qquad (3.4)$$

$$Cost_t^{cu} = SUC_t^{VPPu}.v_t^{VPPu} + SDC_t^{VPPu}.k_t^{VPPu} + c_0.s_t^{VPPu}$$
$$+ \sum_{j=1}^{J} \Omega_j^{VPP} P_{tj}^{VPPu}, \tag{3.5}$$
$$P_{Min}^{VPPu}.s_t^{VPPu} \leq P_t^{VPPu} \leq P_{Max}^{VPPu}.s_t^{VPPu}, \tag{3.6}$$
$$-RD^{VPPu} \leq P_t^{VPPu} - P_{t-1}^{VPPu} \leq RU^{VPPu}. \tag{3.7}$$

The binary variable s_t^{VPPu} will have a value of 1 if the conventional unit is on at time t, otherwise it will have a value of 0. The binary variable v_t^{VPPu} will have a value of 1 if the conventional unit is started in time t, otherwise it will have a value of 0. Constraint (3.3) shows the logical relation between these binary variables. Constraint (3.4) expresses the equality of the output power of a conventional unit with the sum of the production capacities in each block of j blocks of a conventional unit. Eq. (3.5) shows the cost of the conventional unit of a virtual power plant (start-up and shut-down cost and operating cost as a linear approximation). Constraint (3.6) expresses the output limits of a conventional unit with respect to the maximum and minimum capacity, and constraint (3.7) represents the ramp rate of these units. Further, we need

$$\sum_{t=1}^{MUP} s_t^{VPPu} - 1 \geq MUP^{VPPu}, \forall v_t^{VPPu} = 1, \tag{3.8}$$
$$\sum_{t=1}^{MDN} 1 - s_t^{VPPu} \geq MDN^{VPPu}, \forall k_t^{VPPu} = 1. \tag{3.9}$$

Constraints (3.8) and (3.9) indicate the minimum up- and down-time of the unit, respectively.

3.4.4 Energy storage equations

Storages act as consumers in the charge mode and as producers in the discharge mode. Their equations are Eqs. (3.10)–(3.14). Eq. (3.10) shows the amount of energy stored in storages, which depends on the energy of the previous hour and the amount of charge and discharge of the storage. The charging and discharging limits of storages are modeled according to Eqs. (3.11) and (3.12). Eq. (3.13) shows the minimum and maximum charging limits of storages. Eq. (3.14) indicates that charging and discharging do not occur simultaneously [21]:

$$E_t^S = E_{t-1}^S + P_t^{Ch} - P_t^{Dch}, \tag{3.10}$$
$$P_t^{Ch} \leq P^{ChMax}.\alpha_t^{Ch}, \tag{3.11}$$
$$P_t^{Dch} \leq P^{DchMax}.\alpha_t^{Dch}, \tag{3.12}$$
$$E_t^{SMin} \leq E_t^S \leq E_t^{SMax}, \tag{3.13}$$
$$\alpha_t^{Ch} + \alpha_t^{Dch} \leq 1. \tag{3.14}$$

The lower-level problem is introduced by Eqs. (3.15)–(3.19), and the objective function of the lower-level problem is to maximize the social welfare or reduce the overall costs of the system. In other words, the lower-level problem describes the market clearing process. The lower-level problem is linear because from the market operator's point of view, λ_{ti}^{I} is considered constant. The lower-level objective function, which should minimize system costs, includes the following:

$$\text{Minimize} \sum_{t=1}^{T} \left[\sum_{r=1}^{R} \sum_{b=1}^{B} \Omega_{trb}^{R} P_{trb}^{R} + \sum_{b_v=1}^{B_v} \Omega_{tb_v}^{I} P_{tb_v}^{DA} - \sum_{l=1}^{L} \sum_{bl=1}^{BL} \Omega_{tlbl}^{L} P_{tlbl}^{L} \right] \quad (3.15)$$

subject to:

$$0 \leq P_{trb}^{R} \leq P_{trb}^{R\,max} : \partial_{trb}^{R\,min}, \partial_{trb}^{R\,max}, \quad (3.16)$$

$$0 \leq P_{tb_v}^{DA} \leq P_{tb_v}^{I\,max} : \mu_{tb_v}^{I\,min}, \mu_{tb_v}^{I\,max}, \quad (3.17)$$

$$0 \leq P_{tlbl}^{L} \leq P_{tlbl}^{L\,max} : \eta_{tlbl}^{L\,min}, \eta_{tlbl}^{L\,max}, \quad (3.18)$$

$$\sum_{r=1}^{R} \sum_{b=1}^{B} P_{trb}^{R} + \sum_{b_v=1}^{B_v} P_{tb_v}^{DA} - \sum_{l=1}^{L} \sum_{bl=1}^{BL} P_{tlbl}^{L} = 0 : \lambda_{to}^{DA}. \quad (3.19)$$

The term $\Omega_{trb}^{R} P_{trb}^{R}$ refers to bidding of rivals in the day-ahead market. In this regard, P_{trb}^{R} is the cleared power for each producer in the day-ahead market, which is one of the problem variables; Ω_{trb}^{R} is the parameter related to the price bid by the producers to the day-ahead market. In this case, the bid price is equal to the marginal costs of each producer. The term $\Omega_{tb_v}^{I} P_{tb_v}^{DA}$ refers to the VPP biddings in the day-ahead market, which is strategic; $P_{tb_v}^{DA}$ is the cleared power in the day-ahead market for the virtual power plant. The term $\Omega_{tb_v}^{I}$ specifies the price bid by the virtual power plant to the day-ahead market. The term $\Omega_{tlbl}^{L} P_{tlbl}^{L}$ refers to the offer prices of the demand in the network in the day-ahead market. In this regard, P_{tlbl}^{L} is a variable related to the network load's power; Ω_{tlbl}^{L} is the parameter related to the price offered by demand to the day-ahead market. In this case, the offer price is equal to the marginal costs of each demand. Eqs. (3.16)–(3.18) show the power limits of the rivals, VPP, and demand in the network, respectively. Eq. (3.19) shows the power balance limit in the system. This equation guarantees that the injected power minus the powers taken from the system is zero. The dual variable related to the power balance equation of the whole system, $\Omega_{tb_v}^{DA}$, is the marginal price of the system, which indicates an increase in the cost ($/h) of hourly performance of the system to provide 1 MW of additional load consumption. Since the increase in demand must be provided by the marginal producer, which is also located in the system reference bus, the marginal price of the system is equal to the marginal price of the reference bus or $\Omega_{tb_v}^{DA}$.

3.4.5 Formulation of mathematical programming problem with equilibrium Constraint (MPEC)

One common way to solve bi-level optimization problems is to turn it into a single-level problem. Due to the linearity and convexity of the objective function of the lower-level (market clearing problem), the lower problem can be replaced with the first-order KKT optimally conditions:

Maximize $UL \cup LL$ (3.20)

subject to:

(3.2)–(3.15) and (3.17)–(3.20), and (3.21)

$$\Omega_{trb}^{R} - \Omega_{to}^{DA} + \partial_{trb}^{Rmax} - \partial_{trb}^{Rmin} = 0, \quad (3.22)$$

$$\Omega_{tb_v}^{I} - \Omega_{to}^{DA} + \mu_{tb_v}^{Imax} - \mu_{tb_v}^{Imin} = 0, \quad (3.23)$$

$$-\Omega_{tlbl}^{L} + \Omega_{to}^{DA} + \eta_{tlbl}^{Lmax} - \eta_{tlbl}^{Lmin} = 0, \quad (3.24)$$

$$0 \leq \partial_{trb}^{Rmax} \perp P_{trb}^{Rmax} - P_{trb}^{P} \geq 0, \quad (3.25)$$

$$0 \leq \partial_{trb}^{Rmin} \perp P_{trb}^{R} \geq 0, \quad (3.26)$$

$$0 \leq \mu_{tb_v}^{Imax} \perp P_{tb_v}^{Imax} - P_{tb_v}^{DA} \geq 0, \quad (3.27)$$

$$0 \leq \mu_{tb_v}^{Imin} \perp P_{tb_v}^{DA} \geq 0, \quad (3.28)$$

$$0 \leq \eta_{tlbl}^{Lmax} \perp P_{tlbl}^{Lmax} - P_{tlbl}^{L} \geq 0, \quad (3.29)$$

$$0 \leq \eta_{tlbl}^{Lmin} \perp P_{tlbl}^{L} \geq 0. \quad (3.30)$$

3.4.6 Mixed-integer linear programming

The problem of mathematical programming with equilibrium constraints expressed in Eqs. (3.20)–(3.30) has two nonlinear parts:

1. The expression $\Omega_{tb_v}^{DA} P_{tb_v}^{DA}$ in the objective function (3.20),
2. Constraints for complementary conditions (3.25)–(3.30).

The expression $\Omega_{tb_v}^{DA} P_{tb_v}^{DA}$ is linearized using the strong duality theory and some other mathematical equations. As stated earlier, according to the strong duality theory, the objective functions of the primary and dual problems have equal values at the optimal point. Therefore, for the dual form of the lower-level problem, we have:

$$\sum_{t=1}^{T} \left[\sum_{r=1}^{R} \sum_{b=1}^{B} \Omega_{trb}^{R} P_{trb}^{R} + \sum_{b_v=1}^{B_v} \Omega_{tb_v}^{I} P_{tb_v}^{DA} - \sum_{l=1}^{L} \sum_{bl=1}^{BL} \Omega_{tlbl}^{L} P_{tlbl}^{L} \right]$$

$$= -\sum_{r=1}^{R} \sum_{b=1}^{B} \partial_{trb}^{Rmax} P_{trb}^{Rmax} - \mu_{tb_v}^{Imax} P_{tb_v}^{Imax} - \sum_{l=1}^{L} \sum_{bl=1}^{BL} \eta_{tlbl}^{Lmax} P_{tlbl}^{Lmax}. \quad (3.31)$$

Bidding strategy in the electricity market

Using Eq. (3.24), we have

$$\Omega^{I}_{tb_v} = \Omega^{DA}_{to} - \mu^{I max}_{tb_v} + \mu^{I min}_{tb_v}. \tag{3.32}$$

And then by multiplying the above equation by $P^{DA}_{tb_v}$, we have

$$\Omega^{I}_{tb_v} \cdot P^{DA}_{tb_v} = P^{DA}_{tb_v} \cdot \left(\Omega^{DA}_{to} - \mu^{I max}_{tb_v} + \mu^{I min}_{tb_v} \right). \tag{3.33}$$

Now the above equations can be added and converted as follows to be similar to the expression in the lower-level objective function:

$$\sum_{b_v=1}^{B_v} \Omega^{I}_{tb_v} P^{DA}_{tb_v} = \sum_{b_v=1}^{B_v} P^{DA}_{tb_v} \cdot \left(\Omega^{DA}_{to} - \mu^{I max}_{tb_v} + \mu^{I min}_{tb_v} \right). \tag{3.34}$$

Similarly, the following equations can be obtained using Eqs. (3.27) and (3.28):

$$\sum_{b_v=1}^{B_v} \mu^{I max}_{tb_v} P^{I max}_{tb_v} = \sum_{b_v=1}^{B_v} \mu^{I max}_{tb_v} P^{DA}_{tb_v}. \tag{3.35}$$

And also

$$\sum_{b_v=1}^{B_v} \mu^{I min}_{tb_v} P^{DA}_{tb_v} = 0. \tag{3.36}$$

Using Eqs. (3.35) and (3.36), we can simplify Eq. (3.34), and then

$$\sum_{b_v=1}^{B_v} \Omega^{I}_{tb_v} P^{DA}_{tb_v} = \sum_{b_v=1}^{B_v} \Omega^{DA}_{to} P^{DA}_{tb_v} - \sum_{b_v=1}^{B_v} \mu^{I max}_{tb_v} P^{I max}_{tb_v}. \tag{3.37}$$

Now by inserting the expression $\sum_{b_v=1}^{B_v} \Omega^{DA}_{to} P^{DA}_{tb_v}$ into the lower-level objective function (3.15), we have

$$\sum_{t=1}^{T} \left[\sum_{r=1}^{R} \sum_{b=1}^{B} \Omega^{R}_{trb} P^{R}_{trb} + \sum_{b_v=1}^{B_v} \Omega^{DA}_{tb_v} P^{DA}_{tb_v} - \sum_{b_v=1}^{B_v} \mu^{I max}_{tb_v} P^{I max}_{tb_v} \right.$$

$$\left. - \sum_{l=1}^{L} \sum_{bl=1}^{BL} \Omega^{L}_{tlbl} P^{L}_{tlbl} \right]$$

$$= \sum_{t=1}^{T} \left[-\sum_{r=1}^{R} \sum_{b=1}^{B} \partial^{R min}_{trb} P^{R max}_{trb} - \sum_{b_v=1}^{B_v} \mu^{I max}_{tb_v} P^{I max}_{tb_v} - \sum_{l=1}^{L} \sum_{bl=1}^{BL} \eta^{L max}_{tlbl} P^{L max}_{tlbl} \right]. \tag{3.38}$$

Now, using Eq. (3.38), we can obtain

$$\sum_{t=1}^{T}\sum_{b_v=1}^{B_v} \Omega_{tb_v}^{DA} P_{tb_v}^{DA} = -\sum_{t=1}^{T}\left[\sum_{r=1}^{R}\sum_{b=1}^{B} \Omega_{trb}^{R} P_{trb}^{R} - \sum_{l=1}^{L}\sum_{bl=1}^{BL} \Omega_{tlbl}^{L} P_{tlbl}^{L} \right.$$
$$\left. -\sum_{r=1}^{R}\sum_{b=1}^{B} \partial_{trb}^{R^{min}} P_{trb}^{R^{max}} - \sum_{l=1}^{L}\sum_{bl=1}^{BL} \eta_{tlbl}^{L^{max}} P_{tlbl}^{L^{max}} \right]. \quad (3.39)$$

Eq. (3.39) makes it possible to compute the expression $\sum_{t=1}^{T}\sum_{b_v=1}^{B_v} \Omega_{tb_v}^{DA} P_{tb_v}^{DA}$ as a linear equation, and also the nonlinear expression $\mu_{tb_v}^{I^{max}} P_{tb_v}^{I^{max}}$ in constraints (3.31) and (3.37) was finally removed from this equation. On the other hand, the nonlinear constraints in the complementary terms must be rewritten using the equivalent linear form obtained from the Fortuny-Amat method. The method of linearization of complementary conditions is given in the Appendix. Finally, using the stated linear equations, the problem of optimal bidding strategy of a virtual power plant is expressed in the form of mixed-integer linear programming as follows:

Maximize $UL \cup LL \cup AUX$ \quad (3.40)

$$\sum_{t=1}^{T}\left[-\sum_{r=1}^{R}\sum_{b=1}^{B} \Omega_{trb}^{R} P_{trb}^{R} + \sum_{l=1}^{L}\sum_{bl=1}^{BL} \Omega_{tlbl}^{L} P_{tlbl}^{L} - \sum_{r=1}^{R}\sum_{b=1}^{B} \partial_{trb}^{R^{min}} P_{trb}^{R^{max}} \right.$$
$$\left. -\sum_{l=1}^{L}\sum_{bl=1}^{BL} \mu_{tlbl}^{L^{max}} P_{tlbl}^{L^{max}} + \Omega_t^{Bal}\left(dn_{reg}.P_t^{dn} - up_{reg}.P_t^{up}\right) - Cost_t^{cu}\right]$$

subject to:

(3.2)–(3.15), (3.17)–(3.20) and (3.23)–(3.25), as well as \quad (3.41)

$$0 \leq \partial_{trb}^{R^{max}} \leq M_\partial^{max} u_{trb}^{max}, \quad (3.42)$$

$$P_{trb}^{R^{max}} - P_{trb}^{R} \leq M_\partial^{max}\left(1 - u_{trb}^{max}\right), \quad (3.43)$$

$$0 \leq \partial_{trb}^{R^{min}} \leq M_\partial^{min} u_{trb}^{min}, \quad (3.44)$$

$$P_{trb}^{R} \leq M_\partial^{min}\left(1 - u_{trb}^{min}\right), \quad (3.45)$$

$$0 \leq \mu_{tb_v}^{I^{max}} \leq M_\mu^{max} u_{tb_v}^{max}, \quad (3.46)$$

$$P_{tb_v}^{I^{max}} - P_{tb_v}^{DA} \leq M_\mu^{max}\left(1 - u_{tb_v}^{max}\right), \quad (3.47)$$

$$0 \leq \mu_{tb_v}^{I^{min}} \leq M_\mu^{min} u_{tb_v}^{min}, \quad (3.48)$$

$$P_{tb_v}^{DA} \leq M_\mu^{min}\left(1 - u_{tb_v}^{min}\right), \quad (3.49)$$

$$0 \leq \eta_{tlbl}^{L^{max}} \leq M_\eta^{max} u_{tlbl}^{max}, \quad (3.50)$$

$$P_{tlbl}^{R^{max}} - P_{tlbl}^{L} \leq M_\eta^{max}\left(1 - u_{tlbl}^{max}\right), \quad (3.51)$$

$$0 \leq \eta_{tlbl}^{L^{min}} \leq M_\eta^{min} u_{tlbl}^{min}, \quad (3.52)$$

Bidding strategy in the electricity market

$$P_{tlbl}^{L} \leq M_{\eta}^{min}\left(1 - u_{tlbl}^{min}\right), \tag{3.53}$$

$$u_{trb}^{max}, u_{trb}^{min}, u_{tb_v}^{max}, u_{tb_v}^{min}, u_{tlbl}^{max}, u_{tlbl}^{min}, s_t^{VPPu}, v_t^{VPPu}, k_t^{VPPu} \in \{0, 1\}. \tag{3.54}$$

Constraints on the volume of power exchanged in the balancing market are modeled using constraints (3.55)–(3.56).

$$0 \leq P_t^{dn} \leq P_t^{dn,max}, \tag{3.55}$$

$$0 \leq P_t^{up} \leq P_t^{up,max}, \tag{3.56}$$

$$\sum_{b_v=1}^{B_v} P_{tb_v}^{DA} + P_t^{Ch} = P_t^{wind} + P_t^{solar} + P_t^{Bal} + P_t^{VPPu} + P_t^{Dch}. \tag{3.57}$$

3.5 Case study

This virtual power plant is of commercial type and has a wind farm, solar farm, thermal unit, and storage. There are eight power plants as rivals with the specifications presented in Table 3.1 and four large demands with the specification presented in Tables 3.2 and 3.3. It should be noted that bid prices are marginal cost amounts associated with existing participants [22].

Table 3.1 Rival unit characteristics

Units	Generation Units								
	Offer size (MW)				Offer price ($/MWh)			
	b_1	b_2	b_3	b_4	Ω_{b1}	Ω_{b2}	Ω_{b3}	Ω_{b4}
R_1	2.40	3.40	3.60	2.40	23.41	23.78	26.84	30.40
R_2	54.25	38.75	31	31	9.92	10.25	10.68	11.26
R_3	25	25	20	20	18.6	20.03	21.67	22.72
R_4	15.2	22.8	22.8	15.2	11.46	11.96	13.89	15.97
R_5	140	97.5	52.5	70	19.20	20.32	21.22	22.13
R_6	15	15	10	10	0	0	0	0
R_7	100	100	120	80	5.31	5.38	5.53	5.66
R_8	68.95	49.25	39.4	39.4	10.08	10.66	11.09	11.72

Information about the storage and conventional unit is given in Tables 3.4 and 3.5, respectively.

Fig. 3.1 shows the balancing market, as well as up- and down-regulation prices. The up- and down-regulation market prices are considered to be 20% higher and lower than the balancing market price, respectively.

By solving the bi-level optimal bidding problem of the virtual power plant, outputs will be stated below. Fig. 3.2 shows the day-ahead market clearing price. The price blocks offered by the virtual power plant, rivals, and demands are effective in creating this clearing price in the day-ahead market.

Table 3.2 Demands power blocks.

No.	Demand				
	Bid size (MW)				
	bl_1	bl_2	bl_3	bl_4	bl_5
L_1	171	4.75	4.75	4.75	4.75
L_2	243	6.75	6.75	6.75	6.75
L_3	243	6.75	6.75	6.75	6.75
L_4	243	6.75	6.75	6.75	6.75

Table 3.3 Demands price blocks.

t	Ω_{bl1}	Ω_{bl2}	Ω_{bl3}	Ω_{bl4}	Ω_{bl5}
1	17.430	17.25	17.216	16.886	16.79
2	17.25	17.216	16.886	16.79	16.38
3	17.216	16.886	16.79	16.38	16.32
4	17.216	16.886	16.79	16.38	16.32
5	16.886	16.79	16.38	16.32	16.13
6	16.886	16.79	16.38	16.32	16.13
7	17.25	17.216	16.886	16.79	16.38
8	17.94	17.612	17.43	17.25	17.216
9	19.232	18.932	18.806	18.344	18.152
10	20.378	19.922	19.532	19.232	18.932
11	24.968	22.628	20.876	20.606	20.378
12	25.00	22.628	20.876	20.606	20.378
13	25	22.628	20.876	20.606	20.378
14	24.968	22.628	20.876	20.606	20.378
15	20.378	19.922	19.532	19.232	18.932
16	20.378	19.922	19.532	19.232	18.932
17	20.876	20.606	20.378	19.922	19.532
18	25.00	22.628	20.876	20.606	20.378
19	25.00	22.628	20.876	20.606	20.378
20	25.00	22.628	20.876	20.606	20.378
21	25.00	22.628	20.876	20.606	20.378
22	24.968	22.628	20.876	20.606	20.378
23	19.532	19.232	18.932	18.806	18.344
24	17.940	17.612	17.43	17.25	17.216

Table 3.4 ESS characteristics.

Maximum/Minimum Energy Storage Limits [MWh]	Discharging/Charging Power Limits [MW]	Discharging and Charging Efficiency	Initial Energy Storage [MWh]
100/10	30/30	90%	10

Bidding strategy in the electricity market

Table 3.5 Conventional unit characteristics.

Max/Min Capacity [MW]	Min Up/Down Time [h]	Ramp Up/Down	Size of the jth production block [MW]	Fixed cost/Startup cost	Marginal cost of jth production block [$/MWh]
100/0	1/1	40/40	6.5–5–4.5–3.5	10/10	15.09–15.42–20.06–20.24

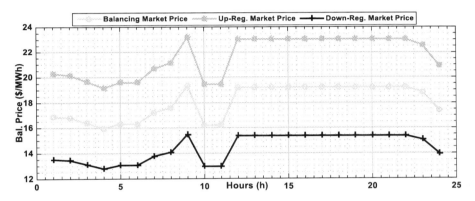

Figure 3.1 Balancing market price.

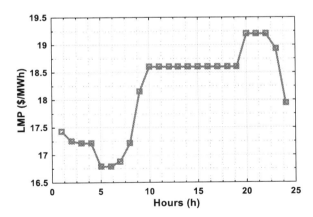

Figure 3.2 Market clearing price.

The generation power of the virtual power plant components is shown in Fig. 3.3. As it is clear from this figure, in the middle hours, the cleared power of the virtual power plant has increased. One of the most important reasons is the increase in production of wind and solar units. In the middle hours, the amount of cleared power is much below the produced power, so there is a lot of down-regulating power during

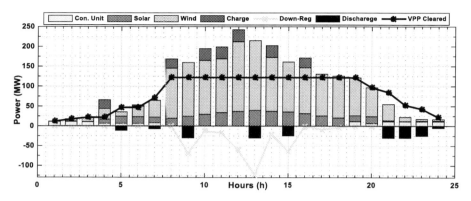

Figure 3.3 VPP power.

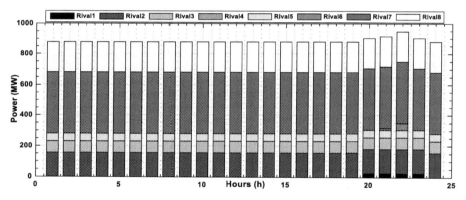

Figure 3.4 Rivals' power.

these hours. Storage has played an important role in the market. Increasing the battery charge reduces the need for down-regulation power, this is evident at hours 8, 10, and 11. In the last hours of the scheduling period, the production of renewable and non-renewable units has sharply decreased. The storage device has discharged a lot so that virtual power plant reaches its cleared power. This discharged power enters the network and compensates for the lack of virtual power plant production. The conventional unit in the virtual power plant generated power in the early and late hours of the scheduling period, and produced almost no power in the middle hours. This is because in the middle hours, wind and solar units have put a lot of power into the network.

The production power of rivals is shown in Fig. 3.4. As mentioned before, there are 8 rivals in this problem who have non-strategic behavior. Rivals 1, 4, and 6 have produced very little or no power. The most important reason is that these have high marginal cost blocks. If the marginal cost is higher than the market cleared price, it is impossible to produce power. As can be seen from the Fig. 3.2, the maximum cleared

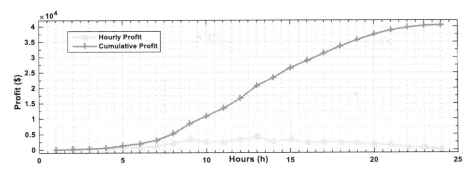

Figure 3.5 VPP profit.

price of the market is 19.2 $/MWh. The cheapest blocks of the fourth and sixth units are equal to this price, so they produce almost no power.

Fig. 3.5 shows the amount of profit of the virtual power plant on an hourly and cumulative basis. As it is clear, the amount of hourly profit is closely related to the market cleared price and the amount of cleared power of the virtual power plant. In the middle hours, the market clearing price and the amount of cleared power of the virtual power plant are high, so the profit has increased in these hours compared to others.

3.6 Conclusion

The main purpose of this chapter is to determine the optimal bidding strategy of the commercial virtual power plant to the day-ahead and balancing market, so that the profit is maximized. To achieve this goal, a bi-level model is proposed. The upper level of the model maximizes the profits of the virtual power plant and the lower level models the problem of social welfare. This bi-level model has been transformed into a MILP model using the KKT optimality conditions. The virtual power plant has renewable and non-renewable production units, storage and loads. There are a number of rivals and demands that they bid/offer to the market. The outputs show that the marginal bid/offer prices by the participants are very effective on the market clearing price and profit of the virtual power plant.

Appendix 3.A

Linearization of complementary conditions

The linearization of complementary conditions was proposed in 1981 by Fortuny-Amat and McCarl [23]. This method replaces the mathematical problem with equilibrium and evolutionary constraints with the programming model with integer constraints, and the problem becomes a mixed-integer linear programming problem or a mixed-integer nonlinear programming problem.

The form of complementary constraints is as follows:

$$0 \leq x \perp y \geq 0. \tag{a}$$

By introducing the binary variable u and parameter M, the complementary conditions are replaced by the linearized constraints as follows:

$$0 \leq x \leq Mu, \tag{b}$$
$$0 \leq y \leq M(1 - u). \tag{c}$$

This method has two main problems because, firstly, it is computationally expensive for large models, and secondly, the constraints of this approach can be sensitive to the M value. However, this problem has been largely solved by improving the model solvers of the mixed-integer linear programming problem and the mixed-integer nonlinear programming problem, such as CPLEX and DICOPT.

It should be noted that the selection of small values for M causes tight formulation and error in formulating the problem. Selecting large values for M increases the number in the condition. Therefore, an improper selection of M causes the answer value to change from the correct value. Despite the said shortcomings, this method has been very practical and noteworthy.

Nomenclature

cu Conventional unit
S Storage
VPP Virtual power plant
DA Day-ahead market
Bal Balancing market

Indices
T Time
J Set of power blocks related to VPP conventional unit
BL Power blocks of demand lth
B$_v$ Virtual power plants' power blocks

Parameters
Ω_t^{Bal} Balancing market price at hour t
dn_{reg} Down-regulation coefficient in the balancing market
up_{reg} Up-regulation coefficient in the balancing market
P_t^{wind} Power of wind unit at hour t
P_t^{solar} Power of solar unit at hour t
SUC_t^{VPPu} Start-up cost of VPP conventional unit
SDC_t^{VPPu} Shut-down cost of VPP conventional unit

c_0 Fixed operation cost of VPP conventional unit
Ω_j^{VPP} Marginal cost of block j of VPP conventional unit
P_{Min}^{VPPu} Minimum power of VPP conventional unit
P_{Max}^{VPPu} Maximum power of VPP conventional unit
RU^{VPPu} Ramp-up rate of VPP conventional unit
RD^{VPPu} Ramp-down rate of VPP conventional unit
E_t^{SMax} Maximum capacity of storage
Ω_{trb}^{R} Bid price of block b of rival r at hour t
Ω_{tlbl}^{L} Offer price of block bl of demand l at hour t
P_{trb}^{Rmax} Maximum power of block b of rival r at hour t
P_{tlbl}^{Lmax} Maximum power of block bl of demand l at hour t

Variables

Ω_t^{DA} DA market clearing price at hour t
$P_{tb_v}^{DA}$ Cleared power for block b_v of VPP at hour t
P_t^{dn} Power sold in balancing market by VPP at hour t
P_t^{up} Power purchased in balancing market by VPP at hour t
$Cost_t^{cu}$ Total cost of VPP conventional unit at hour t
P_t^{VPPu} Sum of produced power by blocks of VPP conventional unit at hour t
P_{tj}^{VPPu} Produced power of block j of VPP conventional unit
P_t^{Ch} Power charged by storage at hour t
P_t^{Dch} Power discharged by storage at hour t
s_t^{VPPu} On/off status of VPP conventional unit
v_t^{VPPu} Binary decision variable of start-up status of VPP conventional unit
k_t^{VPPu} Binary decision variable of shut-down status of VPP conventional unit
α_t^{Dch} Binary decision variable of discharging status of storage at hour t
α_t^{Ch} Binary decision variable of charging status of storage at hour t
E_t^{S} Energy stored in storage at hour t
P_{trb}^{R} Cleared power of block b of rival r at hour t
P_{tlbl}^{L} Cleared power of block bl of demand l at hour t
$P_{tb_v}^{Imax}$ Maximum limit of bidding power of block b_v of VPP conventional unit at hour t
$\Omega_{tb_v}^{I}$ Bid price of block b_v of VPP conventional unit at hour t

References

[1] J.M. Morales, A.J. Conejo, H. Madsen, P. Pinson, M. Zugno, Integrating Renewables in Electricity Markets, Springer, New York, 2014.

[2] A. Alahyari, M. Ehsan, M. Mousavizadeh, A hybrid storage-wind virtual power plant (VPP) participation in the electricity markets: a self-scheduling optimization considering price, renewable generation, and electric vehicles uncertainties, Journal of Energy Storage 25 (2019) 100812.

[3] M. Shafiekhani, A. Badri, A risk-based gaming framework for VPP bidding strategy in a joint energy and regulation market, Iranian Journal of Science and Technology, Transactions of Electrical Engineering 43 (2019) 545–558.

[4] F. Luo, Z.Y. Dong, K. Meng, J. Qiu, J. Yang, K.P. Wong, Bidding strategy of virtual power plant for participating in energy and spinning reserve markets, IET Renewable Power Generation 10 (5) (2016) 623–633.

[5] Y. Wang, X. Ai, Z. Tan, L. Yan, S. Liu, Interactive dispatch modes and bidding strategy of multiple virtual power plants based on demand response and game theory, IEEE Transactions on Power Systems 7 (0) (2016) 1510–1519.

[6] S.M. Nosratabadi, R. Hooshmand, E. Gholipour, A comprehensive review on microgrid and virtual power plant concepts employed for distributed energy resources scheduling in power systems, Renewable & Sustainable Energy Reviews 67 (2017) 341–363.

[7] M. Giuntoli, D. Poli, Optimized thermal and electrical scheduling of a large scale virtual power plant in the presence of energy storages, IEEE Transactions on Smart Grid 4 (2) (2014) 942–955.

[8] M. Peik-Herfeh, H. Seifi, M.K. Sheikh-El-Eslami, Decision making of a virtual power plant under uncertainties for bidding in a day-ahead market using point estimate method, International Journal of Electrical Power & Energy Systems 44 (1) (2013) 88–98.

[9] M. Rahimi, F. Jahanbani Ardakani, A. Jahanbani Ardakani, Optimal stochastic scheduling of electrical and thermal renewable and non-renewable resources in virtual power plant, International Journal of Electrical Power & Energy Systems 127 (2021).

[10] N. Naval, R. Sánchez, J.M. Yusta, A virtual power plant optimal dispatch model with large and small-scale distributed renewable generation, Renewable Energy 151 (2020) 57–69.

[11] S. Sadeghi, H. Jahangir, B. Vatandoust, M.A. Golkar, A. Ahmadian, A. Elkamel, Optimal bidding strategy of a virtual power plant in day-ahead energy and frequency regulation markets: a deep learning-based approach, International Journal of Electrical Power & Energy Systems 127 (2021).

[12] M. Bazaraa, H. Sherali, C.M. Shetty, Nonlinear Programming: Theory and Algorithms, John Wiley & Sons, 2013.

[13] J. Outrata, M. Kocvara, J. Zowe, Nonsmooth Approach to Optimization Problems With Equilibrium Constraints: Theory, Applications and Numerical Results, Springer Science & Business Media, 2013.

[14] E.G. Kardakos, C.K. Simoglou, A.G. Bakirtzis, Optimal offering strategy of a virtual power plant: a stochastic bi-level approach, IEEE Transactions on Smart Grid 7 (2) (2016).

[15] H. Pandžić, J.M. Morales, A.J. Conejo, I. Kuzle, Offering model for a virtual power plant based on stochastic programming, Applied Energy 105 (2013) 282–292.

[16] M. Shafiekhani, A. Badri, M. Shafie-Khah, J.P.S. Catalão, Strategic bidding of virtual power plant in energy markets: a bi-level multi-objective approach, International Journal of Electrical Power & Energy Systems 113 (2019) 208–219.

[17] M. Shafiekhani, A. Badri, F. Khavari, A bi-level model for strategic bidding of virtual power plant in day-ahead and balancing market, in: 2017 Smart Grid Conference (SGC), IEEE, 2017, pp. 1–6.

[18] S. Yin, Q. Ai, Z. Li, Y. Zhang, T. Lu, Energy management for aggregate prosumers in a virtual power plant: a robust Stackelberg game approach, International Journal of Electrical Power & Energy Systems 117 (2020).

[19] T. Dai, W. Qiao, Finding equilibria in the pool-based electricity market with strategic wind power producers and network constraints, IEEE Transactions on Power Systems 32 (1) (2016) 389–399.

[20] T. Dai, W. Qiao, Optimal bidding strategy of a strategic wind power producer in the short-term market, IEEE Transactions on Sustainable Energy 6 (3) (2015) 707–719.

[21] M. Shafiekhani, A. Zangeneh, Integration of electric vehicles and wind energy in power systems, in: Electric Vehicles in Energy Systems, Springer, Cham, 2020, pp. 165–181.

[22] C. Ruiz, A.J. Conejo, Pool strategy of a producer with endogenous formation of locational marginal prices, IEEE Transactions on Power Systems 24 (4) (2009) 1855–1866.
[23] S.A. Gabriel, A.J. Conejo, J.D. Fuller, B.F.H.C. Ruiz, Complementarity Modeling in Energy Markets, Springer, New York, 2013.

Optimization model of a VPP to provide energy and reserve

Niloofar Pourghaderi[a], Mahmud Fotuhi-Firuzabad[b], Moein Moeini-Aghtaie[c], and Milad Kabirifar[a]
[a]Sharif University of Technology, Tehran, Iran, [b]Faculty of Electrical Engineering Department, Sharif University of Technology, Tehran, Iran, [c]Faculty of Energy Engineering Department, Sharif University of Technology, Tehran, Iran

4.1 Introduction

The penetration of distributed energy resources (DERs) is significantly increasing in modern power systems. DERs may contain renewable energy sources (RESs), conventional distributed generators (DGs), energy storage systems (ESSs), electric vehicles (EVs), flexible demands, etc. [1]. DERs may not have visibility to participate in electricity markets or provide services for power system due to their small size, scattering over the distribution network, or lack/deficiency of information and communication infrastructures [2]. Therefore, in order to utilize the DERs' capabilities, they should be appropriately aggregated and coordinated. In order to coordinate and optimize the utilization of DERs in the network, the concept of a virtual power plant (VPP) is taken into consideration. VPP aggregates the DERs in the network and creates a single operating portfolio through composition of the DERs' characteristics [3]. VPP may disregard the network constraints and DERs' location in the network [4–9]. In this case, the impact of the network on the DERs' aggregated profile is not taken into account and the VPP can provide commercial services to the network operator. On the other hand, VPP can take the operation of the network under its supervision into account, which makes the results feasible for implementing on the network practically, as well as facilitating the provision of network services [4], [10–14]. VPP has the opportunity of participating in wholesale electricity markets as well as generation companies (GenCos). However, unlike the GenCos, VPP can act as both producer and consumer in the market, as well as look after the generation/consumption balancing in the network [14]. It is worth mentioning that the electricity market may have enough liquidity to be competitive, in which the VPP's decisions don't affect the market clearing procedure [8], [9], [13], and [14]. On the contrary, the VPP's performance in the market may affect the market clearing procedure which should be taken into consideration in the VPP's decision making [5], [10], and [15]. VPP can schedule the DERs dispersed in a distribution network to provide energy and reserve services in order to make profit in wholesale markets or provide network services. The authors of [5–7], [10–12], and [16] addressed the DERs energy scheduling by VPP to make profit in the energy market, while references [8], [9], [13], [14], and [17–19] represented the

VPP's and the associated DERs' scheduling for energy and reserve provision. In addition, most of the DERs can provide flexibility services to compensate the variability and uncertainty in the power system [20]. Reference [21] presented the DERs' flexibility scheduling, in which the flexibility product was considered as a type of reserve to be traded in the wholesale market. It should be noted that there are different sources of uncertainty in the network (e.g., renewable sources and customers' demand) which should be considered by the VPP to make accurate decisions. References [13], [16], and [20] considered the existence uncertainties using stochastic approach while [8], [9], and [19] applied robust approaches.

In this chapter, a model of VPP to provide energy and reserve services is investigated. In this regard, diverse DERs' energy and reserve models are addressed and a model of VPP to aggregate and manage the DERs to make profit in markets in addition to satisfying the network operational constraints is presented. Furthermore, modeling the uncertainties is briefly addressed. In addition, utilizing the DERs' flexibility through the VPP framework is addressed in the chapter.

In the remaining parts, Section 4.2 covers the energy model of diverse DERs and optimization model of VPP to aggregate DERs providing energy. Furthermore, considering the DERs' uncertainties in the model is addressed. Section 4.3 considers the DERs' reserve models and VPP's optimization model to provide reserve service. Moreover, the model of DERs and VPP to provide the flexibility products are also taken into consideration in the section. Thereafter, the implementation procedure of examples is addressed in detail in Section 4.4, and finally, the conclusion is provided.

4.2 Optimization model of VPP to provide energy

4.2.1 Diverse distributed energy resources' energy models

In this part, energy models of diverse DERs, including DGs, ESSs, EVs, photovoltaic (PV) units, wind turbines (WTs), responsive loads consisting of heating, ventilation, and air conditioning (HVAC) systems and residential appliances, along with interruptible loads, are presented.

4.2.1.1 DGs

DGs like microturbines and diesel generators are dispatchable units which can be scheduled to supply desirable output powers. The output power of DG i at time t is denoted by $P_{DG}^{i,t}$ which should be placed in the allowable limits given by (4.1). The variable $u_{DG}^{i,t}$ determines the commitment status of DG which is 1 when the DG is on, and 0 otherwise:

$$P_{DG\_min}^{i} u_{DG}^{i,t} \leq P_{DG}^{i,t} \leq P_{DG\_max}^{i} u_{DG}^{i,t}. \tag{4.1}$$

DG unit's operating cost mostly depends on the fuel cost and is modeled as a function of the active output power of DG. The cost function of DG is practically formulated as a quadratic polynomial presented in (4.2). In the equation, the factors a^i_{DG}, b^i_{DG} and c^i_{DG} are positive parameters:

$$C^{i,t}_{DG} = a^i_{DG}(P^{i,t}_{DG})^2 + b^i_{DG} P^{i,t}_{DG} + c^i_{DG}. \tag{4.2}$$

It is worth mentioning that the cost curve of DG can be linearized by utilizing piecewise linear function as in (4.3). In this equation, $s^{i,j}$ represents the slope of the segment j in the piecewise linear curve with J segments. Furthermore, $Ps^{i,j,t}_{DG}$ is the power of segment j at time t which should be inside the limits presented through (4.4); $C^i_{DG\_min}$ is the cost of DG to generate the minimum allowed active power:

$$C^{i,t}_{DG} = C^i_{DG\_min} + \sum_{j \in J} s^{i,j} Ps^{i,j,t}_{DG}, \tag{4.3}$$

$$Ps^{i,j}_{DG\_min} \leq Ps^{i,j,t}_{DG} \leq Ps^{i,j}_{DG\_max}. \tag{4.4}$$

The DG's output power in the linearized form is attained as [16]

$$P^{i,t}_{DG} = P^i_{DG\_min} + \sum_{j \in J} Ps^{i,j,t}_{DG}. \tag{4.5}$$

4.2.1.2 ESSs

ESSs can facilitate the integration of RESs in power systems by mitigating RESs' variability and intermittency. ESSs can store energy during the times of low demand and return the energy back to the network in peak-demand periods. Furthermore, ESSs improve the reliable operation of network and provide ancillary services [22].

In order to model the energy management of an ESS unit, binary variables $u^{s,t}_{ESS\_Ch}$ and $u^{s,t}_{ESS\_Dch}$ are defined to determine the charging and discharging statuses of the ESS, respectively [23]; $u^{s,t}_{ESS\_Ch}/u^{s,t}_{ESS\_Dch}$ is 1 if the ESS is charged/discharged, and 0 otherwise. The charging and discharging statuses of ESS cannot be equal at the same time. This constraint is added in (4.6) which should be satisfied for each time:

$$u^{s,t}_{ESS\_Ch} + u^{s,t}_{ESS\_Dch} \leq 1. \tag{4.6}$$

The charging and discharging powers of an ESS unit should be inside the associated limits given by (4.7) and (4.8), respectively:

$$0 \leq P^{s,t}_{ESS\_Ch} \leq P^s_{ESS\_Ch\_max} u^{s,t}_{ESS\_Ch}, \tag{4.7}$$

$$0 \leq P^{s,t}_{ESS\_Dch} \leq P^s_{ESS\_Dch\_max} u^{s,t}_{ESS\_Dch}. \tag{4.8}$$

The state of charge of the ESS at each time is obtained from the state of charge of the ESS at the previous time and the charging/discharging energy of the ESS at the given time. Eq. (4.9) expresses the state of charge of ESS at each time as

$$SoC_{ESS}^{s,t} = SoC_{ESS}^{s,t-1} + (\eta_{ESS\_Ch}^{s} P_{ESS\_Ch}^{s,t} - \frac{1}{\eta_{ESS\_Dch}^{s}} P_{ESS\_Dch}^{s,t})\Delta t, \quad \forall t > 1. \tag{4.9}$$

In the above equation, $\eta_{ESS\_Ch}^{s}$ and $\eta_{ESS\_Dch}^{s}$ are the charging and discharging efficiencies of the ESS unit.

The state of charge of ESS should be within the allowable limits at each time, namely

$$SoC_{ESS\_min}^{s} \leq SoC_{ESS}^{s,t} \leq SoC_{ESS\_max}^{s}. \tag{4.10}$$

In the ESS's energy management model, it is usually assumed that the state of charge of ESS at the end of the day is higher or equal to the ESS's state of charge at the beginning time of scheduling, see (4.11). Furthermore, the ESS's output power is considered as an initial value at the beginning of scheduling time through (4.12) [8]. It should be noted that $P_{ESS}^{s,1}$ is equal to $P_{ESS\_Ch}^{s,1}$ if it is positive and $P_{ESS\_Dch}^{s,1}$ if it is negative:

$$SoC_{ESS}^{s,24} \geq SoC_{ESS}^{s,1}, \tag{4.11}$$

$$P_{ESS}^{s,1} = P_{ESS\_init}^{s}. \tag{4.12}$$

It is worth nothing that the rate of usage of electric vehicles is increasing due to environmental issues. EVs with vehicle-to-grid (V2G) capabilities are considered as portable storage systems. The management of EVs in the VPP's energy scheduling problem should satisfy technical and customers' welfare constraints which are expressed in the following:

Variables $u_{ch\_EV}^{n,t}$ and $u_{Dch\_EV}^{n,t}$ are charging and discharging statuses of EV which cannot be equal simultaneously. Furthermore, the EVs can be charged and discharged just when they are plugged into a charger. Relations (4.13) and (4.14) indicate these constraints, respectively [16]. Parameter α_{EV}^{n} represents the arrival time of the nth EV to home and parameter β_{EV}^{n} expresses the departure time of the EV from home; α_{EV}^{n} and β_{EV}^{n} can be forecasted by VPP utilizing historical data associated with EVs' owners behaviors [24]:

$$u_{EV\_ch}^{n,t}, u_{EV\_disch}^{n,t} = 0 \qquad \forall t \in T - [\alpha_{EV}^{n}, \beta_{EV}^{n}], \tag{4.13}$$

$$u_{EV\_ch}^{n,t} + u_{EV\_disch}^{n,t} \leq 1 \qquad \forall t \in [\alpha_{EV}^{n}, \beta_{EV}^{n}]. \tag{4.14}$$

Constraints (4.15) and (4.16) ensure that the charging and discharging powers of EV are within the minimum and maximum limits:

$$0 \leq P_{EV\_Ch}^{n,t} \leq P_{EV\_Ch\_max}^{n} u_{EV\_Ch}^{n,t}, \tag{4.15}$$

$$0 \leq P_{EV\_Dch}^{n,t} \leq P_{EV\_Dch\_max}^{n} u_{EV\_Dch}^{n,t}. \tag{4.16}$$

The state of charge of EV is attained using (4.17). Variable $SoC_{EV}^{n,0}$ indicates the initial state of charge of EV at the time α_{EV}^n; E_{EV} is the battery capacity of the nth EV, while $\eta_{EV\_Ch}^n$ and $\eta_{EV\_Dch}^n$ represent the charging and discharging efficiencies associated with the EV:

$$SoC_{EV}^{n,t} = SoC_{EV}^{n,t-1} + (\eta_{EV\_Ch}^n P_{EV\_Ch}^{n,t} - \frac{1}{\eta_{EV\_Dch}^n} P_{EV\_Dch}^{n,t})\Delta t. \tag{4.17}$$

The EV should be charged till the required level using the relation (4.18). Furthermore, the state of charge of EV should be within the limits at each time:

$$SoC_{EV}^{n,\beta_{EV}^n} \geq SoC_{EV\_REQ}^n, \tag{4.18}$$

$$SoC_{EV\_min}^n \leq SoC_{EV}^{n,t} \leq SoC_{EV\_max}^n. \tag{4.19}$$

4.2.1.3 Flexible loads

Demand-side management and demand-response programs are receiving attention in modern power systems. One of the most important demand-response programs is exploiting the customers that have the ability of shifting their loads in response to direct or indirect signals. Customers with shiftable loads can reduce their loads at the times of peak load or when the market price is high and consume more in low-demand or low-price periods. HVAC systems and residential appliances like dish washers, washing machines, and dryers are two important shiftable load examples, and their model is explained in this part. Furthermore, the other important responsive customers are those with interruptible loads that sign contracts with the system operator or VPP to interrupt the specific portion of their loads for limited time with a specified price. The model of interruptible loads is also addressed in the following.

- **HVAC systems**

HVAC systems are utilized in residential, commercial, and other buildings to bring the building temperature to the desired level. The HVAC system includes about 47% of the building's demand which is appropriate for utilizing in demand-response programs [25]. The consumed power of an HVAC system should be placed within the limits expressed in (4.20). Furthermore, the temperature of the associated building should be in the customer's desirable range, see (4.21). It should be noted that since the building temperature has inertia to change immediately, the HVAC system's consumption profile can be managed to build the appropriate temperature while contributing the VPP to achieve the optimal consumption pattern [26]:

$$0 \leq P_{HVAC}^{h,t} \leq P_{HVAC\_max}^h, \tag{4.20}$$

$$\theta_{Buil\_min}^c \leq \theta_{Buil}^{c,t} \leq \theta_{Buil\_max}^c. \tag{4.21}$$

When modeling the residential and commercial HVAC systems, an equivalent thermal model (ETP) can be applied [27]. The ETP model can be simplified by matching the measured turn-on and turn-off time under a range of ambient temperatures [27]. In this way, an equivalent model of an HVAC system is obtained, in which the indoor temperature of a building when the HVAC system is turned on/off is depicted in (4.22) and (4.23), respectively:

$$\theta_{Buil}^{c,t} = \theta_O^t + R_{Buil}^c \sum_{h \in H_{Buil}} P_{HVAC}^{h,t}$$
$$- (\theta_O^t + R_{Buil}^c \sum_{h \in H_{Buil}} P_{HVAC}^{h,t} - \theta_{Buil}^{c,t_1}) e^{-\Delta t / R_{Buil}^c C_{Buil}^c}, \quad (4.22)$$

$$\theta_{Buil}^{c,t} = \theta_O^t - (\theta_O^t - \theta_{Buil}^{c,t_1}) e^{-\Delta t / R_{Buil}^c C_{Buil}^c}, \quad (4.23)$$

where θ_O^t, R_{Buil}^c, and C_{Buil}^c are the ambient temperature, equivalent thermal resistance, and heat capacity of space associated with the cth customers' building, respectively.

It should be noted that since the desired temperature of buildings is in a relatively low range, the relations (4.22) and (4.23) can be linearized by utilizing the Taylor's series for linearizing the term $e^{-\Delta t / R_{Buil}^c C_{Buil}^c}$ and ignoring the second- and higher-order terms (4.24). The linearized forms of the abovementioned relations are addressed in (4.25) and (4.26), which is related to turning on/off the HVAC system, respectively:

$$e^{-\Delta t / R_{Buil}^c C_{Buil}^c} = (e^{-\Delta t / R_{Buil}^c C_{Buil}^c})\Big|_{\Delta t = 0} + \frac{d(e^{-\Delta t / R_{Buil}^c C_{Buil}^c})}{d(\Delta t)}\Big|_{\Delta t = 0}$$
$$= 1 - \frac{\Delta t}{R_{Euil}^c C_{Buil}^c}, \quad (4.24)$$

$$C_{Buil}^c (\theta_{Buil}^{c,t} - \theta_{Buil}^{c,t_1}) / \Delta t = (1 / R_{Buil}^c)(\theta_O^t - \theta_{Buil}^{c,t_1}), \quad (4.25)$$
$$t_1 = t - 1,$$
$$C_{Buil}^c (\theta_{Buil}^{c,t} - \theta_{Buil}^{c,t_1}) / \Delta t = (1 / R_{Buil}^c)(\theta_O^t - \theta_{Buil}^{c,t_1}) - \sum_{h \in H_{Buil}} P_{HVAC}^{h,t},$$
$$t_1 = t - 1. \quad (4.26)$$

- *Residential responsive appliances*

The operation of residential responsive appliances, like dishwasher, washing machine, and dryer, can be deferred in a limited time interval in which the customers' welfare is provided [28]. In this regard, the desirable time period for customer c to operate appliance a is denoted by $[\alpha_{AP}^{c,a}, \beta_{AP}^{c,a}]$. The constraint (4.27) ensures that the appliance would be operated during the desired time period. In this relation, $u_{AP}^{c,a,t}$ illustrates the binary variable that shows the start-up status of the appliance and $k_{AP}^{c,a}$ is the number of time intervals required for complete application of the appliance:

$$u_{AP}^{c,a,t} = 0 \quad t \in T - [\alpha_{AP}^{c,a}, \beta_{AP}^{c,a} - k_{AP}^{c,a}]. \quad (4.27)$$

The continuous operation of the appliance should be guaranteed through the constraint

$$\sum_{t \in T} u_{AP}^{c,a,t} = 1. \tag{4.28}$$

The consumed power of the appliance is calculated through equation (4.29), in which $P_{AP}^{c,a,t}$ is the consumed power of the appliance at each time. This relation ensures that the required energy for complete operation of the appliance is provided; $p_{AP}^{c,a,t}$ is the nominal required power of the appliance at each time, which is obtained from the appliance's energy consumption profile (ECP):

$$P_{AP}^{c,a,t} = \sum_{n=1}^{k_{AP}^{c,a}} u_{AP}^{c,a,t-k_{AP}^{c,a}+n} p_{AP}^{c,a,k_{AP}^{c,a}-n+1}. \tag{4.29}$$

- **Interruptible loads**

Load interruption is a voluntary, demand-side management program which is an agreement between the customers and the utility to interrupt a part of the power demand under different specific conditions when the system reliability is jeopardized. The interruptible loads can sign contracts with the VPP to participate in demand-response programs or electricity markets. The contracted prices for utilizing interruptible loads are much lower than the value of lost load (VOLL) [29].

Constraint (4.30) limits the customer's interrupted power between the minimum and maximum amounts; $u_{IL}^{d,t}$ is the status of utilizing the interruptible load which is 1 if the load is interrupted, and 0 otherwise:

$$0 \leq P_{IL}^{l,t} \leq P_{IL\_max}^{l} u_{IL}^{l,t}. \tag{4.30}$$

The total time of load interruption should be lower than the agreement time T_{IL}^{d} using the following equation:

$$\sum_{k=1}^{T} u_{IL}^{l,k} = T_{IL}^{l}. \tag{4.31}$$

Constraint (4.32) guaranties that total energy requirement of the customer is provided by considering the interrupted load through (1.1) [8],

$$\sum_{t \in T} (P_L^{l,t} - P_{IL}^{l,t}) \Delta t \geq E_{L\_min}^{l}. \tag{4.32}$$

The cost of load interruption which should be paid to the customers with interrupted loads can be formulated as a function of the interrupted power as follows [14]:

$$C_{IL}^{l,t}(P_{IL}^{l,t}) = a_{IL}^{l}(P_{IL}^{l,t})^2 + b_{IL}^{l} P_{IL}^{d,t} + c_{IL}^{l}. \tag{4.33}$$

4.2.1.4 RESs

RESs integration with power systems is dramatically increasing due to their various advantages. RESs have uncertain and intermittent output power which should be considered to optimally utilize the resources. PV units and wind turbines are the most common RESs utilized in power systems that we describe in the following.

- *Wind turbines*

Wind turbine output power depends on the wind speed on the turbine blades according to Fig. 4.1. As it is shown in the figure, at very low wind speeds, the torque on the turbine blades is not enough to make the turbine rotate. When the wind speed reaches the cut-in speed $v_{WT\_Cut\_in}$, the wind turbine starts to rotate and generates electric power. By increasing the wind speed, the generation power of WT is increased until its rated output power. The wind speed in which the WT reaches the rated output power is called rated output speed v_{WT_Rated}. Increasing the wind speed above the rated output power, the wind turbine generates its rated output power until the wind speed reaches the cut-out speed $v_{WT\_Cut\_out}$. In this situation the wind turbine is brought to a standstill because of the risk of turbine damage [30].

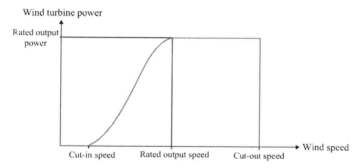

Figure 4.1 WT generated power with respect wind speed.

The output power of a wind turbine is represented in (4.34). In the relation, P_v is the output power of WT when the wind speed is between the cut-in and cut-out speeds and is a function of wind speed; P_v is approximated by fitting different mathematical functions like a polynomial function in (4.35). Furthermore, P_n is the rated output power of WT:

$$P_{WT} = \begin{cases} 0, & v < v_{WT\_Cut\_in}, v > v_{WT\_Cut\_out}, \\ P_v, & v_{WT\_Cut\_in} \leq v \leq v_{WT\_Rated}, \\ P_n, & v_{WT\_Rated} < v \leq v_{WT\_Cut\_out}, \end{cases} \quad (4.34)$$

$$P_v = P_n \left(\frac{v - v_{WT\_Cut\_in}}{v_{WT\_Rated} - v_{WT\_Cut\_in}} \right)^3. \quad (4.35)$$

The equivalent output power of a wind farm containing N WTs with efficiency η_{WT} is obtained using

$$P_{WF} = N P_{WT} \eta_{WT}. \quad (4.36)$$

- *PV units*

The output power of a PV unit depends on the solar radiation and the operating temperature at its mounting location as described in [31]:

$$P_{PV} = N\eta_{PV} A_{PV} G_T (1 + k(T_C - T_{avg})). \tag{4.37}$$

In the above equation, N, A_{PV}, and G_T represent the number of PV arrays, the total area of PV arrays, and the solar radiation incident on the PV panels, respectively; η_{PV} represents the efficiency of PV unit's photoelectric conversion which is assumed to be a constant parameter during the PV life time; T_C is the operating temperature at PV panels' mounting location; T_{avg} indicates the mean temperature during daytime, and k is the temperature coefficient.

4.2.2 Model of VPP to aggregate DERs to provide energy

VPP aggregates the DERs' capacities in the network under its supervision and coordinates the DERs' operation to obtain the optimal energy management and maximize its profit in the wholesale energy market. In this regard, VPP manages the dispatchable resources in addition to taking the load consumption and RESs' production and their associated uncertainties into account to design its optimal performance in the market. VPP is responsible for satisfying the production/consumption power balance in the network and can take the network operating characteristics into account in its decision making. Incorporating the operation of the network in VPP's performance, two main categories of VPP, including commercial VPP (CVPP) and technical VPP (TVPP), are investigated in Section 4.2.2.1. VPP has the opportunity of participating in wholesale electricity market as both an energy consumer and producer. In other words, VPP can sell energy to the wholesale market to gain revenue and also purchase energy from the market by paying the energy cost. The wholesale market may have enough liquidity to be competitive, in which the TVPP has no market power [32]. In this case, VPP is a price-taker agent which receives the market price and manages the DERs according to the price of market. In contrast, the VPP may have market power and act as a price-maker agent in the market [5]. The VPP's scheduling models as either a price-taker or a price-maker agent are addressed in Section 4.2.2.2.

The VPP's energy scheduling framework is illustrated in Fig. 4.2. Diverse DERs containing conventional DGs, ESSs and EVs, responsive customers, RESs, etc., may be dispersed in the network under VPP's supervision. Network operating constraints are also taken into consideration in the case of TVPP.

It is worth noting that if the VPP's decisions affect the wholesale market's clearing procedure, VPP should consider the market model in its scheduling in order to have feedback about the market clearing price. Fig. 4.3 shows the framework of VPP's operation model in this case. As it is shown in the figure, different agents participate in the wholesale market whose performance in the market should be forecasted and the market clearing procedure should be modeled by VPP.

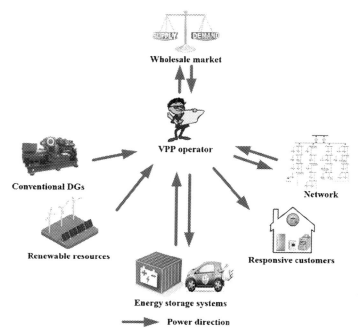

Figure 4.2 Virtual power plant's energy scheduling framework.

4.2.2.1 Commercial vs. technical VPP

CVPP considers the DERs in the network as aggregated resources in which DERs' locations are not important. In the case of CVPP, the impact of the network on the DERs' aggregated profile is not taken into account. CVPP is mostly like an aggregator that combines and manages DERs for participating in the wholesale electricity markets to maximize its profit in addition to offering services to the transmission system operator. In this regard, CVPP collects the required data including DERs' operating parameters, their costs, and associated forecasted data, as well as market's price and procedure, and optimizes its revenue by determining the DERs' optimal scheduling and optimal performance in the market. There may be more than one CVPP in a distribution network region that aggregates the DERs dispersed around the network [3], [33].

CVPP considers the aggregated production–consumption power balance at each time, which is written as

$$P_{VPP}^{t} + \sum_{i \in DG} P_{DG}^{i,t} + \sum_{s \in ESS} P_{ESS\_Dch}^{s,t} + \sum_{r \in RES} P_{RES}^{r,t}$$
$$= \sum_{l \in L} P_{L}^{l,t} + \sum_{s \in ESS} P_{ESS\_Ch}^{s,t}. \tag{4.38}$$

Optimization model of a VPP to provide energy and reserve

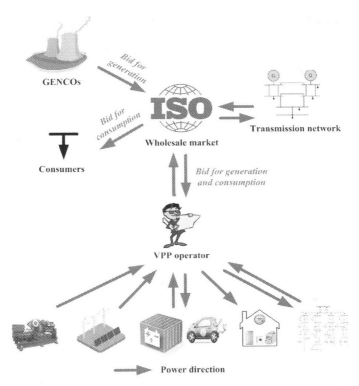

Figure 4.3 VPP's energy scheduling framework considering the wholesale market model.

In the relation (4.38), P_{VPP}^t are the active/reactive powers exchanged between the VPP and the energy market which are positive if the VPP purchases power from the market and negative if the VPP sells power to the market. The terms $\sum_{i \in DG} P_{DG}^{i,t}$, $\sum_{s \in ESS} P_{ESS\_Dch}^{s,t}$, and $\sum_{r \in RES} P_{RES}^{r,t}$ are the aggregated production powers of DGs, ESSs' discharging, and RESs, respectively. Furthermore, $\sum_{l \in L} P_L^{l,t}$ and $\sum_{s \in ESS} P_{ESS\_Ch}^{s,t}$ are the total consumption powers of loads and ESSs' charging. It should be noted that the load consumption $P_L^{l,t}$ contains both the fixed and responsive loads. Furthermore, electric vehicles are considered as a part of ESSs in the formulation.

In the case of TVPP, the impact of network operation is considered on TVPP's operation and scheduling. The DERs' geographic locations in the network are taken into account in the TVPP's portfolio. TVPP can provide network services like the distribution network congestion management along with the transmission system balancing and ancillary services. In order to participate in the market, the TVPP gathers the required data about DERs as well as the network characteristics and manages the network and the DERs to maximize its profit in the market in addition to providing network services.

TVPP requires detailed information about the network under its supervision in order to take the operation of network into consideration. The network operating constraints are addressed in the following.

- *Network operating constraints*

In order to consider the underlying network constraints in TVPP's operation model, the production–consumption power balance at each bus of the network should be satisfied for each time instant, which is addressed in (4.39). In the equation, $S_{Gen}^{b,t}$ represents the complex generating power injected to bus b due to the generation of DGs, RESs, and discharging of ESSs, which are connected to bus b, as well as the power purchased from the market if b is the bus of common coupling with market; $S_{Dem}^{b,t}$ is the complex demand power flowing out of bus b through the ESSs' charging power and load consumption of customers connected to bus b, as well as the power sold to the market at the bus of common coupling with the market. Similarly, $S_{Line}^{b,b',t}$ is the complex power flowing out of bus b through the lines which connect bus b to bus b'; B' is the number of buses connected to bus b:

$$S_{Gen}^{b,t} - S_{Dem}^{b,t} = \sum_{b'=1}^{B'} S_{Line}^{b,b',t}. \tag{4.39}$$

The power flow of the line connected to the buses b and b' is calculated using (4.40), in which V^b and $V^{b'}$ represent the voltage of busses b and b', and $Y^{b,b}$ $Y^{b,b'}$ are the associated components in the admittance matrix:

$$S_{Line}^{b,b',t} = V^b Y^{b,b'} (V^b)^* + V^b (V^b)^* (Y^{b,b})^* - V^b (V^{b'})^* (Y^{b,b'})^*. \tag{4.40}$$

The power flows of the lines should be within the allowable limits:

$$-S_{Line\_max}^{b,b'} \leq S^{b,b',t} \leq S_{Line\_max}^{b,b'}. \tag{4.41}$$

Moreover, (4.42) considers the limits of the network buses' voltages:

$$V_{min}^b \leq V^{b,t} \leq V_{max}^b. \tag{4.42}$$

The relations (4.39)–(4.42) are based on the complex powers. The relations of the active power flows can be obtained from the above relations. In order to model the reactive powers, it should be noted that the reactive powers of DERs are attained based on their active power and characteristics. In order to reduce the computational volume and solving time of the network's power flow calculation, some approximations may be applied to simplify the power flow constraints which have different degrees of accuracy [34].

4.2.2.2 Presence of VPP in energy market as a price-taker vs. price-maker agent

a) VPP as a price-taker agent

VPP determines its optimal energy scheduling and biddings to the energy market with the goal of maximizing its profit. The market may have enough liquidity to be competitive, in which the market participants' performances do not affect the price in the market [32]. In this case, the participants receive or forecast the market price as a parameter and determine their optimal performance in the market. The participants are called price-taker agents in this case [35]. The VPP's framework participating in the energy market as a price-taker agent is represented in Fig. 4.4. As it is shown in the figure, VPP receives the market price λ_E^t and finds the optimal dispatch of DGs, charging/discharging cycles of ESSs, responsive loads scheduling, and network operation, as well as optimal power exchange with the market.

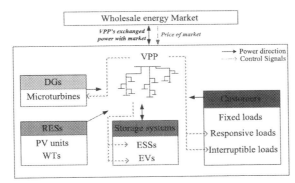

Figure 4.4 VPP's framework to participate in energy market as a price-taker agent.

The optimization problem of VPP's scheduling in the energy market as a price-taker agent is addressed in the following.

- *Objective*

The goal of VPP is to maximize its profit which is expressed as

$$\max \sum_{t \in T} \left[\begin{array}{c} -P_{VPP}^t \lambda_E^t + \lambda_{Ls}^t Load_E^t - \sum_{l \in IL} C_{IL}^{l,t}\left(P_{IL}^{l,t}\right) \\ - \sum_{i \in DG} C_{DG}^{i,t}(P_{DG}^{i,t}) - \lambda_{Lp}^t \sum_{s \in ESS} P_{ESS\_Dch}^{s,t} \end{array} \right]. \quad (4.43)$$

In the relation, the term $P_{VPP}^t \lambda_E^t$ represents the cost of purchasing power from the energy market which is imposed to VPP. It should be noted that if the VPP sells power to the market, P_{VPP}^t is negative and the term $P_{VPP}^t \lambda_E^t$ is a part of VPP's revenue. The terms $\sum_{i \in DG} C_{DG}^{i,t}(P_{DG}^{i,t})$ and $\lambda_{Lp}^t \sum_{s \in ESS} P_{ESS\_Dch}^{s,t}$ are the cost of DGs' power generation and the cost of purchasing the discharging power of ESS units; λ_{Lp}^t is the price of purchasing the ESSs' discharging power which in the case of EVs, it is called the V2G price [35]; $\lambda_{Ls}^{t} Load_{E}^{t}$ shows the VPP's revenue obtained from selling power to customers providing their loads; λ_{Ls}^{t} is the price of selling energy to the end users which can be designed as time of use price, real-time price, etc. Moreover, $Load_{E}^{t}$ which is addressed in (4.44) is the customers' total demand which comprises customers' fixed loads and responsive loads minus the interrupted loads along with charging power of ESSs; $P_{L\_Responsive}^{l,t}$ may contain the loads of HVAC systems, shiftable appliances, and other responsive loads. The term $\sum_{l \in IL} C_{IL}^{l,t} \left(P_{IL}^{l,t} \right)$ shows the cost of load interruption that VPP pays to interrupted customers:

$$Load_{E}^{t} = \sum_{s \in ESS} P_{ESS\_Ch}^{s,t} + \sum_{l \in L}(P_{L\_Fixed}^{l,t} + P_{L\_Responsive}^{l,t} - P_{IL}^{l,t}). \quad (4.44)$$

- **DERs constraints**

Operational constraints of DERs available in the system must be addressed in the optimization problem. Therefore, (4.1)–(4.5) for DGs, (4.6)–(4.19) for ESSs and EVs, (4.20)–(4.33) for different types of flexible loads, and (4.34)–(4.37) for RESs, if available, are considered in the optimization problem.

- **Network constraints**

In the case of CVPP, the production–consumption power balance is considered for aggregated DERs through (4.38). However, CVPP does not take the network operational constraints into account. In the case of TVPP, the production–consumption power balance at each bus of the network, as well as the operational constraints of the network, can be considered through (4.39)–(4.42).

b) VPP as a price-maker agent

The VPP's performance in the wholesale energy market may affect the market clearing procedure. In this case, the VPP acts as a strategic bidder in the market and takes the model of market clearing into account to have feedback about the market price. It should be noted that the wholesale energy market is cleared by an independent system operator (ISO). Different agents participate in the wholesale market and offer their production/consumption bids to the market. After receiving the bids, the ISO clears the market and determines the agents' accepted bids [15]. In order to model the wholesale energy market, the agents' performance in the market would be forecasted by VPP. In this regard, VPP can use the historical data about the rivals' biddings in the market. One of the reputed methods to model the VPP's energy scheduling problem as a price-maker agent is to use a bi-level approach depicted in Fig. 4.5.

According to Fig. 4.5, VPP gathers the required data containing DERs' and network's characteristics, as well as forecasted performances of the rivals, and executes the proposed bi-level problem. The upper-level problem contains energy scheduling and operation of VPP with the goal of maximizing its profit and considering the DERs' and network operational constraints. The VPP's optimal energy scheduling is obtained from this level. The lower-level problem includes the market clearing model from the viewpoint of the ISO with the aim of social welfare maximization while considering

Optimization model of a VPP to provide energy and reserve

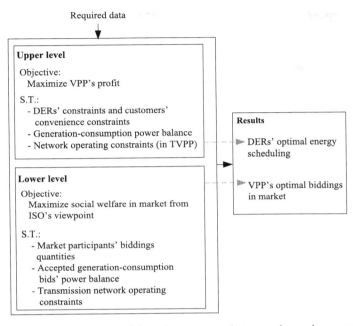

Figure 4.5 VPP's framework to participate in energy market as a price-maker agent.

the biddings of the agents and the accepted bids' power balance. The VPP's power transactions with the market result from this level. The proposed model is mathematically formulated in the following.

- *Upper-level problem*

In the upper-level problem, VPP determines its optimal energy scheduling with the aim of maximizing its profit. The VPP's profit is represented as

$$Max \sum_{t \in T} \left[\begin{array}{l} -(\phi^{VPP,t} - \omega^{VPP,t})\tilde{\lambda}_E^t + \lambda_{Ls}^t Load^t - \sum_{l \in IL} C_{IL}^{d,t}\left(P_{IL}^{l,t}\right) \\ - \sum_{i \in DG} C_{DG}^{i,t}(P_{DG}^{i,t}) - \lambda_{Lp}^t \sum_{s \in ESS} P_{ESS\_Dch}^{s,t} \end{array} \right]. \quad (4.45)$$

In the relation, $\phi^{VPP,t}$ and $\omega^{VPP,t}$ are the production and consumption bidding powers of VPP which are accepted in the market. Furthermore, $\tilde{\lambda}_E^t$ is a variable which indicates the market clearing price and which is obtained from the lower-level problem.

In order to participate in the market, agent m, which has generation capacity, submits the production bids to the market as pairs of power and price in the form of $(\Phi^{m,t}, \pi_\Phi^{m,t})$ for each time instant. Furthermore, an agent with consumption requirement submits consumption offers as pairs of $(\Omega^{m,t}, \pi_\Omega^{m,t})$ to the market. For VPP, the production and consumption bids should be lower than the total generation and consumption in the network, which is presented in (4.46) and (4.47) for each time instant, respectively:

$$\Phi^{VPP,t} \leq \sum_{i \in DG} P_{DG}^{i,t} + \sum_{r \in RES} P_{RES}^{r,t} + \sum_{s \in ESS} P_{ESS\_Dch}^{s,t}, \quad (4.46)$$

$$\Omega^{VPP,t} \leq \sum_{s \in ESS} P_{ESS\_Ch}^{s,t} + \sum_{l \in L}(P_{L\_Fixed}^{l,t} + P_{L\_Responsive}^{l,t} - P_{L\_Responsive}^{l,t}). \quad (4.47)$$

In the upper-level problem, the operational constraints of DERs which are available in the network should be modeled through the constraints determined previously in the price-taker framework. In addition, the aggregated production–consumption power balance is considered in the case of CVPP and the network operational constraints are taken into account by TVPP, as it is explained previously in the price-taker framework.

- *Lower-level problem*

In the lower level, the market clearing problem is modeled from the ISO's viewpoint. The objective in this case is to maximize the social welfare which is expressed as

$$Min \sum_{m \in M} \phi^{m,t} \pi_\Phi^{m,t} - \sum_{m \in M} \omega^{m,t} \pi_\Omega^{m,t}. \quad (4.48)$$

As mentioned, the participants submit their production and consumption bids to the market as pairs of $(\Phi^{m,t}, \pi_\Phi^{m,t})$ and $(\Omega^{m,t}, \pi_\Omega^{m,t})$. Afterwards, the market is cleared in the pay-as-bid form and the participants' accepted production and consumption bidding powers are determined as $\phi^{m,t}$ and $\omega^{m,t}$, respectively. The participants' accepted bids should be lower than their submitted bids to the market as follows:

$$0 \leq \phi^{m,t} \leq \Phi^{m,t}, \quad (4.49)$$
$$0 \leq \omega^{m,t} \leq \Omega^{m,t}. \quad (4.50)$$

Furthermore, the accepted bids' power balance should be satisfied for each time instant:

$$\sum_{m \in M} \phi^{m,t} = \sum_{m \in M} \omega^{m,t}. \quad (4.51)$$

In addition, the operation of the transmission network can be considered in the formulation. The detailed formulation and solution method can be found in [15].

It is worth noting that in order to make accurate and efficient decisions by the VPP, the existence uncertainties should be modeled effectively. In Section 4.2.3 important methods for addressing uncertainties are presented.

4.2.3 Uncertainty modeling

Renewable energy sources like PV units and wind turbines are the main sources of uncertainties in the network. It is due to the dependence of RES's output power on the uncertain, variable, and intermittence parameters like solar radiation on the PV panels and the speed of wind across the turbines blades. Furthermore, the load consumption of the customers is another important source of uncertainty in the power system which should be taken into account. Other parameters like the price of markets, the behavior of EVs' owners, outside temperature of buildings used for HVAC systems scheduling, occurrence of unexpected events in the network, etc., have uncertainty. The existence of sources of uncertainty in the network creates challenges for VPP's operation and scheduling which should be precisely modeled and taken into consideration for VPP's performance. Different methods are applied for modeling the uncertain features of the parameters [34], [36].

One of the most common methods is applying the stochastic programming which is based on the probability density function (PDF) of the uncertain parameter. PDF can be extracted from the statistical and historical data about the occurrence of parameter with uncertainty. For utilizing the stochastic method, different scenarios are considered for realization of uncertain parameter based on the associated PDF [37]. Different methods, like Monte Carlo simulation and scenario tree methods, can be applied to generate the scenarios. After generating the scenarios, scenario reduction approaches are applied for finding a compromise between the computational problems and the accuracy of uncertainty modeling. Specifying the scenarios, the optimization problem is solved for each scenario and the optimal result of the problem is achieved by considering the results associated with the scenarios as well as the probability of scenarios occurrences [34].

Another important approach to consider the existence uncertainties is utilizing robust optimization methods. In the robust approach, confidence bounds are considered for realization of parameters with uncertainty [38]. The uncertain parameters can take value within the confidence bounds. The robustness of the model can be controlled by considering a budget of uncertainties [36]. If the budget of uncertainty is considered to be equal to 0, the uncertainties are disregarded in the model and the solution is not robust against the uncertain parameters. On the other hand, the maximum value of the uncertainty budget is equal to the total number of uncertain parameters in which all the uncertain parameters are taken into account [36]. The advantage of robust optimization over the stochastic programming is that obtaining the confidence bounds is generally easier than generating accurate scenarios. However, the disadvantage of robust optimization is that this method leads to conservative solutions in some cases. Other methods, like fuzzy c-mean clustering method and chance-constrained programming, are also applied for uncertainty modeling in VPP's scheduling problem [35], [39]. Considering existence uncertainties in VPP's energy and reserve provision framework is presented in an example in Section 4.4.2 [26].

4.3 Optimization model of VPP to provide reserve

4.3.1 Diverse distributed energy resources' reserve models

VPP utilizes the capabilities of DERs to provide reserve for network services or market trading. In the following, the reserve management models of main categories of DERs containing DGs, ESSs, flexible loads, which are mainly utilized to provide reserve services, are addressed and applied in VPP's operation framework.

4.3.1.1 DGs

DGs can be dispatched to increase/decrease their generation power to provide up- and down-reserve capacity, respectively. Binary variables $u_{DG}^{i,t}$, $u_{DG\_SU}^{i,t}$, and $u_{DG\_SD}^{i,t}$ are considered to represent DG's commitment, start-up, and shut-down statuses, respectively, which are equal to 1 if the DG is committed for power generation, started up, or shut down at time t:

$$u_{DG}^{i,t}, u_{DG\_SU}^{i,t}, u_{DG\_SD}^{i,t} \in \{0, 1\}. \tag{4.52}$$

The relations between the abovementioned binary variables are presented in (4.53)–(4.55) for each DG at each time instant [14], [17]:

$$u_{DG}^{i,t} - u_{DG}^{i,t-1} \leq u_{DG\_SU}^{i,t}, \tag{4.53}$$

$$u_{DG}^{i,t-1} - u_{DG}^{i,t} \leq u_{DG\_SD}^{i,t}, \tag{4.54}$$

$$u_{DG}^{i,t} - u_{DG}^{i,t-1} \leq u_{DG\_SU}^{i,t} - u_{DG\_SD}^{i,t}. \tag{4.55}$$

Constraint (4.53) guarantees that the output power of DG after allocating the reserve capacity is within the limits [9], [17]; $P_{DG}^{i,t}$ and $R_{DG}^{i,t}$ represent the DG's output power in the energy market and DG's reserve capacity, respectively. Also

$$P_{DG\_min}^{i} u_{DG}^{i,t} \leq P_{DG}^{i,t} + R_{DG}^{i,t} \leq P_{DG\_max}^{i} u_{DG}^{i,t}. \tag{4.56}$$

The reserve capacity R_{DG}^{t} can be divided into up- and down-reserve capacities. In this case, the constraints associated with the output power of DG's limits are expressed as follows [17], [18]:

$$P_{DG}^{i,t} + R_{DG\_U}^{i,t} \leq P_{DG\_max}^{i} u_{DG}^{i,t}, \tag{4.57}$$

$$P_{DG}^{i,t} - R_{DG\_D}^{i,t} \geq P_{DG\_min}^{i} u_{DG}^{i,t}. \tag{4.58}$$

The difference between output powers of DG in two consecutive periods should be lower than up-ramping limit of DG if it is committed previously or DG's start-up ramp rate if the DG is started up at this time. This constraint is provided in (4.59). In a similar way, constraint (4.60) ensures that the difference between output powers of DG in two consecutive periods is greater than down-ramping limit of DG if it is committed previously or DG's shut-down ramp rate if the DG is shut down in this time [9]:

$$(P_{DG}^{i,t} + R_{DG\_U}^{i,t}) - (P_{DG}^{i,t-1} - R_{DG\_D}^{i,t-1}) \leq (RU_{DG}^i u_{DG}^{i,t-1} + RSU_{DG}^i u_{DG\_SU}^{i,t}), \quad (4.59)$$

$$(P_{DG}^{i,t-1} + R_{DG\_U}^{i,t-1}) - (P_{DG}^{i,t} - R_{DG\_D}^{i,t}) \leq (RD_{DG}^i u_{DG}^{i,t-1} + RSD_{DG}^i u_{DG\_SD}^{i,t}). \quad (4.60)$$

Mean up-time (MUT) and mean down-time (MDT) associated with DG should be considered through constraints (4.61) and (4.62) [14]; $T_{DG\_ON}^t$ and $T_{DG\_OFF}^t$ are the numbers of hours that the DG unit has been on/off before hour t, respectively:

$$(T_{DG\_ON}^{i,t} - MUT_{DG}^i)(u_{DG}^{i,t} - u_{DG}^{i,t+1}) \geq 0, \quad (4.61)$$

$$(T_{DG\_OFF}^{i,t} - MDT_{DG}^i)(u_{DG}^{i,t} - u_{DG}^{i,t+1}) \geq 0. \quad (4.62)$$

It should be noted that the linear formulation to consider DG's MUT and MDT constraints can be considered as follows:

$$\sum_{k=t}^{t+MUT_{DG}^i-1} u_{DG}^{i,k} \geq MUT_{DG}^i u_{DG\_SU}^{i,t}, \quad \forall k \in \{1, 2, ..., T - MUT_{DG}^i + 1\}, \quad (4.63)$$

$$\sum_{k=t}^{T} (u_{DG}^{i,k} - u_{DG\_SU}^{i,t}) \geq 0, \quad \forall k \in \{T - MUT_{DG}^i + 2, ..., T\}, \quad (4.64)$$

$$\sum_{k=t}^{t+MDT_{DG}^i-1} (1 - u_{DG}^{i,k}) \geq MDT_{DG}^i u_{DG\_SD}^{i,t},$$

$$\forall k \in \{1, 2, ..., T - MDT_{DG}^i + 1\}, \quad (4.65)$$

$$\sum_{k=t}^{T} (1 - u_{DG}^{i,k} - u_{DG\_SD}^{i,t}) \geq 0, \quad \forall k \in \{T - MDT_{DG}^i + 2, ..., T\}. \quad (4.66)$$

The cost function of DG is formulated as in (4.67), considering the allocated reserve capacities:

$$C_{DG}^{i,t} = a_{DG}^i (P_{DG}^{i,t} + R_{DG\_U}^{i,t} - R_{DG\_D}^{i,t})^2 + b_{DG}^i (P_{DG}^{i,t} + R_{DG\_U}^{i,t} - R_{DG\_D}^{i,t}) + c_{DG}^i. \quad (4.67)$$

4.3.1.2 ESSs

ESS is a significant resource to contribute in providing reserve service. When allocating the ESS's reserve capacities at each time, the charging and discharging powers of the ESS should be lower than the associated maximum charging and discharging capacities by considering the constraints (4.68) and (4.69), respectively [8]. In the relations, $P_{ESS\_Ch}^{s,t}$ and $P_{ESS\_Dch}^{s,t}$ represent the ESS's charging and discharging powers in the energy market, respectively. Moreover, $R_{ESS\_D}^{s,t}$ is the down-reserve capacity of ESS: its positive value represents increasing of ESS's charging power, while the negative value indicates reducing the ESS's charging power. Likewise, $R_{ESS\_U}^{s,t}$ is the ESS's up-reserve capacity, the positive/negative values of which represent the increasing/decreasing the ESS's discharging power:

$$P_{ESS\_Ch}^{s,t} + R_{ESS\_D}^{s,t} \leq P_{ESS\_Ch\_max}^{s} u_{ESS\_Ch}^{s}, \quad (4.68)$$

$$P_{ESS\_Dch}^{s,t} + R_{ESS\_U}^{s,t} \leq P_{ESS\_Dch\_max}^{s} u_{ESS\_Dch}^{s}. \quad (4.69)$$

The state of charge of ESS after allocation of reserve capacity is calculated using equation (4.70) for each time instant,

$$SoC_{ESS}^{s,t} = SoC_{ESS}^{s,t-1} + \left(\eta_{ESS\_Ch}^{s} (P_{ESS\_Ch}^{s,t} + R_{ESS\_D}^{s,t}) - \frac{1}{\eta_{ESS\_Dch}^{s}} (P_{ESS\_Dch}^{s,t} + R_{ESS\_U}^{s,t}) \right) \Delta t. \quad (4.70)$$

The maximum state of charge of ESSs should be taken into account through (4.71) and (4.72), respectively [8], [13]:

$$\eta_{ESS\_Ch}^{s} (P_{ESS\_Ch}^{s,t} + R_{ESS\_D}^{s,t}) \Delta t \leq SoC_{ESS\_max}^{s} - SoC_{ESS}^{s,t-1}, \quad (4.71)$$

$$\frac{1}{\eta_{ESS\_Dch}^{s}} (P_{ESS\_Dch}^{s,t} + R_{ESS\_U}^{s,t}) \Delta t \leq SoC_{ESS}^{s,t-1} - SoC_{ESS\_min}^{s}. \quad (4.72)$$

In order to consider the reserve management of ESS for a representative day, it is assumed that the state of charge of ESS at the end of the day is higher or equal to the ESS's state of charge at the beginning time of scheduling. Furthermore, there is an initial value for the ESS's power considered as in (4.73) [8]:

$$SoC_{ESS}^{s,24} \geq SoC_{ESS}^{s,1}, \quad (4.73)$$

$$P_{ESS}^{s,1} = P_{ESS\_init}^{s}. \quad (4.74)$$

4.3.1.3 Flexible loads

Flexible loads have the capability of increasing, decreasing, or shifting their loads to provide reserve service. The flexible customer's demand should be within the limits after utilizing the reserve capacity. The relations (4.75) and (4.76) depict the above-mentioned constraints, in which $P_L^{l,t}$, $R_{L\_D}^{l,t}$, and $R_{L\_U}^{l,t}$ are the consumed power, down-reserve capacity caused by increasing the loads, and up-reserve capacity caused by decreasing the loads [8]:

$$P_L^{l,t} + R_{L\_D}^{l,t} \leq P_{L\_max}^{l}, \quad (4.75)$$

$$P_L^{l,t} - R_{L\_U}^{l,t} \leq P_{L\_min}^{l}. \quad (4.76)$$

There may be limitations on the amount of increasing and decreasing the loads at each time which are depicted in constraints (4.77) and (4.78), respectively [18]:

$$0 \leq R_{L\_U}^{l,t} \leq R_{L\_U\_max}^{l}, \tag{4.77}$$

$$0 \leq R_{L\_D}^{l,t} \leq R_{L\_D\_max}^{l}. \tag{4.78}$$

The up- and down-ramping limits of the loads should be considered through (4.79) and (4.80), respectively. In this regard, the difference between the customer's consumption powers in two consecutive time instants should be limited to the maximum reserve ramping [8]:

$$(P_L^{l,t} + R_{L\_D}^{l,t}) - (P_L^{l,t-1} - R_{L\_U}^{l,t-1}) \leq RU_L^l, \tag{4.79}$$

$$(P_L^{l,t-1} + R_{L\_D}^{l,t-1}) - (P_L^{l,t} - R_{L\_U}^{l,t}) \leq RD_L^l. \tag{4.80}$$

Applying the reserve management should guarantee that total energy requirement of the customer be provided by considering the constraint [8]

$$\sum_{t \in T} (P_L^{l,t} + R_{L\_D}^{l,t} - R_{L\_U}^{l,t}) \Delta t \leq E_{L\_min}^{l}. \tag{4.81}$$

There may be time limitation for utilizing the reserve capability of flexible load through the contract between the flexible customer and the VPP. In this case, binary variable $u_{LR}^{l,t}$ is defined to illustrate the load's utilization status in the reserve management problem, which is limited to specific time $T_{LR\_max}^{l}$. The constraints (4.82)–(4.84) indicate the abovementioned limit [17]:

$$P_L^{l,t} + R_{L\_D}^{l,t} \leq R_{L\_max}^{l} u_L^{l,t}, \tag{4.82}$$

$$P_L^{l,t} - R_{L\_U}^{l,t} \leq P_{L\_min}^{l} u_L^{l,t}, \tag{4.83}$$

$$\sum_{t=1}^{24} u_{LR}^{l,t} \leq T_{LR\_max}^{l}. \tag{4.84}$$

4.3.2 Model of VPP to aggregate DERs to provide reserve

VPP aggregates the DERs and manages their reserve capabilities to make profit by offering reserve services in the wholesale market, as well as provides the reserve requirements of the network under its supervision. The framework of VPP's energy and reserve scheduling in the market is illustrated in Fig. 4.7. In the following, the formulation of VPP's optimization problem providing energy and reserve is addressed.

a) Objective

VPP aims at maximizing its profit considering the energy and reserve capacities of DERs under its supervision, along with the energy and reserve offered to the wholesale market. The objective of the VPP is given as

$$\max \sum_{t \in T} \begin{bmatrix} -P_{VPP}^t \lambda_E^t + (\lambda_{R\_U}^t + k_{R\_U} \hat{\lambda}_{R\_U}^t \Delta t) R_{VPP\_U}^t \\ +(\lambda_{R\_D}^t - k_{R\_D} \hat{\lambda}_{R\_D}^t \Delta t) R_{VPP\_D}^t \\ - \sum_{i \in DG} \begin{pmatrix} C_{DG}^{i,t}(P_{DG}^{i,t} + R_{DG\_U}^{i,t} - R_{DG\_D}^{i,t}) \\ + C_{DG\_SU}^i u_{DG\_SU}^t + C_{DG\_SD}^i u_{DG\_SD}^t \end{pmatrix} \\ + \lambda_{Ls}^t Load_{ER}^t - \sum_{l \in IL} C_{IL}^{l,t} \left(P_{IL}^{l,t} + R_{IL}^{l,t} \right) \\ - \lambda_{Lp}^t \sum_{s \in ESS} \frac{1}{\eta_{ESS\_Dch}^s} \left(P_{ESS\_Dch}^{s,t} + R_{ESS\_U}^{s,t} \right) \end{bmatrix}. \quad (4.85)$$

In relation (4.85), the term P_{VPP}^t represents the VPP's exchanged power with the energy market, which is positive/negative if the VPP purchases/sells power from/to the energy market. Furthermore, λ_E^t is the price of the energy in the market.

The terms $\lambda_{R\_U}^t R_{VPP\_U}^t$ and $\lambda_{R\_D}^t R_{VPP\_D}^t$ represent the VPP's revenue due to allocating up- and down-reserve capacities to the reserve market, respectively. Moreover, the energy values of the reserve capacities which are called on to produce are presented in the objective function through the terms $(k_{R\_U} \hat{\lambda}_{R\_U}^t \Delta t) R_{VPP\_U}^t$ and $(k_{R\_D} \hat{\lambda}_{R\_D}^t \Delta t) R_{VPP\_D}^t$; k_{R_U} and k_{R_D} are the coefficients in the interval [0,1] which determine the portion of up- and down-reserve capacities which are called on to produce, respectively.

The costs associated with DGs' energy and reserve procurement, as well as DGs' start-up and shut-down costs, are also considered in the objective function and subtracted from the VPP's revenue. Furthermore, the term $\lambda_{Ls}^t Load_{ER}^t$ shows the VPP's revenue due to providing the customers' loads while considering their energy and reserve scheduling; $Load_{ER}^t$ comprises the charging schedules of ESSs, as well as other loads of the customers. The cost of utilizing the customers reserve capabilities is imposed to VPP through the term $\sum_{l \in L} C_L^{l,t} \left(R_{L\_D}^{l,t} + R_{L\_U}^{l,t} \right)$, and the cost of load interruption is represented by the term $\sum_{l \in IL} C_{IL}^{l,t} \left(P_{IL}^{l,t} + R_{IL}^{l,t} \right)$, which should be paid to interrupted customers; $R_{IL}^{l,t}$ represents the increasing in interrupted power of customers during the reserve scheduling problem if it is positive and shows the amount of reduction in the interrupted load if it is negative. Besides, the cost of purchasing the discharging power of ESS is also considered in the objective function:

$$Load_{ER}^t = \sum_{s \in ESS} \eta_{ESS\_Ch}^s \begin{pmatrix} P_{ESS\_Ch}^{s,t} \\ + R_{ESS\_D}^{s,t} \end{pmatrix} + \sum_{l \in L} \begin{pmatrix} P_{L\_Fixed}^{l,t} + P_{L\_Responsive}^{l,t} \\ + R_{L\_D}^{l,t} - R_{L\_U}^{l,t} \\ -(P_{IL}^{l,t} + R_{IL}^{l,t}) \end{pmatrix}.$$

(4.86)

b) Constraints

The objective function described above is maximized considering the constraints associated with DERs and network operational constraints, along with power exchange limits with market. The constrains are addressed in the following.

- *DERs constraints*

DERs' energy and reserve constraints are taken into account in the model. In this regard, the energy management models of DERs and their reserve models should be considered as (4.1)–(4.38) and (4.52)–(4.84), respectively.

- *Network constraints*

The production–consumption power balance should be considered by taking the reserve capacities into account. The CVPP considers the DERs as aggregated resources and applies the aggregated power balance according to (4.87). However, TVPP considers the actual locations of DERs and considers the production–consumption power balance at each bus of the network through (4.88), in which the reactive power balance can be taken into account [13]. The term $(P_{Gen}^{b,t} + R_{Gen\_U}^{b,t} - R_{Gen\_D}^{b,t})$ is the total production power injected to bus b due to the generation sources connected to the bus. Furthermore, $(P_{Dem}^{b,t} - R_{Dem\_U}^{b,t} + R_{Dem\_D}^{b,t})$ is the total consumption power flowing out of bus b due to the consuming DERs connected to the bus. The exchanged energy and reserve powers of VPP with the market are also considered for the bus of common coupling; $\sum_{b'=1}^{B'} P_{Line\_ER}^{b,b',t}$ is the power flow of lines connected to bus b in the energy and reserve management problem:

$$P_{VPP}^t - k_{R\_U}^t R_{VPP\_U}^t + k_{R\_D}^t R_{VPP\_D}^t + \sum_{i \in DG}(P_{DG}^{i,t} + R_{DG\_U}^{i,t} - R_{DG\_D}^{i,t})$$
$$- Load_{ER}^t + \sum_{r \in RES}(P_{RES}^{r,t}) + \sum_{s \in ESS} \frac{1}{\eta_{ESS\_Dch}^s}(P_{ESS\_Dch}^{s,t} + R_{ESS\_U}^{s,t}) = 0, \tag{4.87}$$

$$(P_{Gen}^{b,t} + R_{Gen\_U}^{b,t} - R_{Gen\_D}^{b,t}) - (P_{Dem}^{b,t} - R_{Dem\_U}^{b,t} + R_{Dem\_D}^{b,t}) = \sum_{b'=1}^{B'} P_{Line\_ER}^{b,b',t}. \tag{4.88}$$

The network operational constraints are considered in the energy and reserve scheduling of VPP by adding the reserve capacities to (4.40)–(4.42).

- *Power exchange limits*

There may be limitations for VPP to exchange reserve capacities which are addressed in

$$0 \leq R_{VPP\_U}^t \leq R_{VPP\_U\_max}, \tag{4.89}$$
$$0 \leq R_{VPP\_D}^t \leq R_{VPP\_D\_max}. \tag{4.90}$$

Furthermore, the power transfer between the VPP and market should be inside the allowable limits as represented in

$$0 \leq \left| P_{VPP}^{t} + k_{R\_U}^{t} R_{VPP\_U}^{t} - k_{R\_D}^{t} R_{VPP\_D}^{t} \right| \leq P_{Exch\_max}. \tag{4.91}$$

- *VPP adequacy constraints to meet reserve margin*

VPP must ensure that the reserve capacities of the network are greater than the network's reserve margin and sufficient to weather the unexpected events. The reserve capacities of the network consist of the committed capacities of DGs which are not dispatched yet and are ready to utilize, the amount of loads which can be decreased or interrupted, the contribution of the ESS to provide reserve margin, as well as the amount of power which can be purchased from the market. The associated constraint is provided by

$$\sum_{i \in DG} (P_{DG\_max}^{i} u_{DG}^{i,t} - (P_{DG}^{i,t} + R_{DG\_U}^{i,t} - R_{DG\_D}^{i,t}))$$
$$+ \sum_{s \in ESS} \left(SoC_{ESS}^{s,t-1} / \Delta t + \left(\begin{array}{c} \eta_{ESS\_Ch}^{s} (P_{ESS\_Ch}^{s,t} + R_{ESS\_D}^{s,t}) \\ - \frac{1}{\eta_{ESS\_Dch}^{s}} (P_{ESS\_Dch}^{s,t} + R_{ESS\_U}^{s,t}) \end{array} \right) \right)$$
$$+ \sum_{l \in IL} (P_{IL\_max}^{l} - (P_{IL}^{l,t} + R_{IL}^{l,t})) + P_{Exch\_max}$$
$$- (P_{VPP}^{t} - R_{VPP\_U}^{t} + R_{VPP\_D}^{t}) \geq RM^{t}. \tag{4.92}$$

4.3.3 Model of DERs and VPP to provide flexibility

In order to compensate the increasing uncertainty and variability of the net load,[1] utilizing the flexibility capabilities of the power system has come into attention. There are different definitions for the flexibility concept. Flexibility of the network refers to the ability of the network to preserve the production–consumption power balance during expected/unexpected variability [40], [41]. Furthermore, flexibility capabilities of the resources refer to the ability of changing resources' output profiles from their initial dispatch in response to direct or indirect signals [2]. The important difference of the flexibility and reserve services is that the reserve product occupies the resources' capacities and the reserved capacity is barred from participating in the energy market. In the case of net load deviation from the predicted level in a time step, the reserve product is available as unused standby capacity to be utilized in the same time interval. However, the flexibility product refers to the ability of the resources to change their output power to compensate the net load deviation for the next time interval and does not occupy any capacity of resources for the energy market participation. Moreover, the flexibility products are traded in the markets closer to real-time [41]. DERs have significant flexibility capabilities which can be exploited through VPP's coordination

---
[1] Net load is defined as the total load consumption minus the RESs' generation, which accommodates most of the uncertainty sources.

and management. The VPP's energy and scheduling framework to participate in the energy and flexibility markets is considered as a two-stage problem in [20] which is depicted in Fig. 4.6.

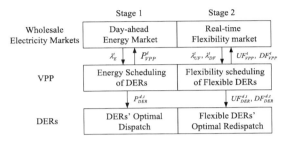

Figure 4.6 VPP's energy and flexibility scheduling framework.

The model of VPP to provide DERs' energy and flexibility scheduling for market participation is formulated in the following.

a) Objective

The VPP aims at maximizing its profit in the energy and flexibility market as given in

$$\text{Profit} = \max \sum_{t \in T} \begin{bmatrix} \left(-P_{VPP}^t \lambda_E^t + \lambda_{Ls}^t Load_E^t - \sum_{l \in IL} C_{IL}^{l,t}\left(P_{IL}^{l,t}\right) \right. \\ - \sum_{i \in DG} C_{DG}^{i,t}(P_{DG}^{i,t}) - \lambda_{Lp}^t \sum_{s \in ESS} P_{ESS\_Dch}^{s,t} \\ \left. + \left(UF_{VPP}^t \lambda_{UF}^t + DF_{VPP}^t \lambda_{DF}^t\right) \right) \end{bmatrix}. \quad (4.93)$$

According to relation (4.93), the VPP's profit comprises two main parts, including the profit attained from energy scheduling in the energy market and the profit earned though the flexibility scheduling in the flexibility market. It should be noted that the flexibility product is considered as tertiary (non-spinning) reserve in the form of upward and downward flexibility powers [21]. The VPP's upward and downward flexibility power is represented by UF_{VPP}^t and DF_{VPP}^t which are obtained through (4.94) and (4.95), respectively:

$$UF_{VPP}^t = \sum_{d \in DERs} UF_{DER}^{d,t}, \quad (4.94)$$

$$DF_{VPP}^t = \sum_{d \in DERs} DF_{DER}^{d,t}. \quad (4.95)$$

In the above relations, $UF_{DER}^{d,t}$ represents each DER's upward flexibility power which is obtained by increasing the DER's production power or decreasing its consumption power. In contrast, the DER's down flexibility power $DF_{DER}^{d,t}$ is obtained due to decreasing the DER's generation power or increasing its consumption power. The set of DERs in the relations contains all available DERs in the model like DGs, ESSs, RESs, and flexible loads.

b) Constraints

- *DERs' constraints*

The general flexibility model of flexible DERs is presented in the following. The detailed models of diverse DERs providing flexibility are addressed in [20] and [21].

Exploiting DERs' upward and downward flexibility at the same time can be prevented through (4.96)–(4.98). $y_{UF\_DER}^{d,t}$ and $y_{DF\_DER}^{d,t}$ are binary variables considered to indicate the status of upward and downward flexibility utilization of each DER. It should be noted that if the model does not address the real-time flexibility deployment (e.g. model for day-ahead flexibility scheduling), the capability of providing both upward and downward flexibility is considered at each time.

$$y_{UF\_DER}^{d,t} + y_{DF\_DER}^{d,t} \leq 1, \tag{4.96}$$

$$DF_{DER}^{d,t} \leq P_{\max\_DER}^{d} y_{DF\_DER}^{d,t}, \tag{4.97}$$

$$UF_{DER}^{d,t} \leq P_{\max\_DER}^{d} y_{UF\_DER}^{d,t}. \tag{4.98}$$

Constraints (4.99) and (4.100) define the allowable limits of upward and downward flexibility powers for generation units, respectively:

$$0 \leq UF_{DER}^{d,t} \leq P_{\max\_DER}^{d,t} - P_{DER}^{d,t}, \tag{4.99}$$

$$0 \leq DF_{DER}^{d,t} \leq P_{DER}^{d,t}. \tag{4.100}$$

Furthermore, the limits of upward and downward flexibility powers for consumption units are presented in (4.101) and (4.102), respectively:

$$0 \leq UF_{DER}^{d,t} \leq P_{DER}^{d,t}, \tag{4.101}$$

$$0 \leq DF_{DER}^{d,t} \leq P_{\max\_DER}^{d,t} - P_{DER}^{d,t}. \tag{4.102}$$

Besides the constraints (4.96)–(4.102), DERs' constraints in the VPP's energy scheduling problem should be considered through (4.1)–(4.38). Furthermore, constraints (4.1)–(4.38) should be satisfied by considering the DERs' upward and downward flexibility powers.

- *Network constraints*

The aggregated production–consumption power balance for CVPP should be considered through (4.38), and the complete operational constraints of the network for TVPP should be taken into account utilizing (4.39)–(4.42) in the energy as well as flexibility market by adding the DERs' flexibility powers to the relations.

4.4 Examples for VPP optimization model providing energy and reserve

In this section, the role of VPP in the energy and reserve provision is investigated through two examples. The first example considers the optimization problem of a

TVPP that aggregates DERs to provide energy and spinning reserve service in the wholesale market [14]. The second example proposes a self-scheduling model of CVPP for providing both energy and reserve in a day-ahead market. In the model, the uncertainties associated with wind power, reserve deployment request, and electricity price are considered [9]. The proposed results demonstrate the effective role of VPP in aggregating DERs for energy and reserve provision.

4.4.1 VPP's energy and spinning reserve provision

In this example, the TVPP creates a single operation portfolio from a composite of diverse DERs characteristics and takes the network constraints into account. The DERs are considered as DGs, ESSs, and customers with flexible loads. The TVPP manages the DERs centrally via a control coordination center to achieve their optimal dispatch for providing energy and reserve. Furthermore, TVPP considers the available reserve of the system to preserve the adequacy of the network. The TVPP participates in the energy and spinning reserve market as a price-taker agent and determines its optimal performance in markets, as well as optimal energy and reserve scheduling of the DERs.

In the following, the optimization model of TVPP proposed in the example is addressed. Thereafter, the mathematical formulation is presented. Then the model is implemented on two test TVPP, and the results are presented and investigated [42].

a) System model

In the proposed model, a TVPP aggregates DERs in the network and defines their optimal energy and reserve management to make profit in the energy and reserve markets. The DERs in the network are considered as DGs, ESSs, and customers with flexible loads. The DGs and ESSs are owned and directly managed by TVPP. The customers are supplied with the local network price. The TVPP signs contracts with the customers to have permission to interrupt their loads. The maximum amount of load interruption, number of hours of interruption, and the associated interruption cost are specified in the contracts. The interruption cost should be paid to the interrupted customer if the TVPP utilizes the interruptible loads. In this model, it is supposed that the DGs and customers are utilized to provide both energy and spinning reserve while the ESSs are considered just in the energy scheduling problem. In order to achieve the optimal scheduling of DERs, TVPP centrally manages the DERs through a control coordination center to maximize its profit. The TVPP participates in day-ahead joint energy and spinning reserve markets and optimizes its performance in the markets to maximize its profit. The TVPP acts as a price-taker agent, which receives the energy and reserve prices of associated markets and determines the optimal scheduling of DGs, optimal charge and discharge patterns of ESSs, and utilization of interruptible loads to make profit while operational constraints of network are preserved. The framework of the proposed model is depicted in Fig. 4.7.

b) Mathematical formulation

The TVPP aims at maximizing its profit which consists of the revenue attained from the energy and reserve markets, the revenue of selling energy to customers, and the

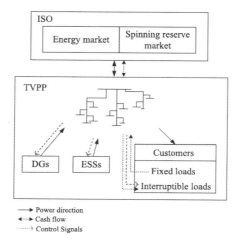

Figure 4.7 Relations between TVPP, wholesale markets, and DERs.

costs of DGs, ESSs, and utilization of interruptible loads. The objective is as follows:

$$\max \sum_t \begin{bmatrix} -P_{VPP}^t \lambda_E^t + R_{VPP}^t \lambda_R^t \\ -\sum_{i \in DG} \begin{pmatrix} C_{DG}^{i,t}(P_{DG}^{i,t} + R_{DG}^{i,t})u_{DG}^{i,t} \\ +C_{DG\_SU}^i u_{DG\_SU}^{i,t} + C_{DG\_SD}^i u_{DG\_SD}^{i,t} \end{pmatrix} \\ -\sum_{s \in ESS} C_{ESS}^{s,t}(P_{ESS}^{s,t}) \\ +\lambda_{Lp}^t \sum_{d \in D} \left(Load^{d,t} - P_{IL}^{d,t} - R_{IL}^{d,t} \right) - \sum_{l \in L} C_{IL}^{l,t} \left(P_{IL}^{l,t} + R_{IL}^{l,t} \right) \end{bmatrix}.$$

(4.103)

It should be noted that in some markets, the agents whose bids have been accepted are paid additionally based on the energy called on to produce in reserve markets [43]. Since in this model the TVPP does not predict which amount of energy would be called on to produce in the reserve market, the TVPP ignores the profit attained in this way and considers the forecasted price of the reserve market which is paid for the reserve capacity. In other words, the TVPP maximizes the minimum expected profit based on [44].

The presented objective function would be maximized considering the DERs' constraints, as well as the network operational constraints addressed in the following.

- *DGs constraints*

In the proposed model, DGs are owned and directly dispatched by the TVPP. The cost function of a DG unit is considered to be a quadratic polynomial with respect to the active production power:

$$C_{DG}^{i,t}(P_{DG}^{i,t}) = a_{DG}^i (P_{DG}^{i,t})^2 + b_{DG}^i P_{DG}^{i,t} + c_{DG}^i. \quad (4.104)$$

The coefficients a_{DG}^i and b_{DG}^i are positive parameters. The start-up and shut-down costs of DGs are also taken into account.

The DGs' technical constraints are given in (4.105)–(4.114). Eq. (4.105) guarantees that the output power of DG is within its limits. The upper bound of DG's spinning reserve power is addressed in (4.106). The difference between output power of DG in two consecutive periods limits is based on the up-ramping limit (if the output power of DG is increased) and down-ramping limit (if the output power of DG is decreased) through (4.107) and (4.108), respectively. MUT and MDT of DG unit is taken into account through (4.109) and (4.110). The relations between the on/off, start-up and shut-down statuses of DG units are considered in (4.111)–(4.114):

$$P_{DG\_min}^i \leq P_{DG}^{i,t} u_{DG}^{i,t} + R_{DG}^{i,t} u_{DG}^{i,t} \leq P_{DG\_max}^{i,t}, \tag{4.105}$$

$$R_{DG}^{i,t} \leq \min\{R_{DG\_max}^i, (P_{DG\_max}^i - P_{DG}^{i,t})u_{DG}^{i,t}\}, \tag{4.106}$$

$$P_{DG}^{i,t+1} - P_{DG}^{i,t} \leq RU_{DG}^i, \tag{4.107}$$

$$P_{DG}^{i,t} - P_{DG}^{i,t+1} \leq RD_{DG}^i, \tag{4.108}$$

$$(T_{DG\_ON}^{t-1} - MUT_{DG}^i)(u_{DG}^{i,t-1} - u_{DG}^{i,t}) \geq 0, \tag{4.109}$$

$$(T_{DG\_OFF}^{t-1} - MDT_{DG}^i)(u_{DG}^{i,t-1} - u_{DG}^{i,t}) \geq 0, \tag{4.110}$$

$$u_{DG}^{i,t}, u_{DG\_SU}^{i,t}, u_{DG\_SD}^{i,t} \in \{0,1\}, \quad \forall i \in I, t \in T, \tag{4.111}$$

$$u_{DG}^{i,t} - u_{DG}^{i,t-1} \leq u_{DG\_SU}^{i,t}, \quad \forall i \in I, t \in T, \tag{4.112}$$

$$u_{DG}^{i,t-1} - u_{DG}^{i,t} \leq u_{DG\_SD}^{i,t}, \quad \forall i \in I, t \in T, \tag{4.113}$$

$$u_{DG}^{i,t} - u_{DG}^{i,t-1} \leq u_{DG\_SU}^{i,t} - u_{DG\_SD}^{i,t}, \quad \forall i \in I, t \in T. \tag{4.114}$$

- **ESSs constraints**

In the model, it is supposed that the ESSs are owned by TVPP and, therefore, their associated costs should be modeled in the TVPP's energy and reserve management problem. The operational cost of an ESS is mainly due to the maintenance costs which are considered as a linear function with respect to the absolute of ESSs power [45]. Eq. (4.115) represents the ESS cost function in which the coefficients a_{ESS}^s and b_{ESS}^s are positive parameters:

$$C_{ESS}^{s,t}(P_{ESS}^{s,t}) = a_{ESS}^s |P_{ESS}^{s,t}| + b_{ESS}^s. \tag{4.115}$$

The technical constraints of ESSs in the VPP's energy and reserve scheduling problem are considered as (4.116)–(4.118). The first constraint is

$$-(SoC_{ESS\_0}^s - SoC_{ESS\_min}^s)/\Delta t \leq \sum_{k=1}^t P_{ESS}^{s,k} \leq (SoC_{ESS\_max}^s - SoC_{ESS\_0}^s)/\Delta t. \tag{4.116}$$

The output power of ESS in each time interval limits is based on maximum charging power (if the ESS is charged) and maximum discharging power (if the ESS is discharged):

$$P_{ESS}^{s,t} \leq P_{ESS\_Ch\_max}^{s,t}, \quad (4.117)$$

$$-P_{ESS}^{s,t} \leq P_{ESS\_Dch\_max}^{s,t}. \quad (4.118)$$

Although in this example the ESSs are utilized in the energy market, it should be noted that the ESSs have the capability to provide reserve which is addressed in the next example.

- *Interruptible loads' constraints*

Customers are supplied with the local network price. Customers' interruptible loads are utilized by the TVPP by paying the interruption cost. The cost of utilizing the interruptible load is considered as a function of interrupted power and considered as a quadratic polynomial function as follows [46]:

$$C_{IL}^{l,t}(P_{IL}^{l,t}) = a_{IL}^{l}(P_{IL}^{l,t})^2 + b_{IL}^{l} P_{IL}^{l,t} + c_{IL}^{l}. \quad (4.119)$$

The constraints associated with the interruptible loads' energy and reserve scheduling are considered as in (4.120) and (4.121), respectively:

$$0 \leq P_{IL}^{l,t} \leq P_{IL\_max}^{l}, \quad \forall t \in T_{IL}^{l}, \quad (4.120)$$

$$0 \leq P_{IL}^{l,t} + R_{IL}^{l,t} \leq P_{IL\_max}^{l}, \quad \forall t \in T_{IL}^{l}. \quad (4.121)$$

- *Network constraints*

The supply–demand power balance constraint is considered through (4.122) or (4.123), when the accepted spinning reserve is or is not called on to produce. Variables $Loss_E^t$ and $Loss_{ER}^t$ are the network losses when the accepted spinning reserve is or is not called on to produce:

$$P_{VPP}^t + \sum_{i \in DG} P_{DG}^{i,t} - \eta_{ESS\_Ch}^s P_{ESS\_Ch}^{s,t} + \frac{1}{\eta_{ESS\_Dch}^s} P_{ESS\_Dch}^{s,t}$$
$$= \sum_{l \in L}(Load^{l,t} - P_{IL}^{l,t}) + Loss_E^t, \quad (4.122)$$

$$P_{VPP}^t + R_{VPP}^t + \sum_{i \in DG}(P_{DG}^{i,t} + R_{DG}^{i,t}) - \eta_{ESS\_Ch}^s P_{ESS\_Ch}^{s,t}$$
$$+ \frac{1}{\eta_{ESS\_Dch}^s} P_{ESS\_Dch}^{s,t} = \sum_{l \in L}(Load^{l,t} - P_{IL}^{l,t} - R_{IL}^{l,t}) + Loss_{ER}^t. \quad (4.123)$$

The constraints associated with the network operation consist of Kirchhoff's laws, power flow limits of lines, and buses' voltage limits are taken into consideration according to the relations:

$$P_{Gen}^{b,t} - P_{Dem}^{b,t} = \sum_{b'=1}^{B'} P_{Line}^{b,b',t}, \quad (4.124)$$

$$|P^{b,b',t}| \leq P^{b,b'}_{Line\_max}, \tag{4.125}$$

$$V^b_{min} \leq V^{b,t} \leq V^b_{max}. \tag{4.126}$$

Furthermore, the exchanged power between TVPP and the market should be lower than the limit P_{Exch_max}. Inequality (4.127) is for the case that the spinning reserve is not called on to produce and (4.128) is applied otherwise:

$$|P^t_{VPP}| \leq P_{Exch\_max}, \tag{4.127}$$

$$|P^t_{VPP} + R^t_{VPP}| \leq P_{Exch\_max}. \tag{4.128}$$

Moreover, the network adequacy should be guaranteed during the TVPP's energy and reserve management problem. Therefore, the committed DGs, ESSs, interruptible loads, and power that can be purchased from the market should satisfy the network reserve margin based on (4.129) and (4.130), when the spinning reserve is not and is called on to produce, respectively.

$$\sum_{i \in DG}(P^i_{DG\_max} - P^{i,t}_{DG})u^{i,t}_{DG}$$

$$+ \sum_{s \in ESS}\left(SoC^{s,t-1}_{ESS}/\Delta t + \left(\eta^s_{ESS\_Ch}P^{s,t}_{ESS\_Ch} - \frac{1}{\eta^s_{ESS\_Dch}}P^{s,t}_{ESS\_Dch}\right)\right)$$

$$+ \sum_{l \in L}(P^l_{IL\_max} - P^{l,t}_{IL}) + P_{Exch\_max} - P^t_{VPP} \geq RM^t, \tag{4.129}$$

$$\sum_{i \in DG}(P^i_{DG\_max} - P^{i,t}_{DG} - R^{i,t}_{DG})u^{i,t}_{DG}$$

$$+ \sum_{s \in ESS}\left(SoC^{s,t-1}_{ESS}/\Delta t + \left(\eta^s_{ESS\_Ch}P^{s,t}_{ESS\_Ch} - \frac{1}{\eta^s_{ESS\_Dch}}P^{s,t}_{ESS\_Dch}\right)\right)$$

$$+ \sum_{l \in L}(P^l_{IL\_max} - P^{l,t}_{IL} - R^{l,t}_{IL}) + P_{Exch\_max} - P^t_{VPP} - R^t_{VPP} \geq RM^t. \tag{4.130}$$

c) Solution method

The problem is a nonlinear mixed-integer programming which is solved by applying the genetic algorithm (GA). The advantages of utilizing GA to solve this optimization problem other than utilizing the mathematical methods are explained in [42].

To apply the GA, the optimization problem variables, including the DGs' output power, interrupted loads of interruptible customers, and charge/discharge powers of ESSs are considered as chromosomes. For each chromosome, the backward/forward power flow is used to calculate the power losses in the network and TVPP's exchanged power with the energy market. During this procedure, the network adequacy and operating constraints to be satisfied are investigated. Afterward, the power flow is checked again, by considering DGs' total production, interrupted loads in the energy and reserve markets, and the output power of ESSs. The TVPP's energy and reserve scheduling is defined in this step. Thereafter, the objective function is calculated and satisfaction of all the constraints is investigated. In the case of violations in the constraints, a penalty term is taken into account. For each infeasible chromosome, the penalty is subtracted from the objective function value of the worst feasible chromosome in the current population. Consequently, infeasible chromosomes are not discarded and their information can be used for improving the algorithm search. Moreover, the roulette-wheel selection is applied for creating the next generation. The detailed information about the algorithm can be found in [42].

d) Case study

- *Test system and required data*

The presented method is implemented for two test TVPP. TVPP1 supervises a small distribution network to comprehensively investigate the proposed model. TVPP2 supervises a larger distribution network to show the effectiveness of the model. At first, TVPP1 just participates in the energy market and it is supposed that the TVPP cannot participate in the spinning reserve market in this case. After that, participation of TVPPs to provide reserve is investigated. In the following, at first the structure of TVPPs is defined and then the results of TVPPs' scheduling problem to provide energy and reserve are addressed.

TVPP1 supervises the network depicted in Fig. 4.8(a) with a DG, an ESS, and two loads. The technical characteristics of DG and ESS are presented in (4.131) and (4.132), respectively. It is supposed that the DG has been off for 2 hours before the beginning period of the scheduling problem:

$$P_{DG\_min} = 20 \text{ kW}, \quad P_{DG\_max} = 100 \text{ kW},$$
$$MUT_{DG} = 3 \text{ hr}, \quad MDT_{DG} = 2 \text{ hr},$$
$$R_{DG\_max} = 3 \text{ kW/min}, \quad (4.131)$$
$$C_{DG} = 0.01 P_{DG}^2 + 8.5 P_{DG},$$
$$C_{DG\_SU} = 70 \text{ MonetaryUnit}, \quad C_{DG\_SD} = 20 \text{ MonetaryUnit},$$

$$E_{ESS\_min} = 5 \text{ kWh}, \quad E_{ESS\_max} = 65 \text{ kWh},$$
$$\eta_{ESS} = 90\%, \quad SOC_{ESS\_0} = 5 \text{ kWh}. \quad (4.132)$$

The forecasted load of TVPP1 is shown in Fig. 4.8(b). It is supposed that a portion of the load (up to 25 kW) is interruptible which can be interrupted during hours 7–8 and 11–18, if necessary. The cost of load interruption which should be paid to interrupted customers is defined in

$$C_{IL}^{l,t}(P_{IL}^{l,t}) = 0.01(P_{IL}^{l,t})^2 + P_{IL}^{l,t}. \quad (4.133)$$

In summary, the main information about the DERs under supervision of TVPP1 is provided in Table 4.1.

Table 4.1 Associated parameters of DERs under supervision of TVPP1.

DGs	P_{DG_min}	P_{DG_max}	MUT_{DG}/MDT_{DG}	R_{DG_max}	C_{DG}	C_{DG_SU}/C_{DG_SD}
DG1	20 kW	100 kW	3/2 hr	3 kW/min	$0.01P_{DG}^2 + 8.5P_{DG}$	70/20 Monetary Unit
ESSs	E_{ESS_min}	E_{ESS_max}	η_{ESS}			SOC_{ESS_0}
ESS1	5 kWh	65 kWh	90%			5 kWh
Interruptible Loads	$P_{IL\_min}^l$	$P_{IL\_max}^l$	Interruption hours			$C_{IL}^{l,t}$
Load1, Load 2	0	25 kW	7–8 and 11–18			$0.01(P_{IL}^{l,t})^2 + P_{IL}^{l,t}$

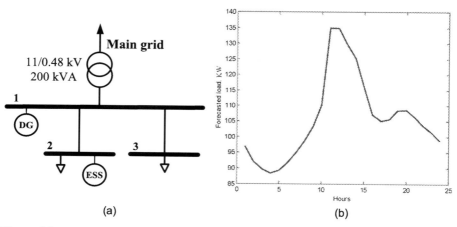

Figure 4.8 (a) Network under supervision, and (b) forecasted daily load for TVPP1 [42].

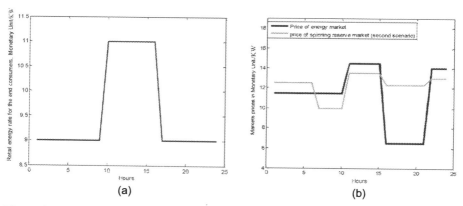

Figure 4.9 (a) Local network price and (b) energy and reserve markets prices for TVPP1 [42].

The local network price for end users is shown in Fig. 4.9(a). The market prices associated with the energy trading and spinning reserve services are shown in Fig. 4.9(b).

TVPP2 supervises the network shown in Fig. 4.10. There are 8 DGs in the network for which the technical characteristics are taken from [42] and presented in Table 4.2. It is assumed that DG1, DG4, and DG8 are utilized in both the energy and reserve markets for which their ramp rates given as follows:

$$R^1_{DG\_max} = 3 \text{ kW/min},$$
$$R^4_{DG\_max} = 2.5 \text{ kW/min}, \quad (4.134)$$
$$R^8_{DG\_max} = 3.5 \text{ kW/min}.$$

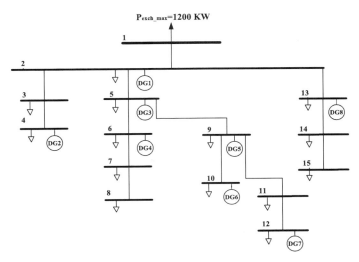

Figure 4.10 Network under supervision of TVPP2 [42].

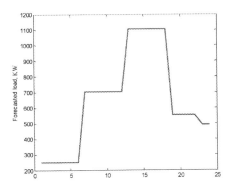

Figure 4.11 Forecasted daily load for TVPP2 [42].

The forecasted daily load curve in the case of TVPP2 is depicted in Fig. 4.11. Moreover, it is assumed that the loads connected to busses 4 and 7 of TVPP2 are interruptible, which can be interrupted up to 30 and 40 MW during the hours 7–22, respectively. The interruption cost functions of the loads connected to busses 4 and 7 are addressed in (4.135) and (4.136), respectively:

$$C_{IL}^{l,t}(P_{IL}^{l,t}) = 0.01(P_{IL}^{l,t})^2 + 3P_{IL}^{l,t}, \qquad (4.135)$$

$$C_{IL}^{l,t}(P_{IL}^{l,t}) = 0.01(P_{IL}^{l,t})^2 + 1.5P_{IL}^{l,t}. \qquad (4.136)$$

The information about DERs supervised by TVPP2 is summarized in Table 4.2.

The local network energy price for the end users and the energy and reserve market prices are shown in Fig. 4.12, (a) and (b), respectively.

Table 4.2 Associated parameters of DERs under supervision of TVPP2.

DGs	P_{DG_min}	P_{DG_max}	MUT_{DG}	MDT_{DG}	R_{DG_max}	C_{DG}
DG1	20 kW	85 kW	-	-	3 kW/min	$0.03P_{DG}^2 + 10.3P_{DG}$
DG2	35 kW	115 kW	4 hr	4 hr	-	$0.01P_{DG}^2 + 6.5P_{DG}$
DG3	30 kW	110 kW	3 hr	3 hr	-	$0.01P_{DG}^2 + 9.2P_{DG}$
DG4	20 kW	25 kW	-	-	2.5 kW/min	$0.03P_{DG}^2 + 12.6P_{DG}$
DG5	25 kW	80 kW	2 hr	2 hr	-	$0.01P_{DG}^2 + 7.2P_{DG}$
DG6	30 kW	80 kW	3 hr	3 hr	-	$0.01P_{DG}^2 + 7P_{DG}$
DG7	30 kW	100 kW	-	-	-	$0.01P_{DG}^2 + 30.1P_{DG}$
DG8	30 kW	90 kW	-	-	3.5 kW/min	$0.05P_{DG}^2 + 12.5P_{DG}$
Interruptible Loads	$P_{IL\_min}^l$	$P_{IL\_max}^l$	\multicolumn{2}{c}{Interruption hours}		$C_{IL}^{l,t}$
Load1	0	30 kW	\multicolumn{2}{c}{7–22}		$0.01(P_{IL}^{l,t})^2 + 3P_{IL}^{l,t}$
Load 2	0	40 kW	\multicolumn{2}{c}{7–22}		$0.01(P_{IL}^{l,t})^2 + 1.5P_{IL}^{l,t}$

Optimization model of a VPP to provide energy and reserve

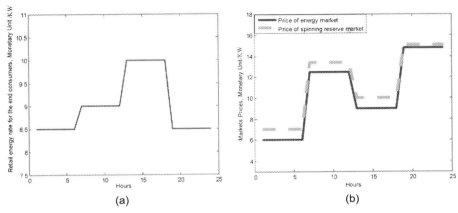

Figure 4.12 (a) Local network price and (b) energy and reserve markets prices for TVPP2 [42].

- *Numerical results*
- **Energy scheduling of TVPP1**

In this case, TVPP1 participates in the energy market and determines the DERs optimal energy scheduling. It is supposed that the TVPP cannot provide the spinning reserve service in this case. The result of TVPP1's energy scheduling is shown in Figs. 4.13, 4.14, and 4.15.

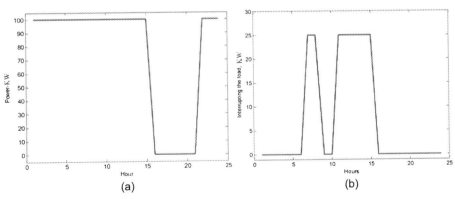

Figure 4.13 (a) Energy dispatch of DG and (b) load interruption utilization for TVPP1 [42].

As it is shown in the above figures, in the time intervals 7–8 and 11–15, when the energy market price is higher than the local network price, TVPP utilizes the load interruption options to sell power to the energy market or reduce the purchased power from the energy market. In the time interval 16–18, when the energy market price is lower than the local network price, DG is off and the total energy requirement of TVPP1 is bought from the market.

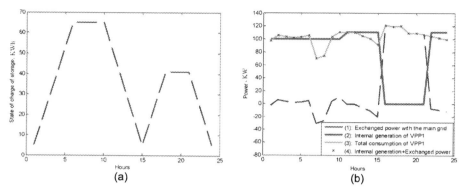

Figure 4.14 (a) ESS's state of charge and (b) total generation, consumption, and exchanged power of TVPP1 [42].

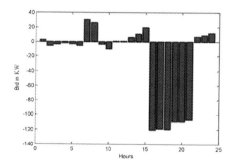

Figure 4.15 Energy bid in the energy market for TVPP1 [42].

According to Fig. 4.14(a), during low-demand time intervals the ESS is charged to store energy and return the energy back to the network at the times of peak load. In this regard, the ESS is completely charged during the time interval 1–8, when the price of market is 11.5 monetary units/kW. Afterward, the ESS is discharged at the time interval 11–14, when the price of market is 14.5 monetary units/kW. In the same way, the ESS is charged again at time interval 16–18 and discharged at 21–24.

Fig. 4.14(b) shows the optimal consumption, generation, and the optimal exchanged power of TVPP1 with the main grid. The consumption of TVPP is due to the customers' load consumption, the charging power of ESS units, and the network losses. The TVPP's generation capabilities are due to generation of DG units and discharge of ESSs. The exchanged power of TVPP into the market is positive when the TVPP purchases energy from the market, and otherwise it is negative. The sum of the power that TVPP purchases from the market and the internal generating power of TVPP is equal to the TVPP's total consumption at each time.

Fig. 4.15 represents the TVPP's bidding to the energy market in which, for better understanding, the energy sold to the market is shown by the positive sign, while the energy purchased from the market is shown by the negative sign. The maximum expected profit of the TVPP in this case is obtained as 2015.20 monetary units.

– Energy and reserve scheduling of TVPP1

In this case, participation of TVPP1 in the joint energy and reserve markets is investigated and the optimal energy and reserve management of DERs, as well as TVPP's performance in the markets, is addressed in Figs. 4.16 and 4.17.

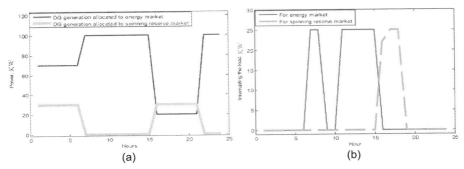

Figure 4.16 (a) Energy and reserve dispatch of DGs and (b) interruption load utilization in energy and reserve markets for TVPP1 [42].

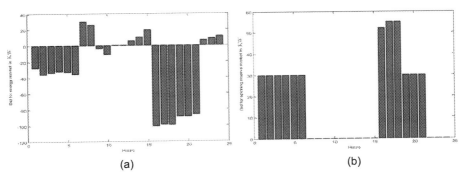

Figure 4.17 (a) Energy bid and (b) spinning reserve bid of TVPP 1 [42].

As it can be seen from the figures, the DG is turned on at the time intervals 1–15 and 22–24, since the price of energy market is higher than the DG's generation price at these times. During the time period 1–6, the price of the reserve market is higher than the price of the energy market; therefore, the DG is scheduled to allocate its maximum reserve capacity equal to 30 kW for the reserve market and the rest of its capacity equal to 70 kW for the energy market.

Moreover, at the times 16–21, when the price of energy market is low and the price of reserve market is high, the DG is committed and dispatching at the minimum allowable power (20 kW) to have the opportunity to allocate its maximum reserve capacity equal to 30 kW for spinning the reserve market. It should be noted that the cost of DG to produce 20 kW in the energy market and 30 kW in reserve market is equal to 450 monetary units for each hour of this time period. On the other hand, the value of buying 20 kW from the energy market with the price of 6.5 monetary units/kW is equal to 130 monetary units, and the value for buying 30 kW from the reserve market is equal to 369 monetary units. Therefore, the utilization of DG brings the profit of 49 monetary units in each hour of this time period for TVPP1.

Fig. 4.16(b) illustrates the interruptible load scheduling. It is worth mentioning that at the time intervals 7–8 and 11–15, the price of the energy market is higher than the price of the reserve market. Furthermore, the local price of end users is lower that the market prices. The interruption of the loads is beneficial in these times considering the interruption cost of the load. The total profit of TVPP is equal to 2689.3 monetary units in this case.

– Energy and reserve scheduling of TVPP2

In this case, the energy and reserve scheduling of TVPP2 for the optimal performance in the energy and reserve markets is considered. Figs. 4.18, 4.19, and 4.20 represent the optimal energy and reserve dispatch of the DERs, as well as optimal TVPP's bidding in market.

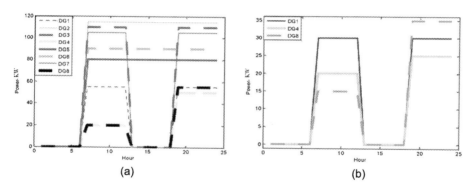

Figure 4.18 (a) Energy dispatch of DGs and (b) reserve dispatch of DGs in TVPP2 [42].

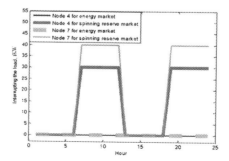

Figure 4.19 Interruption load utilization in energy and reserve markets for TVPP2 [42].

Optimization model of a VPP to provide energy and reserve

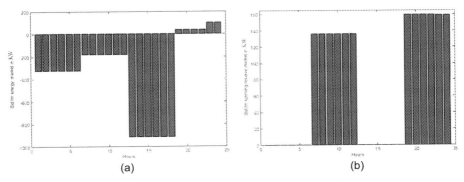

Figure 4.20 (a) Energy bid and (b) spinning reserve bid of TVPP2 [42].

It should be noted that in this case, the price of the spinning reserve market is higher than the price of the energy market at all the time intervals. During the time interval 1–6, the prices of the energy and spinning reserve markets are low. Therefore, all the DGs are off and the TVPP's consumption is provided from the market. During the time interval 7–18, DGs are committed and the TVPP's consumption is provided by DGs generation as well as purchasing from the market. During the interval 19–24, the price of market is high and it is profitable for TVPP to dispatch the DGs to sell power to the market. It is worth mentioning that at the times 7–12, when the price of energy market is high, all the DGs, except for DGs 1, 4, and 8, which have the opportunity to participate in the reserve market, are generated at their maximum capacity in the energy markets. The price of the spinning reserve market in these time intervals is high enough for DG1 to allocate its maximum reserve capacity of 30 kW and produce the rest of its capacity (55 kW) in the energy market. However, the generation costs of DGs 4 and 8 are higher than that of DG 1, and they are committed and scheduled to produce their minimum capacities (20 and 15 kW, respectively) in the energy market to have the opportunity to use their capabilities for providing spinning reserve service. During the time interval 13–18, the price of market decreases and DGs 1, 3, 4, 7, and 8 are turned off. During the interval 19–24, the energy and reserve markets' prices are high. Therefore, DGs 2, 3, 5, 6, and 7 are committed and dispatched at their maximum capacity in the energy market. DGs 1, 4, and 8 provide their maximum reserve capacity (30, 25, and 35 kW, respectively) in the spinning reserve market and the rest of their capacity in the energy market.

As it is found from Fig. 4.19, since the reserve market price is higher than the energy market price, the interruptible loads are used in the reserve market. The total profit of TVPP in this case is equal to 7009.21 monetary units.

4.4.2 VPP's scheduling in energy and reserve markets considering uncertainty

In this example, a VPP aggregates a DG, an ESS, a wind turbine, and a flexible load. The VPP optimizes the use of these energy resources to participate in the energy and reserve markets as a single entity. In this model, the uncertainty in market prices is considered through the stochastic programming approach. Furthermore, the uncertainties associated with wind turbine's output power, as well as requests for reserve deployment, are considered through confidence bounds and intervals, respectively.

In the deterministic formulation of the proposed model, the VPP aims at maximizing its profit in the energy and reserve markets considering the objective function (4.137). In this relation, $\lambda_{R\_U}^t R_{VPP\_U}^t$ and $\lambda_{R\_D}^t R_{VPP\_D}^t$ represent the revenue from provision of up- and down-reserve capacity, while the terms $k_{R\_U} \hat{\lambda}_{R\_U}^t R_{VPP\_U}^t \Delta t$ and $k_{R\_D} \hat{\lambda}_{R\_D}^t R_{VPP\_D}^t \Delta t$ express the revenue associated with the portion of reserve service which is requested to be deployed. Furthermore, the terms k_{R_U} and k_{R_D} are the parameters in the range of [0,1] which determine the fraction of requested reserve; C_{F_DG} and C_{V_DG} represent the fixed and variable costs associated with a DG unit. It should be noted that in the following formulation, P_{DG}^t, P_{WT}^t, $P_{ESS\_Ch}^t$, $P_{ESS\_Dch}^t$, and P_L^t contain the energy resources' power in both the energy and reserve markets:

$$\text{Max} \sum_{t \in T} \begin{bmatrix} -P_{VPP}^t \lambda_E^t + (\lambda_{R\_U}^t + k_{R\_U} \hat{\lambda}_{R\_U}^t \Delta t) R_{VPP\_U}^t \\ +(\lambda_{R\_D}^t - k_{R\_D} \hat{\lambda}_{R\_D}^t \Delta t) R_{VPP\_D}^t \\ -\begin{pmatrix} C_{F\_DG} u_{DG}^t + C_{V\_DG} P_{DG}^t \Delta t \\ +C_{DG\_SU}^i u_{DG\_SU}^t + C_{DG\_SD}^i u_{DG\_SD}^t \end{pmatrix} \end{bmatrix}. \quad (4.137)$$

The VPP considers the total production–consumption power balance in the network through constraint (4.138) by considering the energy and reserve powers. Inequalities (4.139)–(4.141) are considered to model the power trading limits within the energy and reserve markets, respectively. To model the VPP's energy resources, constraints (4.142)–(4.146) are associated with the generator unit, while constraints (4.147) and (4.148) model the flexible load. Moreover, the ESS is modeled through (4.149)–(4.152). Inequality (4.153) represents that the wind turbine's output power can take values lower than the available power:

$$P_{VPP}^t - k_{R\_U}^t R_{VPP\_U}^t + k_{R\_D}^t R_{VPP\_D}^t + P_{DG}^t - P_L^t + P_{WT}^t$$
$$- \left(P_{ESS\_Ch}^t - P_{ESS\_Dch}^t \right) = 0, \quad (4.138)$$

$$P_{VPP\_min} \leq P_{VPP}^t \leq P_{VPP\_max}, \quad (4.139)$$

$$0 \leq R_{VPP\_U}^t \leq R_{VPP\_U\_max}, \quad (4.140)$$

$$0 \leq R_{VPP\_D}^t \leq R_{VPP\_D\_max}, \quad (4.141)$$

$$u_{DG}^{i,t}, u_{DG\_SU}^{i,t}, u_{DG\_SD}^{i,t} \in \{0, 1\}, \quad (4.142)$$

$$u_{DG}^{i,t} - u_{DG}^{i,t-1} = u_{DG\_SU}^{i,t} - u_{DG\_SD}^{i,t}, \quad (4.143)$$

$$P_{DG\_min}^i u_{DG}^{i,t} \leq P_{DG}^{i,t} \leq P_{DG\_max}^i u_{DG}^{i,t}, \quad (4.144)$$

$$P_{DG}^{i,t} - P_{DG}^{i,t-1} \leq (RU_{DG}^i u_{DG}^{i,t-1} + RSU_{DG}^i u_{DG\_SU}^{i,t}), \quad (4.145)$$

$$P_{DG}^{i,t-1} - P_{DG}^{i,t} \leq (RD_{DG}^i u_{DG}^{i,t-1} + RSD_{DG}^i u_{DG\_SD}^{i,t}), \quad (4.146)$$

$$\sum_{t \in T} P_L^{l,t} \Delta t \geq E_{L\_min}^l, \qquad (4.147)$$

$$P_{L\_min}^l \leq P_L^{l,t} \leq P_{L\_max}^l, \qquad (4.148)$$

$$0 \leq P_{ESS\_Ch}^{s,t} \leq P_{ESS\_Ch\_max}^s, \qquad (4.149)$$

$$0 \leq P_{ESS\_Dch}^{s,t} \leq P_{ESS\_Dch\_max}^s, \qquad (4.150)$$

$$SoC_{ESS}^{s,t} = SoC_{ESS}^{s,t-1} + \left(\eta_{ESS\_Ch}^s P_{ESS\_Ch}^{s,t} - \frac{1}{\eta_{ESS\_Dch}^s} P_{ESS\_Dch}^{s,t} \right) \Delta t, \qquad (4.151)$$

$$SoC_{ESS\_min}^s \leq SoC_{ESS}^{s,t} \leq SoC_{ESS\_max}^s, \qquad (4.152)$$

$$0 \leq P_{WT} \leq P_{WT\_Available}. \qquad (4.153)$$

In order to consider the abovementioned uncertainties, a combination of scenario-based stochastic programming and adaptive robust optimization is used, and the resulting stochastic adaptive robust optimization problem is modeled as a tri-level problem. In the upper level, VPP determines its energy and reserve decisions in the markets, as well as the status of the DG. The middle-level problem models the worst case realization of reserve request and available wind power as uncertain parameters while the lower level identifies the optimal operation of VPP. The resulting problem is solved using column-and-constraint generation algorithm and the Karush–Kuhn–Tucker optimality conditions. The detailed explanations and associated mathematical formulation of the tri-level problem can be found in [9].

In order to demonstrate the proposed model by numerical results, it is supposed that the time interval Δt is one hour and the scheduling is performed for 24 hours. The information about electricity market is extracted from Spain electricity market and its associated uncertainty is considered using 34 scenarios. The confidence bounds and average amount of available wind power are depicted in Fig. 4.21. The associated parameters of energy resources are summarized in Table 4.3. In addition, the power exchange limits in the energy market are considered to be equal to -110 and 110 MW, while the participation limit in reserve market is 70 MW.

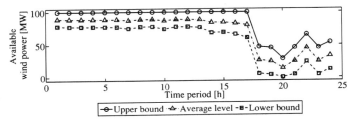

Figure 4.21 Average amount and confidence bounds of available wind power [9].

After solving the proposed problem by considering the abovementioned information, the main results demonstrate the VPP's power exchange in the energy and reserve markets which are depicted in Figs. 4.23, 4.24, and 4.25. In these figures, the VPP's

Table 4.3 Associated parameters of the resources under VPP's supervision

Generators	P_{DG_min}	P_{DG_max}	R_{DG_max}	C_{DG}	C_{DG_SU}	C_{DG_SD}
Generator1	20 MW	80 MW	50 MW/h	$40 P_{DG} + 50$	100 \$	100 \$
ESSs	E_{ESS_min}	E_{ESS_max}	$P^s_{ESS\_Ch\ max}$	$P^s_{ESS\_Dch\ max}$	η_{ESS}	$SOC_{ESS\ 0}$
ESS1	0	100 MWh	50 MW	50 MW	90%	50 MWh
Interruptible Loads	$E^l_{L\_min}$			Hourly Consumption limit		
Load1	700 MWh			In Fig. 4.22		

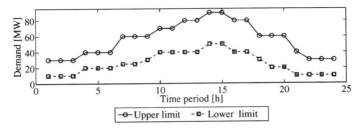

Figure 4.22 Hourly consumption limits of flexible load [9].

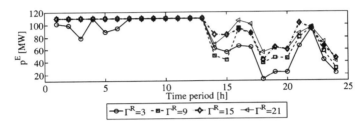

Figure 4.23 VPP's energy schedule [9].

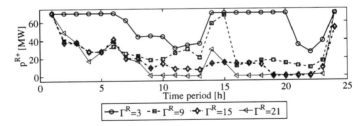

Figure 4.24 VPP's up-reserve schedule [9].

power exchange is presented for different uncertainty budgets. In this regard, the uncertainty budget of wind power (i.e., Γ^W) is assumed to be 12 and the robustness of the model is analyzed for different uncertainty budgets of reserve deployment request (i.e., Γ^R can be changed in the range from 0 to 24). It should be noted that a lower uncertainty budget leads to less conservative solutions while higher values yield more robust solutions.

Fig. 4.23 represents the VPP's schedule for participation in the energy market. As it can be seen in the figure, at the first hours of the day, VPP sells more power in the energy market because of high wind power generation and less load consumption. On the contrary, at the last hours of the day, the wind power generation is decreased and VPP sells less power in the energy market. The schedule of VPP in the up-reserve market is depicted in Fig. 4.24. It can be noticed that VPP has more participation schedule for lower amounts of uncertainty budget. Fig. 4.25 demonstrates the VPP participation in the down-reserve market. In this case, the most scheduled participation of VPP is equal to 70 MW for $\Gamma^R = 3$.

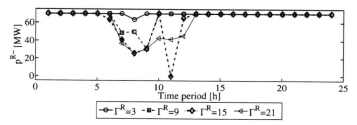

Figure 4.25 VPP's down-reserve schedule [9].

4.5 Conclusion

In this chapter, energy and reserve scheduling and operation models of VPP were put under investigation. In this regard, energy and reserve management models of DERs under VPP's supervision were addressed. Thereafter, the VPP's optimization models to provide energy and reserve service for market trading and satisfying the network operational constraints were taken into account. Afterward, the model of VPP exploiting DERs' flexibility was addressed. Finally, two examples illustrating VPP's optimal performance in the energy and reserve markets were considered. The results indicated the effective role of VPP to determine DERs' optimal energy and reserve management, as well as VPP's optimal offering in the energy and reserve markets to gain profit while satisfying the network constraints. Furthermore, it was shown that by providing the reserve service, the VPP can make more profit besides other advantages of reserve provision.

Nomenclature

Indices (sets)
i (*DG*) Distributed generation units
s (*ESS*) Energy storage systems
n (*EV*) Electric vehicles
h (*HVAC*) Heating, ventilation, and air conditioning systems
c (*C*) Customers
a (*AP*) Responsive appliances
l (*L/IL*) Loads/interruptible loads
r (*RES*) Renewable energy sources
w (*WT*) Wind turbines
p (*PV*) Photovoltaic units
t, t_1 (*T/T_{IL}*) Time steps
b, b' (*B, B'*) Network buses
Li (*Line*) Network lines
m (*M*) Market participants
v Wind speed

Parameters
$P^i_{DG\_min}$, $P^i_{DG\_max}$ Minimum and maximum limits of DGs' output power
$a^i_{DG}, b^i_{DG}, c^i_{DG}$ Coefficients of DG units' cost function

$C_{DG\_min}^i$ Cost of DGs to produce minimum allowable power
$C_{F\_DG}^i, C_{V\_DG}^i$ Fixed and variable costs associated with DGs
$s^{i,j}$ Slope of the segments of DG's piecewise linear cost curve
$Ps_{DG\_min}^{i,j,t}, Ps_{DG\_max}^{i,j,t}$ Minimum and maximum allowed power of segments of DG's piecewise linear cost curve
RU_{DG}^i, RD_{DG}^i Up-ramping and down-ramping limits of DGs
RSU_{DG}^i, RSD_{DG}^i Start-up and shut-down ramp rates of DGs
MUT_{DG}^i, MDT_{DG}^i Mean up-time and mean down-time of DGs
$C_{DG\_SU}^i\ C_{DG\_SD}^i$ Start-up and shut-down costs of DGs
$P_{ESS\_Ch\_max}^s$ Maximum allowed charging power of ESSs
$P_{ESS\_Dch\_max}^s$ Maximum allowed discharging power of ESSs
$\eta_{ESS\_Ch}^s, \eta_{ESS\_Dch}^s$ Charging and discharging efficiency of ESSs
$SoC_{ESS\_min}^s$ Minimum limit of ESSs' state of charge
$SoC_{ESS\_max}^s$ Maximum limit of ESSs' state of charge
$P_{ESS\_init}^s$ Initial power of ESSs
a_{ESS}^s, b_{ESS}^s Coefficients of ESSs' cost function
$\alpha_{EV}^n, \beta_{EV}^n$ Arrival and departure time of EVs to/from their homes
$P_{EV\_Ch\_max}^s$ Maximum allowed charging power of EVs
$P_{EV\_Dch\_max}^s$ Maximum allowed discharging power of EVs
$\eta_{EV\_Ch}^s, \eta_{EV\_Dch}^s$ Charging and discharging efficiency of EVs
$SoC_{EV\_min}^s$ Minimum limit of EVs' state of charge
$SoC_{EV\_max}^s$ Maximum limit of EVs' state of charge
$SoC_{EV\_REQ}^s$ Required level of ESSs' state of charge at the end of scheduling program
$P_{HVAC\_max}^h$ Maximum limit of consumed power of HVAC systems
$\theta_{Buil\_min}^c, \theta_{Buil\_max}^c$ Minimum and maximum desired limits of buildings' temperature
R_{Buil}^c, C_{Buil}^c Equivalent thermal resistance and heat capacity of space of customers' buildings
θ_O^t Ambient temperature
$\alpha_{AP}^{c,a}, \beta_{AP}^{c,a}$ Desirable time period for customers to start appliance s
$k_{AP}^{c,a}$ Number of time intervals required for complete application of appliances
$p_{AP}^{c,a,t}$ Nominal required power of appliances at each time step
$P_{IL\_max}^l$ Customers' interruption power limit
T_{IL}^l Customers' interruption time limit
$E_{L\_min}^l$ Minimum required energy of customers
$a_{IL}^l, b_{IL}^l, c_{IL}^l$ Coefficients of load interruption cost
$P_{L\_Fixed}^{l,t}$ Fixed loads power
$P_{L\_min}^l, P_{L\_max}^l$ Minimum and maximum demand of responsive customers
$R_{L\_U\_max}^l, R_{L\_D\_max}^l$ Maximum allowed increase and decrease of loads at each time step
$RU_L^l\ RD_L^l$ Up- and down-ramping limits of loads
$T_{LR\_max}^l$ Time limit of loads' utilization in reserve management problem
$P_{RES}^{r,t}$ Production power of RESs
$P_{WT\_Available}^{r,t}$ Available power of wind turbines
Y Admittance matrix of the network
$S_{Line\_max}^{b,b'}$ Maximum power flow of lines
V_{min}^b, V_{max}^b Minimum and maximum voltages of buses
λ_E^t Price of energy market

$\lambda_{Ls}^{t}, \lambda_{Lp}^{t}$ Price of selling/purchasing energy to/from customers
$\lambda_{R\_U}^{t} \lambda_{R\_D}^{t}$ Price of up- and down-reserve capacities
$\hat{\lambda}_{R\_U}^{t} \hat{\lambda}_{R\_D}^{t}$ Energy price of reserve capacities which are called on to produce
$\lambda_{UF}^{t} \lambda_{DF}^{t}$ Price of upward and downward flexibility powers
$k_{R\_U} \, k_{R\_D}$ Coefficients determined the portion of up- and down-reserve capacities which are called on to produce
$R_{VPP\_U\_max}, R_{VPP\_D\_max}$ Limits for VPP to exchange up- and down-reserve capacities
P_{Exch_max} Maximum allowed power transfer between the VPP and market
RM^{t} Network reserve margin
$T, \Delta t$ Number of time steps in scheduling horizon and duration of time steps
Γ^{W}, Γ^{R} Uncertainty budgets associated with wind power uncertainty and reserve deployment request

Variables

$P_{DG}^{i,t}$ Output power of DGs in energy market
$u_{DG}^{i,t}$ Commitment status of DGs
$C_{DG}^{i,t}$ Generation cost of DGs
$Ps_{DG}^{i,j,t}$ Power of segments of DG's in piecewise linear cost curve
$u_{DG\_SU}^{i,t}, u_{DG\_SD}^{i,t}$ Start-up and shut-down statuses of DGs
$R_{DG}^{i,t}$ DGs' reserve capacity
$R_{DG\_U}^{i,t}, R_{DG\_D}^{i,t}$ DGs' up- and down-reserve capacities
$T_{DG\_ON}^{i,t}, T_{DG\_OFF}^{i,t}$ Numbers of hours that DGs has been on/off before hour t
$P_{ESS\_Ch}^{s}, P_{ESS\_Dch}^{s}$ Charging and discharging powers of ESSs in energy market
$u_{ESS\_Ch}^{s,t}, u_{ESS\_Dch}^{s,t}$ Charging and discharging statuses of ESSs
$SoC_{ESS}^{s,t}$ State of charge of ESSs
$R_{ESS\_U}^{s,t}, R_{ESS\_D}^{s,t}$ Up- and down-reserve capacity of ESSs
$P_{EV\_Ch}^{n,t}, P_{EV\_Dch}^{n,t}$ Charging and discharging powers of EVs in energy market
$u_{ch\_EV}^{n,t}, u_{Dch\_EV}^{n,t}$ Charging and discharging statuses of EVs
$SoC_{EV}^{s,t}$ State of charge of EVs
P_{HVAC}^{h} Consumed power of HVAC systems in the energy market
$\theta_{Buil}^{c,t}$ Temperature of customers' buildings
$P_{AP}^{c,a,t}$ Consumed power of appliances in the energy market
$u_{AP}^{c,a,t}$ Start-up status of the appliances
$P_{IL}^{l,t}$ Customers' interrupted loads in the energy market
$u_{IL}^{d,t}$ Status of utilizing the interruptible loads
$C_{IL}^{l,t}$ Cost of load interruption
$P_{L}^{l,t}$ Customers' loads consisting of their responsive and interruptible loads
$R_{L\_U}^{l,t}, R_{L\_D}^{l,t}$ Up- and down-reserve capacities of loads
$u_{LR}^{l,t}$ Loads' utilization status in the reserve management problem
$R_{IL}^{l,t}$ Increase in interrupted power of customers during the reserve scheduling problem
$UF_{DER}^{d,t}, DF_{DER}^{d,t}$ Upward and downward flexibility powers of DERs
$y_{UF\_DER}^{d,t}, y_{UF\_DER}^{d,t}$ Statuses of upward and downward flexibility utilization of DERs
$S_{Gen}^{b,t}$ Generating power injected to buses
$S_{Dem}^{b,t}$ Demand power flowing out of buses

$S_{Line}^{b,b',t}$ Power flowing through the lines
V^b Voltage of network buses
P_{VPP}^t VPP's exchanged power with the energy market
$(\Phi^{m,t}, \pi_\Phi^{m,t})$ Agents' production bids to the market as pairs of power and price
$(\Omega^{m,t}, \pi_\Omega^{m,t})$ Agents' consumption offers to the market as pairs of power and price
$\phi^{VPP,t}, \omega^{VPP,t}$ Production and consumption bid powers of VPP which are accepted in the market
$\tilde{\lambda}_E^t$ Energy market clearing price
$R_{VPP\_U}^t, R_{VPP\_D}^t$ Upward and downward flexibility powers of VPP in the flexibility market

References

[1] Ontario's power system (IESO), [Online]. Available: http://www.ieso.ca/en/Learn/Ontario-Power-System/A-Smarter-Grid/Distributed-Energy-Resources.
[2] Simone Minniti, et al., Local markets for flexibility trading: key stages and enablers, Energies 11 (11) (2018) 3074.
[3] D. Pudjianto, et al., The virtual power plant: enabling integration of distributed generation and demand, Fenix Bulletin 2 (2008) 10–16.
[4] Hedayat Saboori, M. Mohammadi, R. Taghe, Virtual power plant (VPP), definition, concept, components and types, in: 2011 Asia–Pacific Power and Energy Engineering Conference, IEEE, 2011, pp. 1–4.
[5] Evaggelos G. Kardakos, Christos K. Simoglou, Anastasios G. Bakirtzis, Optimal offering strategy of a virtual power plant: a stochastic bi-level approach, IEEE Transactions on Smart Grid 7 (2) (2016) 794–806.
[6] Mehrdad Tahmasebi, Jagadeesh Pasupuleti, Self-scheduling of wind power generation with direct load control demand response as a virtual power plant, Indian Journal of Science and Technology 6 (11) (2013) 5443–5449.
[7] Matteo Vasirani, et al., An agent-based approach to virtual power plants of wind power generators and electric vehicles, IEEE Transactions on Smart Grid 4 (3) (2013) 1314–1322.
[8] Yizhou Zhou, et al., Four-level robust model for a virtual power plant in energy and reserve markets, IET Generation, Transmission & Distribution 13 (11) (2019) 2036–2043.
[9] Ana Baringo, Luis Baringo, Jose M. Arroyo, Day-ahead self-scheduling of a virtual power plant in energy and reserve electricity markets under uncertainty, IEEE Transactions on Power Systems 34 (3) (2018) 1881–1894.
[10] Niloofar Pourghaderi, et al., Reliability-based optimal bidding strategy of a technical virtual power plant, IEEE Systems Journal (2021).
[11] Marco Giuntoli, Davide Poli, Optimized thermal and electrical scheduling of a large scale virtual power plant in the presence of energy storages, IEEE Transactions on Smart Grid 4 (2) (2013) 942–955.
[12] Malahat Peik-Herfeh, H. Seifi, M.K. Sheikh-El-Eslami, Decision making of a virtual power plant under uncertainties for bidding in a day-ahead market using point estimate method, International Journal of Electrical Power & Energy Systems 44 (1) (2013) 88–98.
[13] Peyman Karimyan, et al., Stochastic approach to represent distributed energy resources in the form of a virtual power plant in energy and reserve markets, IET Generation, Transmission & Distribution 10 (8) (2016) 1792–1804.

[14] Elaheh Mashhour, Seyed Masoud Moghaddas-Tafreshi, Bidding strategy of virtual power plant for participating in energy and spinning reserve markets—part I: problem formulation, IEEE Transactions on Power Systems 26 (2) (2010) 949–956.
[15] Carlos Ruiz, Antonio J. Conejo, Pool strategy of a producer with endogenous formation of locational marginal prices, IEEE Transactions on Power Systems 24 (4) (2009) 1855–1866.
[16] Niloofar Pourghaderi, et al., Energy scheduling of a technical virtual power plant in presence of electric vehicles, in: Electrical Engineering (ICEE), 2017 Iranian Conference on, IEEE, 2017.
[17] Hossein Nezamabadi, Mehrdad Setayesh Nazar, Arbitrage strategy of virtual power plants in energy, spinning reserve and reactive power markets, IET Generation, Transmission & Distribution 10 (3) (2016) 750–763.
[18] Ali Ghahgharaee Zamani, Alireza Zakariazadeh, Shahram Jadid, Day-ahead resource scheduling of a renewable energy based virtual power plant, Applied Energy 169 (2016) 324–340.
[19] Huy Nguyen-Duc, Nhung Nguyen-Hong, A study on the bidding strategy of the Virtual Power Plant in energy and reserve market, Energy Reports 6 (2020) 622–626.
[20] Niloofar Pourghaderi, et al., Energy and flexibility scheduling of DERs under TVPP's supervision using market-based framework, in: 2020 IEEE 4th International Conference on Intelligent Energy and Power Systems (IEPS), IEEE, 2020.
[21] Jose P. Iria, Filipe Joel Soares, Manuel A. Matos, Trading small prosumers flexibility in the energy and tertiary reserve markets, IEEE Transactions on Smart Grid 10 (3) (2018) 2371–2382.
[22] Zhen Shu, Panida Jirutitijaroen, Optimal operation strategy of energy storage system for grid-connected wind power plants, IEEE Transactions on Sustainable Energy 5 (1) (2013) 190–199.
[23] Wenjun Tang, Hong-Tzer Yang, Optimal operation and bidding strategy of a virtual power plant integrated with energy storage systems and elasticity demand response, IEEE Access 7 (2019) 79798–79809.
[24] 2017 National Household Travel Survey, U.S. transportation department, [Online]. Available: https://nhts.ornl.gov/assets/2017_nhts_summary_travel_trends.pdf.
[25] Terry Cousins, Using time of use (TOU) tariffs in industrial, commercial and residential applications effectively, TLC Engineering Solutions (2009).
[26] Niloofar Pourghaderi, et al., Commercial demand response programs in bidding of a technical virtual power plant, IEEE Transactions on Industrial Informatics 14 (11) (2018) 5100–5111.
[27] Ning Lu, An evaluation of the HVAC load potential for providing load balancing service, IEEE Transactions on Smart Grid 3 (3) (2012) 1263–1270.
[28] Milad Kabirifar, et al., Centralized framework to coordinate residential demand response potentials, in: Proceeding of International Smart Grid Conference, Gwangjo, Korea, 2015, pp. 243–248.
[29] Jingjing Zhang, et al., Two-stage load-scheduling model for the incentive-based demand response of industrial users considering load aggregators, IET Generation, Transmission & Distribution 12 (14) (2018) 3518–3526.
[30] Allen J. Wood, Bruce F. Wollenberg, Gerald B. Sheblé, Power Generation, Operation, and Control, John Wiley & Sons, 2013.
[31] Cai Tao, Duan Shanxu, Chen Changsong, Forecasting power output for grid-connected photovoltaic power system without using solar radiation measurement, in: The 2nd International Symposium on Power Electronics for Distributed Generation Systems, IEEE, 2010.

[32] Rodrigo Henríquez, et al., Participation of demand response aggregators in electricity markets: Optimal portfolio management, IEEE Transactions on Smart Grid 9 (5) (2017) 4861–4871.

[33] Christophe Kieny, et al., On the concept and the interest of virtual power plant: some results from the European project Fenix, in: 2009 IEEE Power & Energy Society General Meeting, IEEE, 2009.

[34] Milad Kabirifar, et al., Deterministic and probabilistic models for energy management in distribution systems, in: Handbook of Optimization in Electric Power Distribution Systems, Springer, Cham, 2020, pp. 343–382.

[35] Niloofar Pourghaderi, et al., Energy management framework for a TVPP in active distribution network with diverse DERs, in: 2019 27th Iranian Conference on Electrical Engineering (ICEE), IEEE, 2019.

[36] Nima Amjady, et al., Adaptive robust expansion planning for a distribution network with DERs, IEEE Transactions on Power Systems 33 (2) (2017) 1698–1715.

[37] Liwei Ju, et al., Multi-objective stochastic scheduling optimization model for connecting a virtual power plant to wind-photovoltaic-electric vehicles considering uncertainties and demand response, Energy Conversion and Management 128 (2016) 160–177.

[38] Morteza Rahimiyan, Luis Baringo, Strategic bidding for a virtual power plant in the day-ahead and real-time markets: a price-taker robust optimization approach, IEEE Transactions on Power Systems 31 (4) (2015) 2676–2687.

[39] Yizhou Zhou, et al., A linear chance constrained model for a virtual power plant in day-ahead, real-time and spinning reserve markets, in: 2019 IEEE Power & Energy Society General Meeting (PESGM), IEEE, 2019.

[40] Andreas Ulbig, Göran Andersson, On operational flexibility in power systems, in: 2012 IEEE Power and Energy Society General Meeting, IEEE, 2012.

[41] José Villar, Ricardo Bessa, Manuel Matos, Flexibility products and markets: literature review, Electric Power Systems Research 154 (2018) 329–340.

[42] Elaheh Mashhour, Seyed Masoud Moghaddas-Tafreshi, Bidding strategy of virtual power plant for participating in energy and spinning reserve markets—part II: numerical analysis, IEEE Transactions on Power Systems 26 (2) (2010) 957–964.

[43] Yann Rebours, Daniel Kirschen, Marc Trotignon, Fundamental design issues in markets for ancillary services, The Electricity Journal 20 (6) (2007) 26–34.

[44] Mohammad Shahidehpour, Hatim Yamin, Zuyi Li, Market Operations in Electric Power Systems: Forecasting, Scheduling, and Risk Management, John Wiley & Sons, 2003.

[45] Elaheh Mashhour, S.M. Moghaddas-Tafreshi, Mathematical modeling of electrochemical storage for incorporation in methods to optimize the operational planning of an interconnected micro grid, Journal of Zhejiang University. Science C 11 (9) (2010) 737–750.

[46] Rodrigo Palma-Behnke, Luis S. Vargas, Alejandro Jofré, A distribution company energy acquisition market model with integration of distributed generation and load curtailment options, IEEE Transactions on Power Systems 20 (4) (2005) 1718–1727.

Provision of ancillary services in the electricity markets

Mehrdad Setayesh Nazar and Kiumars Rahmani
Faculty of Electrical Engineering, Shahid Beheshti University, Tehran, Iran

5.1 Introduction

Distributed energy resources (DERs) are energy conversion and storage facilities that can be dispatched and controlled by the system operators and/or their own local operators. Different forms of DERs commitment strategies were proposed in recent years based on centralized or decentralized dispatching procedures. The virtual power plant concept is one of the most common forms of these conceptual frameworks that utilizes a hierarchical framework to dispatch the DERs. A virtual power plant can be equipped with renewable energy resources, demand response program alternatives, and custom energy conversion technologies [1].

Reference [1] presented an algorithm for the scheduling of VPP considering energy storage degradation. The process used a two-stage stochastic optimization procedure that performed the day-ahead scheduling in the first stage, and the optimal decision variables' values for the balancing market were searched for in the second stage. The model was linearized and the CPLEX solver was employed for optimizing the problem. Reference [2] introduced a two-stage multi-VPP dispatching model that considered the demand response mechanism. The first and second approaches modelled the optimal scheduling of a distribution system energy transaction within the VPP and multi-VPP scheduling problems, respectively. The first solution utilized linear programming and the second-stage optimization was carried out using game theory. Reference [3] assessed a two-level optimization framework for the optimal scheduling of multi-operator VPPs considering the interactions with the distribution system operator. The lower-level problem optimized the economic dispatch of distribution system resources; meanwhile, the upper-level problem determined the optimal bidding strategy of VPPs. The model considered the uncertainties of dispatching variables and used a predictive control model. Reference [4] presented a two-layer optimization model for optimal VPP scheduling considering intermittent electricity generation and its uncertainties. The upper-layer problem predicted the electricity generation of wind and photovoltaic units and maximized the profit of VPP for the day-ahead horizon. The lower-layer revised the predicted values and minimized the costs of energy procurement considering demand response programs. The introduced non-linear problem was linearized, and a robust optimization procedure was carried out to find the optimal solutions. Reference [5] introduced a joint energy and regulating reserve optimization framework for VPP that modelled distributed generation, electric vehicle, and pump storage facilities. The optimization was carried out using the point estimate method.

Reference [6] evaluated a probabilistic optimization method for optimal scheduling of VPP that consisted of heating and electrical energy resources. The model considered the uncertainties of loads, prices, and electricity generation. The model was linearized and the XPRESS solver of GAMS was utilized to solve the problem. Reference [7] presented a conditional value-at-risk (CVaR) model to schedule the distributed energy-based VPP that participated in the demand response program. The risk-averse model maximized the revenue of VPP, and the optimization was carried out using the CPLEX solver for 30-bus IEEE test system. Reference [8] introduced a three-level formulation for multi-energy VPP that used a scenario based optimization process. The upper-level problem minimized the energy procurement costs for the day-ahead horizon that consisted of an energy resource commitment process. The mid-level problem optimized the economic dispatch of energy resources for a 30-minute horizon. The lower-level problem optimized the optimal real-time dispatch of the system for a five-minute interval and considered the penalty costs of deviations. Reference [9] proposed a stochastic optimization framework for the scheduling of VPP. The model considered intermittent electricity generation facilities, market price, and load uncertainties. The objective functions consisted of the VPP profit maximization and emission minimization. The simulation was carried out using particle swarm optimization. The emission minimization problem utilized a two-stage stochastic optimization process. Reference [10] introduced a risk management algorithm for day-ahead and intra-day optimal scheduling of VPP that considered single-line contingencies and demand response programs. The model maximized the profit of VPP using the CVaR minimization operational strategy. The case study was carried out for a real industrial VPP. Reference [11] proposed a multi-objective optimization model for VPP that consisted of wind, photovoltaic, electric vehicle, and conventional distributed generation facilities. The objective functions consisted of compensation costs, abandoned energy costs, and revenue of VPP. The first and second objective functions were minimized, and the third was maximized. A chance-constrained optimization process was employed to solve the problem.

Reference [12] evaluated an online distributed optimization algorithm to schedule the DER-based VPPs considering coupled constraints of VPP and the system. The real-time problem was optimized centrally, and the case study was performed for a 33-bus IEEE test system. Reference [13] proposed a framework for optimal scheduling of VPP that consisted of electric vehicles' parking lots. The optimization was carried out for the energy and reserve markets in day-ahead horizon considering the uncertainties of the electricity market, electricity generation, and stochastic behavior of electric vehicles. The CVaR profit of VPP was used as the objective function of the maximization process. Reference [14] introduced an integrated model for distributed energy resource. The model considered of photovoltaic, wind turbine, and hydroelectric facilities. The optimization was carried out using a mixed-integer linear programming solver, and the case study was performed for Aragon (Spain). Reference [15] developed an optimization framework for scheduling of VPP in the day-ahead energy, reactive, and reserve markets. The proposed model considered the arbitrage strategy in different markets and maximized the profit of the VPP. The DICOPT solver of GAMS optimized the non-linear program; two test systems assessed the proposed algorithm that comprised 6-bus and 30-bus IEEE test systems. Reference [16] assessed a two-level optimization algorithm that consisted of intra-day and real-time optimization processes. The upper-level problem optimized the power exchange of VPP with the market; meanwhile, the lower-level problem minimized the regulation deviations. The model considered the dynamic behavior of thermostatic loads and utilized a mixed-integer non-linear optimization process to solve the problem. Reference [17] presented a CVaR-based optimization process for optimal scheduling of VPP considering demand response, contingencies, and system uncertainties. The objective function maximized the profit of VPP for the energy and reserve markets. The optimization process was performed using the CPLEX solver of GAMS. Reference [18] introduced a bi-level optimization process that optimally dispatched the VPP energy resources for the day-ahead horizon. The upper-level problem maximized the profit of the VPP; meanwhile, the lower-level problem minimized the system energy procurement costs. The case study was performed for the 6-bus IEEE RBTS, and the IPOPT solver was used for the non-linear programming problem. Reference [19] proposed a two-stage optimization procedure for optimal VPP scheduling. The first-stage problem optimized the bidding strategies of VPP for the day-ahead horizon. The second-stage problem minimized the penalty costs of the VPP for the intra-day market. The case study was performed for a real test system in China, and the optimization process was carried out using a heuristic optimization algorithm. Reference [20] presented a CVaR-based optimization process to schedule an integrated wind–thermal energy system considering the uncertainties of wind and market price. The optimization procedure utilized a risk-averse decision parameter to determine the bidding strategy of the energy system. Different scenarios of coordinated and uncoordinated bidding energy system were assessed for the Spanish electricity market. References [1–20] did not consider the simultaneous optimization of bidding strategies in the day-ahead and real-time markets. This book chapter is about the simultaneous optimization of VPP bidding strategy in the day-ahead and real-time market of energy and ancillary services.

5.2 Problem modelling and formulation

The virtual power plant consists of distributed energy resources that are DG units, WT facilities, PV arrays, PHEV parking lots, and demand response programs. As shown in Fig. 5.1, the VPP operator centrally dispatches the distributed energy resources for the day-ahead and real-time horizons and exchanges energy with the electricity market. The VPP sells active and reactive power, as well as reserve, to the day-ahead electricity market; meanwhile, it sells active and reactive power to the real-time electricity market. The VPP should optimize its energy and ancillary service transactions with the upward electricity market for the day-ahead and real-time horizons.

The uncertainties of the day-ahead optimization process are intermittent electricity generation, PHEVs electricity generation/consumption, and the day-ahead energy and ancillary service prices.

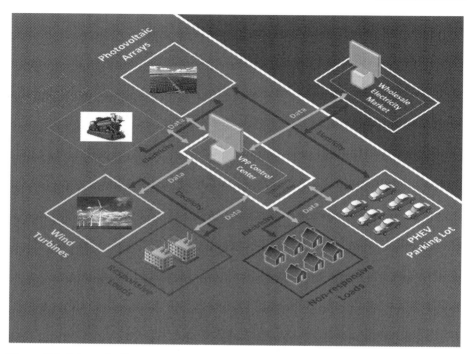

Figure 5.1 Schematic diagram of a virtual power plant and electricity market interactions.

5.2.1 First-stage problem

The day-ahead optimization process maximizes the expected revenue of the virtual power plant for the scheduling horizon. The optimization procedure determines the commitment statuses of distributed energy resources, which are centrally controlled by the virtual power plant operator. The proposed objective function of the first-stage problem can be presented as

$$Max\mathbb{S} = \begin{bmatrix} (\lambda^{DA}_{active} \cdot P^{DA}_{VPP} + \lambda^{DA}_{reactive} \cdot Q^{DA}_{VPP} + \lambda^{DA}_{reserve} \cdot S^{DA}_{VPP}) \\ +(\zeta^{DA}_{active} \cdot P_{Load} + \zeta^{DA}_{reactive} \cdot Q_{Load}) \\ -(C^{DA}_{DG} \cdot \varphi^{DA}_{DG} + C^{DA}_{DRP} \cdot \varphi^{DA}_{DRP} \\ + \sum_{PVAS} prob.C^{DA}_{PVA} + \sum_{WTS} prob.C^{DA}_{WT} \\ + \sum_{PHEVS} prob.C^{DA}_{PHEV} \cdot \varphi^{DA}_{PHEV}) \\ -\beta(\xi - \frac{1}{1-\alpha} \sum_{NOS} prob.\vartheta) \end{bmatrix}. \quad (5.1)$$

The objective function comprises the following terms: (1) the revenue of active and reactive power, and reserve sold to the electricity market; (2) the revenue of active and reactive power sold to the customers; (3) the cost of DG units, DRP, photovoltaic arrays, wind turbine facilities, and PHEVs commitment; and (4) the conditional value-at-risk term.

The β parameter is a weighting factor, and a risk-averse strategy corresponds to $\beta = 1$. The detailed uncertainty model is presented in [20].

The facilities loading constraints and AC load flow constraints are considered in this stage of optimization.

The power balance constraint can be presented as

$$\sum P^{DG} \mp \sum P^{ESSAs} - \sum P^{Loss} - \sum P^{Load} \mp \sum P^{PHEV}$$
$$+ \sum P^{WT} + \sum P^{PVA} \mp \sum P^{DRP} + \sum P^{Upward\ Market} = 0. \quad (5.2)$$

It is assumed that the VPP loads can be controlled by the VPP operator and the DRP constraints can be represented as

$$P_{Load} = P_{Load}^{Critical} + P_{Load}^{Dispatchable}. \quad (5.3)$$

The PHEVs constraints are charge and discharge, and maximum capacity constraints [21]. The charge/discharge of electric vehicles is considered as a stochastic process, and the detailed formulation of this procedure is presented in [21].

5.2.2 Second-stage problem

The second-stage optimization problem maximizes the revenue of the VPP in the real-time market. The VPP should maximize the profit of the sold active and reactive power in the real-time market and minimize the penalty of deviations. Thus, the objective function of the second-stage problem can be presented as

$$Max\mathbb{S} = \begin{bmatrix} (\lambda_{active}^{RT} \cdot P_{VPP}^{RT} + \lambda_{reactive}^{RT} \cdot Q_{VPP}^{RT}) - (\Delta C_{DG}^{RT} + \Delta C_{DRP}^{RT} + \Delta C_{PVA}^{RT} \\ + \Delta C_{WT}^{RT} + \Delta C_{PHEV}^{RT}) - Penalty_{active} - Penalty_{reactive} \\ - Penalty_{reserve} \end{bmatrix}.$$
$$(5.4)$$

Eq. (5.4) comprises of the following terms: (1) the revenue of active and reactive power sold to the real-time electricity market; (2) the mismatches of costs of DG, DRP, photovoltaic arrays, wind turbine facilities, and PHEVs; and (3) the penalty costs of active, reactive, and reserve mismatch.

The ramp-rates of DG units and PHEV charge/discharge limits, maximum and minimum loading of facilities, and AC load flow of system constraints are considered in this stage [21].

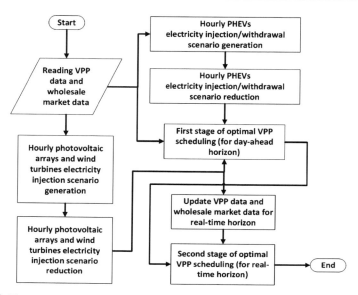

Figure 5.2 The proposed multi-stage optimization procedure.

5.3 Solution algorithm

The first- and second-stage problems are linearized and solved by the CPLEX solver of GAMS. Fig. 5.2 depicts the proposed optimization algorithm.

At first, the VPP and wholesale electricity market data are uploaded. Then, the photovoltaic arrays and wind turbines electricity generation scenarios are generated, and the scenarios are reduced using SCENRED of GAMS library [21]. The MATLAB® and GAMS packages are used for scenario generation/reduction and optimization procedures. The detailed linearization technique is available in [21] and is not presented for the sake of space saving.

The day-ahead optimization determines the 24 optimal dispatch and commitment statuses of the VPP DERs. However, the real-time optimization problem is calculated every 15 minutes and the optimal rescheduling of DERs determined.

5.4 Numerical results

The case study was carried out for a 57-bus IEEE test system. The virtual power plant consisted of six wind farms, seven DGs, two PVAs, four demand response program (provider) groups, and five PHEV parking lots. Table 5.1 presents the parameters of DGs.

The total installed capacity of thermal units was 200 MW, and the wind unit installed capacity was 150 MW. The maximum electricity generation of the first and

Provision of ancillary services in the electricity markets

Table 5.1 Distributed generation units' data.

#DG	P_{min}	P_{max}	Ramp-up	Ramp-down	Start-up cost	a	b	c
1	5	35	12	11	40	93	83	0.009
2	5	25	17	16	36	62	42	0.004
3	5	35	17	17	48	61	42	0.002
4	5	25	16	14	49	86	67	0.006
5	8	25	11	18	41	91	82	0.007
6	5	30	15	16	48	94	82	0.008
7	5	25	12	11	38	83	77	0.009

Table 5.2 The optimization input data.

System parameter	Value
Number of solar irradiation scenarios	200
Number of wind turbine power generation scenarios	400
Number of PHEVs contribution scenarios	500
Number of solar irradiation reduced scenarios	20
Number of wind turbine power generation reduced scenarios	40
Number of PHEVs contribution reduced scenarios	50
α (Confidence level)	0.95

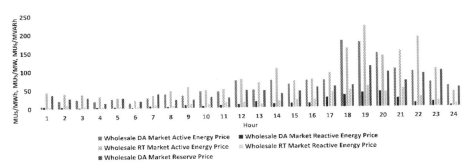

Figure 5.3 The forecasted prices for the active, reactive, and reserve day-ahead markets, as well as active and reactive real-time markets.

second photovoltaic arrays was 10 and 24 MW, respectively. Further, the aggregated capacity of PHEV parking electricity generation was about 30 MW. The aggregated electricity reduction of demand response program provider groups was about 25 MW.

Table 5.2 presents the input parameters of the optimization process.

Fig. 5.3 presents the forecasted prices for the active, reactive, and reserve day-ahead markets, as well as active and reactive real-time markets. The MU stands for monetary unit.

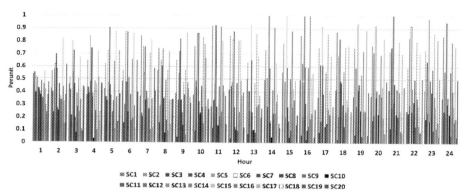

Figure 5.4 The estimated values of per unit electricity generation of wind turbines of the VPP for the 1st through 20th scenarios.

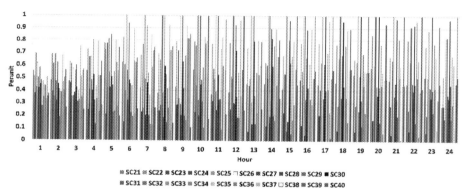

Figure 5.5 The estimated values of per unit electricity generation of wind turbines of the VPP for the 21st through 40th scenarios.

Figs. 5.4 and 5.5 present the estimated values of per unit electricity generation of wind turbines of the VPP. The total number of scenario reductions for wind electricity generation was 40 cases.

Figs. 5.6 and 5.7 depict the estimated values of electricity generation of the first and second PV arrays for different scenarios, respectively. The maximum values of the first and second photovoltaic arrays were about 8.81194 and 22.02825 MW, respectively.

Two cases were considered in our studies:

Case 1: The VPP coordinated its distributed energy resources bidding strategies and participated in the day-ahead and real-time markets,
Case 2: The VPP did not coordinate its distributed energy resources' bidding strategies and it only participated in the day-ahead market.

Fig. 5.8 shows the active energy bidding curves of distributed generation units for the first case, the uncoordinated and coordinated bidding conditions, and different values of risk-averse parameters. As shown in Fig. 5.8, the VPP increased the bidding

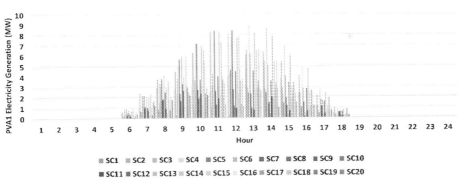

Figure 5.6 The estimated values of electricity generation of the first PV arrays for different scenarios.

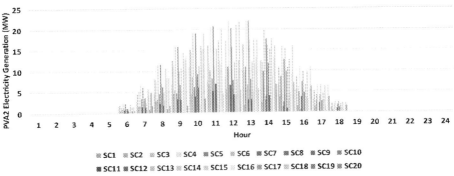

Figure 5.7 The estimated values of electricity generation of the first PV arrays for different scenarios.

price of its distributed generation units when these facilities were coordinately committed based on the fact that the coordinated bidding of distributed generation units increased their market share. Further, the risk-averse strategy of virtual power plant operator ($\beta = 0$) led to an increase in bidding price of distributed generation units.

Fig. 5.9 depicts the active power bidding curves of wind farm facilities for the first case, the uncoordinated and coordinated bidding conditions, and different values of risk-averse parameters.

As shown in Fig. 5.9, the VPP increased the bidding price of its wind farm facilities for uncoordinated and coordinated bidding strategies. The wind farms were increased their bidding prices for the coordinated bidding conditions and $\beta = 0$.

Fig. 5.10 presents the aggregated active power bidding curves of virtual power plant for the first case, the uncoordinated and coordinated bidding conditions, and different values of risk-averse parameters. The uncoordinated bidding price of the active power of VPP for $\beta = 1$ was the lowest value of bidding prices; meanwhile, the bidding price of coordinated bidding strategy for $\beta = 0$ was the highest value of bidding prices based

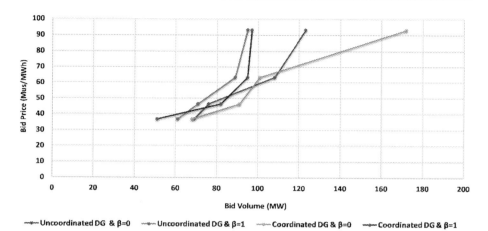

Figure 5.8 Active energy bidding curves of distributed generation units for the first case, the uncoordinated and coordinated bidding conditions, and different values of risk-averse parameters.

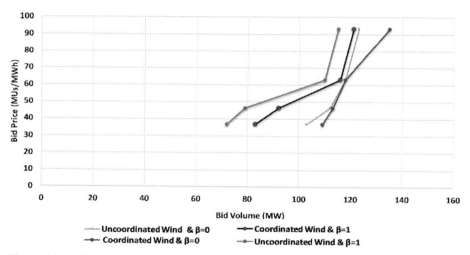

Figure 5.9 Active power bidding curves of wind farm facilities for the first case, the uncoordinated and coordinated bidding conditions, and different values of risk-averse parameters.

on the fact that the VPP operator endeavored to increase its market share and profit for coordinated bidding strategy and risk-averse conditions.

Fig. 5.11 depicts the active power bidding curves of wind farm facilities for the second case, the uncoordinated and coordinated bidding conditions, and different values of β. By comparing Figs. 5.9 and 5.11, it can be concluded that the wind farms of VPP reduced their bidding price and increased their volume of active power for uncoordinated and coordinated bidding strategies for the second case based on the fact that the VPP operator only maximized its profit for the day-ahead market.

Provision of ancillary services in the electricity markets 121

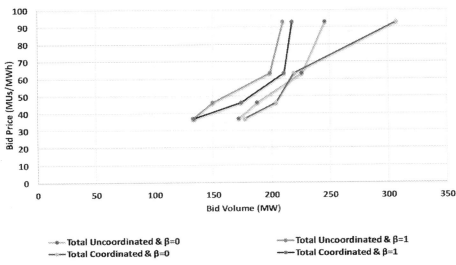

Figure 5.10 Aggregated active power bidding curves of virtual power plant for the first case, the uncoordinated and coordinated bidding conditions, and different values of risk-averse parameters.

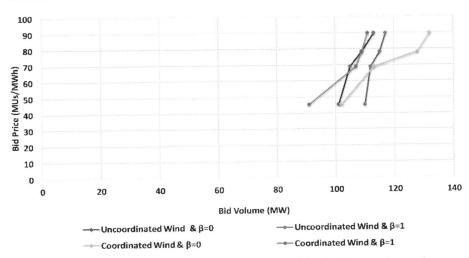

Figure 5.11 Active power bidding curves of wind farm facilities for the second case, the uncoordinated and coordinated bidding conditions, and different values of β.

Fig. 5.12 presents the active energy bidding curves of distributed generation units for the second case, the uncoordinated and coordinated bidding conditions, and different values of β. The distributed generation facilities reduced their bidding price and increased their volume of active power for the second case with respect to the corresponding bidding curves of the first case.

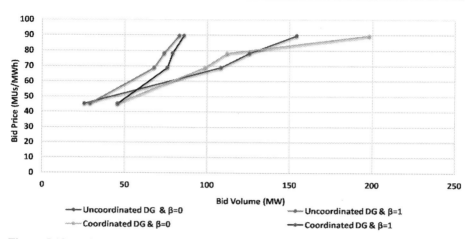

Figure 5.12 Active energy bidding curves of distributed generation units for the second case, the uncoordinated and coordinated bidding conditions, and different values of β.

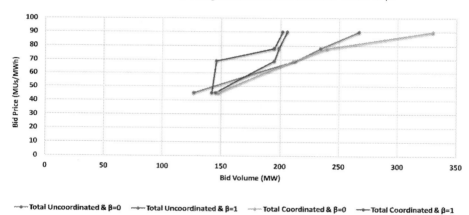

Figure 5.13 Aggregated active power bidding curves of virtual power plant for the second case, the uncoordinated and coordinated bidding conditions, and different values of β.

Fig. 5.13 depicts the aggregated active power bidding curves of virtual power plant for the second case, the uncoordinated and coordinated bidding conditions, and different values of β. By comparing Figs. 5.10 and 5.13, it can be concluded that the VPP increased the volume of active power and reduced the bidding price for the second case based on the fact that the VPP operator endeavors to maximize its profit in only the day-ahead market and did not participate in the real-time market.

Fig. 5.14 presents the estimated values of the sold energy of VPP in the reserve market for the first and second cases. The average values of energy sold in the reserve markets for the first and second cases were 37.8429 and 50.5057 MWh, respectively. The VPP sold less reserve in the day-ahead market for the first case based on the fact that the VPP gained more profit by participating in the real-time market. The

Provision of ancillary services in the electricity markets

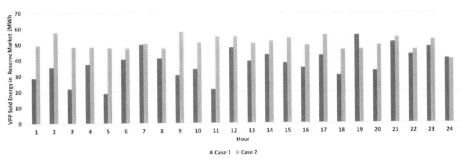

Figure 5.14 Estimated values of the sold energy of VPP in the reserve market for the first and second cases.

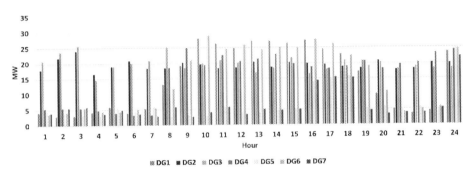

Figure 5.15 Estimated day-ahead market scheduling of distributed generation units.

maximum and minimum values of sold reserve energy for the first case were about 18.85106 and 55.57526 MWh, respectively. Further, the maximum and minimum reserve energy sold for the second case was about 40.46088 and 57.65502 MWh, respectively.

Fig. 5.15 depicts the estimated day-ahead market scheduling of distributed generation units. The aggregated energy production of distributed generation units for the day-ahead market was about 2361.3216 MWh. The maximum and minimum of active energy production of units were about 2.6884 and 28.7818 MWh, respectively.

Fig. 5.16 shows the estimated values of the day-ahead active power production/consumption for plug-in hybrid electric vehicles. The aggregated active energy production/consumption of plug-in hybrid electric vehicles for the day-ahead market was about -3.0417 MWh. The maximum values of active energy consumption and production were about 11.934 and 7.1191 MWh, respectively.

Fig. 5.17 presents the estimated values of the day-ahead reactive power production/consumption for plug-in hybrid electric vehicles. The aggregated reactive energy production/consumption of plug-in hybrid electric vehicles for the day-ahead market was about -2.73573 MVARh. The maximum values of reactive energy consumption and production were about 10.7407 and 6.40719 MVARh, respectively.

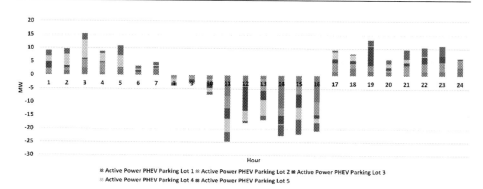

Figure 5.16 Estimated values of the day-ahead active power production/consumption for plug-in hybrid electric vehicles.

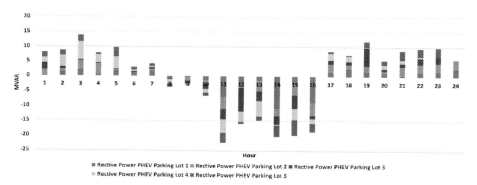

Figure 5.17 Estimated values of the day-ahead reactive power production/consumption for plug-in hybrid electric vehicles.

Fig. 5.18 depicts the estimated values of the allocated reserve of plug-in hybrid electric vehicles. The aggregated the allocated reserve energy of plug-in hybrid electric vehicles for the day-ahead market was about 113.04193 MWh. The maximum value of the allocated reserve was about 4.9329 MW.

Fig. 5.19 presents the estimated values of the day-ahead active power production/consumption for demand response program provider groups. The aggregated active energy consumption of DRP groups for the day-ahead market was about 107.3437 MWh. The maximum values of active energy consumption and production were about 15.896 and 16.9966 MWh, respectively.

Fig. 5.20 depicts the estimated values of the day-ahead reactive power production/consumption for demand response program provider groups. The aggregated reactive energy consumption of DRP groups for the day-ahead market was about 51.9758 MVARh. The maximum values of reactive energy consumption and production were about 7.6966 and 8.2297 MVARh, respectively.

Fig. 5.21 shows the estimated values of the real-time active power production/consumption for plug-in hybrid electric vehicles. The aggregated active energy produc-

Figure 5.18 Estimated values of the allocated reserve of plug-in hybrid electric vehicles.

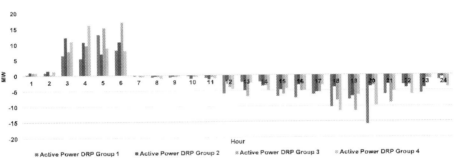

Figure 5.19 Estimated values of the day-ahead active power production/consumption for demand response program provider groups.

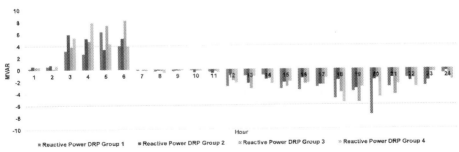

Figure 5.20 Estimated values of the day-ahead reactive power production/consumption for demand response program provider groups.

tion/consumption of plug-in hybrid electric vehicles for the real-time market was about 90.1078 MWh. The maximum values of active energy consumption and production were about 13.56 and 12.60 MWh, respectively.

Fig. 5.22 presents the estimated values of the real-time reactive power production/consumption for plug-in hybrid electric vehicles. The aggregated reactive energy production/consumption of plug-in hybrid electric vehicles for the real-time market

Figure 5.21 Estimated values of the real-time active power production/consumption for plug-in hybrid electric vehicles.

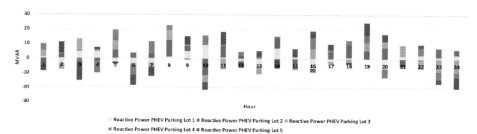

Figure 5.22 Estimated values of the real-time reactive power production/consumption for plug-in hybrid electric vehicles.

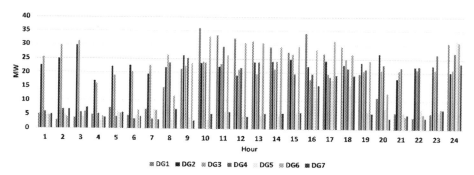

Figure 5.23 Estimated real-time market scheduling of distributed generation units.

was about 81.097 MVARh. The maximum values of reactive energy consumption and production were about 12.204 and 11.342 MVARh, respectively.

Fig. 5.23 shows the estimated real-time market scheduling of distributed generation units. The aggregated energy production of distributed generation units for the real-time market was about 2728.1175 MWh. The maximum and minimum active energy production of units were about 2.81442 and 35.9552 MWh, respectively.

Fig. 5.24 depicts the estimated values of the real-time active power production/consumption for demand response program provider groups. The aggregated active energy consumption of DRP groups for the real-time market was about 11.6417 MWh.

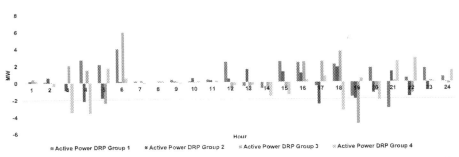

Figure 5.24 Estimated values of the real-time active power production/consumption for demand response program provider groups.

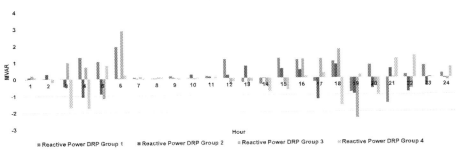

Figure 5.25 Estimated values of the real-time reactive power production/consumption for demand response program provider groups.

The maximum values of active energy consumption and production were about 4.9407 and 5.9303 MWh, respectively.

Fig. 5.25 presents the estimated values of the real-time reactive power production/consumption for demand response program provider groups. The aggregated reactive energy consumption of DRP groups for the real-time market was about 5.6369 MVARh. The maximum values of reactive energy consumption and production were about 2.3923 and 2.8714 MVARh, respectively.

Fig. 5.26 presents the expected values of the day-ahead problem objective function for different values of β. The expected values of the VPP profit for the $\beta = 1$ and $\beta = 0$ strategies were about 332092.5 and 412692 MUs, respectively.

5.5 Conclusions

A day-ahead and real-time optimization framework for the virtual power plant was reviewed in this book chapter. The introduced model considered plug-in hybrid electric vehicles parking lots, wind turbines, photovoltaic arrays, distributed generation units, and demand response programs. The optimization process was carried out in a two-stage procedure that consisted of day-ahead and real-time market horizons. The

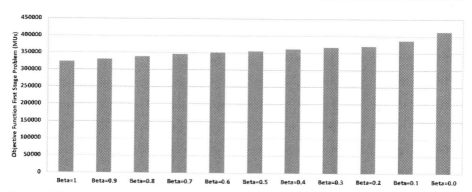

Figure 5.26 Expected values of the day-ahead problem objective function for different values of β.

first-stage problem considered the virtual power plant scheduling in the active energy, reactive energy, and reserve markets. The second-stage problem optimized the virtual power plant scheduling for participating in the active and reactive energy markets. The case study was performed for the 57-bus IEEE test system, and two cases were considered for virtual power plant bidding strategies.

Nomenclature

Abbreviations
CVaR Conditional value-at-risk
DG Distributed generation
DER Distributed energy resource
DRP Demand response program
PHEV Plug-in hybrid electric vehicle
PV Photovoltaic
VPP Virtual power plant
WT Wind turbine

Parameters and variables
λ_{active}^{DA} Price of active energy in day-ahead market
$\lambda_{reactive}^{DA}$ Price of reactive energy in day-ahead market
$\lambda_{reserve}^{DA}$ Price of reserve in day-ahead market
ζ_{active}^{DA} Price of active energy sold to the downward customers in day-ahead market
$\zeta_{reactive}^{DA}$ Price of reactive energy sold to the downward customers in day-ahead market
P_{VPP}^{DA} Active energy sold by VPP in day-ahead market
Q_{VPP}^{DA} Reactive energy sold by VPP in day-ahead market
S_{VPP}^{DA} Reserve sold by VPP in day-ahead market
P_{Load} Active energy of the downward customers in day-ahead market
Q_{Load} Reactive energy of the downward customers in day-ahead market
φ_{DG}^{DA} DG commitment for day-ahead scheduling binary decision variable
φ_{DRP}^{DA} DRP commitment for day-ahead scheduling binary decision variable
φ_{PHEV}^{DA} PHEV commitment for day-ahead scheduling binary decision variable

C_{DG}^{DA} Operational costs of DG for day-ahead scheduling
C_{DRP}^{DA} Operational costs of DRP for day-ahead scheduling
C_{PVA}^{DA} Operational costs of PV array for day-ahead scheduling
C_{WT}^{DA} Operational costs of WT for day-ahead scheduling
C_{PHEV}^{DA} Operational costs of PHEV for day-ahead scheduling
prob Probability of scenario
PVAS Number of PV arrays electricity generation scenarios
WTS Number of WT electricity generation scenarios
PHEVS Number of PHEV electricity generation scenarios
NOS Number of operational scenarios
β Weighting parameter for risk-aversion attitude
ξ, ϑ Auxiliary variable used to compute the CVaR
α Confidence level

References

[1] B. Zhou, X. Liu, Y. Cao, C. Li, C.Y. Chung, K.W. Chan, Optimal scheduling of virtual power plant with battery degradation cost, IET Gener. Transm. Distrib. 10 (2016) 712–725.
[2] Y. Wang, X. Ai, Z. Tan, L. Yan, S. Liu, Interactive dispatch modes and bidding strategy of multiple virtual power plants based on demand response and game theory, IEEE Trans. Smart Grid 7 (2016) 510–519.
[3] X. Kong, J. Xiao, C. Wang, K. Cui, Q. Jin, D. Kong, Bi-level multi-time scale scheduling method based on bidding for multi-operator virtual power plant, Appl. Energy 249 (2019) 178–189.
[4] J. Ju, J. Tan, J. Yuan, Q. Tan, H. Li, F. Dong, A bi-level stochastic scheduling optimization model for a virtual power plant connected to a wind–photovoltaic–energy storage system considering the uncertainty and demand response, Appl. Energy 171 (2016) 184–199.
[5] A. Shayegan-Rad, A. Badri, A. Zanganeh, Day-ahead scheduling of virtual power plant in joint energy and regulation reserve markets under uncertainties, Energy 121 (2017) 114–125.
[6] A. Ghahgharaee Zamani, A. Zakariazadeh, S. Jadid, A. Kazemi, Stochastic operational scheduling of distributed energy resources in a large scale virtual power plant, Int. J. Electr. Power Energy Syst. 82 (2016) 602–620.
[7] Z. Tan, G. Wang, L. Ju, Q. Tan, W. Yang, Application of CVaR risk aversion approach in the dynamical scheduling optimization model for virtual power plant connected with wind–photovoltaic–energy storage system with uncertainties and demand response, Energy 124 (2017) 198–213.
[8] J. Naughton, H. Wang, S. Riaz, M. Cantoni, P. Mancarella, Optimization of multi-energy virtual power plants for providing multiple market and local network services, Electr. Power Syst. Res. 189 (2020) 106775.
[9] S. Hadayeghparast, A. Soltaninejad, H. Shayanfar, Day-ahead stochastic multi-objective economic/emission operational scheduling of a large scale virtual power plant, Energy 172 (2019) 630–646.
[10] R. Hooshmand, S.M. Nosratabadi, E. Gholipour, Event-based scheduling of industrial technical virtual power plant considering wind and market prices stochastic behaviors – a case study in Iran, J. Clean. Prod. 172 (2016) 1748–1764.

[11] L. Ju, H. Li, J. Zhao, K. Chen, Q. Tan, Z. Tan, Multi-objective stochastic scheduling optimization model for connecting a virtual power plant to wind–photovoltaic–electric vehicles considering uncertainties and demand response, Energy Convers. Manag. 128 (2016) 160–177.

[12] S. Fan, J. Liu, Q. Wu, M. Cui, H. Zhou, G. He, Optimal coordination of virtual power plant with photovoltaics and electric vehicles: a temporally coupled distributed online algorithm, Appl. Energy 277 (2020) 115583.

[13] A. Alahyari, M. Ehsan, M. Mousavizadeh, A hybrid storage-wind virtual power plant (VPP) participation in the electricity markets: a self-scheduling optimization considering price, renewable generation, and electric vehicles uncertainties, J. Energy Storage 25 (2019) 100812.

[14] N. Naval, R. Sanchez, J.M. Yusta, A virtual power plant optimal dispatch model with large and small scale distributed renewable generation, Renew. Energy 151 (2020) 57–69.

[15] H. Nezamabadi, M. Setayesh Nazar, Arbitrage strategy of virtual power plants in energy, spinning reserve and reactive power markets, IET Gener. Transm. Distrib. 10 (2016) 750–763.

[16] C. Wei, J. Xu, J. Liao, Y. Sun, Y. Jiang, D. Ke, Z. Zhang, J. Wang, A bi-level scheduling model for virtual power plants with aggregated thermostatically controlled loads and renewable energy, Appl. Energy 224 (2018) 659–670.

[17] M. Vahedipour-Dahraie, H. Rashidizadeh-Kermani, A. Anvari-Moghaddam, P. Siano, Risk-averse probabilistic framework for scheduling of virtual power plants considering demand response and uncertainties, Electr. Power Syst. Res. 121 (2020) 106126.

[18] M. Foroughi, A. Pasban, M. Moeini-Aghtaie, A. Fayaz-Heidari, A bi-level model for optimal bidding of a multi-carrier technical virtual power plant in energy markets, Electr. Power Syst. Res. 125 (2020) 106397.

[19] W. Guo, P. Liu, X. Shu, Optimal dispatching of electric-thermal interconnected virtual power plant considering market trading mechanism, J. Clean. Prod. 279 (2021) 123446.

[20] K. Rahmani, M. Setayesh Nazar, Coordinated bidding of wind and thermal energy in joint energy and reserve markets of Spain by considering the uncertainties, Energy Environ. 28 (2017) 846–869.

[21] A. Bostan, M.S. Nazar, Optimal scheduling of distribution systems considering multiple downward energy hubs and demand response programs, Energy (2020) 116349.

Frequency control and regulating reserves by VPPs

Taulant Kërçi[a], Weilin Zhong[a], Ali Moghassemi[b], Federico Milano[a], and Panayiotis Moutis[c]
[a]School of Electrical & Electronic Engineering, University College Dublin, Dublin, Ireland,
[b]Department of Electrical Engineering, South Tehran Branch, Islamic Azad University, Tehran, Iran, [c]Wilton E. Scott Institute for Energy Innovation, Carnegie Mellon University, Pittsburgh, PA, United States

6.1 Introduction

An intuitive definition of the frequency of an AC circuit is how many times voltage cycles every second [1]. For example, in the European power system the voltage cycles 50 times per second or, equivalently, at 50 Hz. In the North American power systems, the voltage cycles 60 times per second, i.e., at 60 Hz. In mathematical terms, frequency can be defined as the time derivative of the phase angle of bus voltage phasors [2]. This definition implies that the frequency can be constant only in ideal, steady-state conditions.

Keeping the frequency constant and as close to the nominal value as possible everywhere in the system is one of the main objectives of transmission system operators (TSOs). However, this is challenging as frequency fluctuates due to the variations of demand, stochastic noise and harmonics, events such as line connection and disconnections and, last but not least, contingencies [3].

Frequency variations are caused by synchronous machines that inextricably link frequency to power imbalances, as follows:

$$M \frac{d}{dt}\omega(t) \approx p_{\text{gen}}(t) - p_{\text{load}}(t) - p_{\text{loss}}(t), \qquad (6.1)$$

where $\omega(t)$ represents the system frequency, M represents the total inertia of the synchronous machines, $p_{\text{gen}}(t)$ represents the total power generation, $p_{\text{load}}(t)$ represents the total power consumption, and $p_{\text{loss}}(t)$ represents the total losses. Notably, Eq. (6.1) shows that the magnitude of frequency variations depends on the value of M. This means that the greater the system inertia, the smaller the frequency variations, and vice versa. The role of inertia has been the focus of intense research in recent years as the high penetration of large rotating asynchronous renewable sources has significantly reduced the amount of inertia in the system and increased the level of stochastic fluctuations of power [4].

Large frequency deviations may lead to blackouts. For example, they were the main cause of the Italian power system blackout in September 2003 [5]. The blackout of the

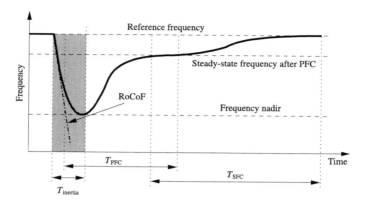

Figure 6.1 Qualitative transient behavior of the frequency in power systems following a loss of a generating unit [9].

European interconnected power system in November 2006 was also caused by high frequency deviations (49 and 51.4 Hz) that led to a split of the system into three islands after cascading events [6]. The interested reader can find other frequency related events that led to power system blackouts in [7].

In conventional power systems, the frequency is controlled hierarchically. After a power imbalance occurs, the inertia of the synchronous machines inherently responds (no control needed) to constrain the magnitude of the frequency deviation. Intuitively, it may be described as the response of the rotating machines to the rate of change of the imbalance. Following this, the primary frequency control (PFC) is realized. PFC is a local actuation process that measures the rotor speed of the generator (as an equivalent of the electrical frequency), compares it to a reference to calculate the deviation, and controls the unit to increase or decrease its output proportionally to that deviation. This control takes place in the time scales of tens of seconds and up to a few minutes, and in most cases is a mandatory service for multiple generating units with an installed capacity greater than a specified threshold. The generating units involved in PFC and the threshold for their eligibility to participate in PFC are defined by TSOs or regulating authorities. Typically, generating units that provide PFC must guarantee an active power reserve greater than $\pm 1.5\%$ of their nominal for interconnected systems and $\pm 10\%$ of their nominal for islanded (or low-inertia) systems [8]. After PFC, the secondary frequency control (SFC), also known as automatic generation control (AGC), follows and reinstates frequency to the nominal value, while it may also restore power exchanges between different areas to their scheduled values, in case they have contributed to PFC. The SFC is planned centrally (although it may be realized by local controllers) and takes place in the time scales of several tens of seconds and up to around 15 minutes. The generating units that provide SFC must guarantee an active power reserve that ranges from $\pm 6\%$ of their nominal for thermal units to $\pm 15\%$ of their nominal for hydroelectric units [8]. A typical frequency response in power systems following a loss of a generating unit is shown in Fig. 6.1.

Traditionally, TSOs have relied on large conventional power plants connected to the transmission network to provide primary and secondary frequency regulation. As these generators are being gradually decommissioned and substituted with small distributed energy resources (DERs), that are mainly connected to the distribution network, it becomes apparent that the latter have to provide these services as well. This chapter discusses the primary and secondary frequency control of VPPs. The chapter also presents examples to illustrate common control approaches for frequency control of VPPs, namely droop and proportional-integral (PI) controller for the primary, and AGC and mixed-integer linear programming (MILP) approach for the secondary.

This chapter is organized as follows. Having already summarized the need for frequency control in power systems, the rest of this Section 6.1 provides a comprehensive review of the most relevant works on VPP frequency regulation. In the review, particular emphasis is given to the role of storage systems, including both large scale batteries and electric vehicles. Section 6.2 expands on the literature review and offers a detailed taxonomy of VPP components, frequency control and reserves frameworks, as well as control and operation strategies. Section 6.3 compares the performance of a few different VPP primary and secondary frequency control approaches through case studies based on the IEEE 9-bus and 39-bus systems. Finally, Section 6.4 summarizes the most relevant aspects of this chapter.

6.1.1 Literature review

While the concept of VPP is relatively recent, there is already a fair number of studies on the implementation and impact of frequency control of VPPs on the performance of power systems. This section is divided into three parts, namely DERs, microgrids (MGs), and energy storage systems (ESSs).

Distributed energy resources

DERs are the main feature of VPPs, as they allow the scheduling of the VPP and defining its capacity. The integration, coordination, and feasibility of the DERs that compose the VPP have thus been extensively studied.

Regarding the integration of DERs and VPPs, reference [10] explores the matter of frequency regulation by wind and photovoltaic energy resources through battery storage systems. The proposed algorithms incorporate large numbers of all aforementioned resources and properly coordinate them for the purpose. The aim is to deliver the service without relying on considerable upgrades of storage capacity. The studies in [11–13] specify that the electric generators of wind turbines that are not directly connected to the grid are largely decoupled from the electric frequency, i.e., they cannot contribute to frequency regulation. Deloading such wind generators to retain frequency reserves for frequency drops is the way they may be enabled to participate in this service. At the same time a short-term release of active power by drawing kinetic energy from the wind turbine rotor is the means to emulate an inertial response by said assets.

The control of power converters used with DERs plays a key role in the potential of VPPs in offering grid services. A control design for an inverter-interfaced photovoltaic system that decouples the dynamics of the generation from those of the grid is presented in [14]. The method ensures proper control across various operating scenarios and is largely unaffected by whether the grid is more resistive or reactive. In [15] the authors specify that the deloading of a photovoltaic plant may be realized by DC switches controlling groups (strings) of photovoltaic modules per each inverter. Frequency regulation signals can accordingly drive more DC breakers to reconnect, thus increasing photovoltaic power at an occurrence of a frequency drop, and vice versa. Field tests have shown negligible effect on PV inverter operation and fast response times. A VSC-interfaced battery storage model is developed in [16] in order to accommodate dynamic and transient behavior studies of such systems. It is a crucial modeling block for the assessment of battery interactions with the grid when the offer of grid services will be considered.

Over-frequency is another problem in modern power systems which can happen every so often. This problem is addressed in [17] that proposes a decision tree (DT) based method to reduce the active power output of VPPs and, consequently, mitigate the over-frequency occurrences.

The impact of the frequency control of DERs connected at the distribution level on the dynamic behavior of high voltage transmission systems is studied in [18]. The paper compares three strategies to generate the input signal used by the DER frequency regulators, namely, decentralized, centralized, and average. It is shown that the centralized approach leads to a better dynamic performance of the system. However, communication delays can significantly impact the noted performance. The impact of VPPs on power system transient response is analyzed in [19]. This work considers two strategies for the frequency control of the DERs included in the VPP, namely, coordinated and uncoordinated. The paper shows that the coordinated control approach leads to a better dynamic response of the system as compared to that without coordination. The paper also shows that communication delays associated with the coordinated control approach have a negligible impact on the dynamic response of the VPP.

For practical implementation purposes, is it desirable that the frequency response is straightforward to apply. Reference [20] proposes a test system for easy and dependable determination of frequency response from a VPP connected to a utility grid and including many distinct power plants. The test system includes a module providing a frequency test sequence that is composed of a set of values and a unit injecting them simultaneously to all nodes of the VPP. Each node includes at least one of either a single generating unit, a single storage unit, or a combination of generating and storage units supervised by a plant controller.

The power output control of VPPs is also important in the context of frequency regulation. The control method proposed in [21] aims to provide primary frequency regulation and, at the same time, carefully adjust the aggregated power output of the VPP comprising photovoltaics and controllable loads (ice machines in this paper). This is done by coordinating the power output of the PVs and power consumption of the ice machines; namely, curtailing a certain amount of PV power and adjusting the number of controllable loads by solving a mixed-integer program (MIP). PFC is achieved with a quadratic interpolation-based active power control strategy. Each PV can operate in a power dispatch mode and simultaneously estimate its maximum available power.

The aggregation of resources within a VPP is another aspect of frequency control because it concerns generation–demand imbalances that the VPP itself may cause. The researchers in [22] propose to operate a VPP with loads having thermal inertia. The control algorithm acts directly on these loads by optimizing their consumption in a specified period, as well as minimizing the difference between generation and demand. In [23], the definition and numerical simulation methods for three aggregation control strategies of distributed generation units in a VPP are proposed. The strategies seek to maximize the efficiency of VPPs as well as minimize the power deviation during dynamic load conditions. Similarly, [24] proposes a VPP model which integrates distributed generators, ESSs, and flexible loads. This VPP performs frequency regulation by decomposing the automatic generation control signal and distributing it to its integrated assets.

The impact of the aggregate response of DERs on the power system dynamic behavior is studied in [25]. The VPP concept is utilized to effectively aggregate the DERs. Two VPP control strategies are proposed, namely AGC approach and MILP. A case study shows that the AGC-based VPP limits the frequency excursions of the system more efficiently as compared to the MILP-based VPP. The work in [26] develops a semi-empirical lifetime model of lithium-ion batteries operated to provide primary frequency regulation in the Danish energy market. The model is proposed as a tool to study the economic profitability of the investment in lithium-ion battery energy storage system. A feasibility study of VPPs that provide ancillary services, including active and reactive power control, for a 50-kV distribution network in Sweden, is presented in [27]. The paper provides a quantification of the economic profits simulated via measuring the variations in the hourly production and consumption at the network nodes. Some authors have emphasized the importance of the configurations, architectures, and the components/model of VPPs like different types of VPPs, the integration of VPP with DERs like PV or wind system, information and communication technology (ICT) systems, etc.

A comprehensive review of the types, architectures, operations, optimization algorithms, communication requirements, and current implementations of VPPs is provided in [28]. A method of operating an MG as VPP, based on artificial intelligence methods, is outlined. The paper anticipates that VPPs will become common solutions for grid operations.

In [29] a distributed generator inverter controller is developed to manage active and reactive power transfer over any type of line, either resistive or reactive. The approach enables complete power control regardless of the type of network (transmission or distribution) to which the generator is connected to. This study, although not explicitly addressing generation–demand imbalances, it describes how dispersed VPP resources can ensure the maximum power transfer when contributing to frequency regulation services.

In [30] the authors present how a commercial VPP can integrate distributed resources and trade them in the energy market, while accounting for the network constraints and other operating constraints imposed by the technical VPP. An overall algorithm to combine these two aspects of a VPP is proposed and the challenges forward are identified. The ICT requirements, in light of enabling virtual power plants to procure, offer, and realize ancillary services, are presented in [31]. The current standards frameworks need to be rethought and/or enhanced according to new directions. Based on the example of IEC 61850, ancillary services from VPPs will be enabled if additional interactions among the various actors are defined and the data are properly determined.

The work in [32] proposes a capability-coordinated frequency control (CCFC) method for a VPP, including an adjustable-speed pumped storage hydroelectric plant, a wind power plant, and an ESS, to improve the controllability of the VPP. The CCFC method can reduce the frequency nadir, with a signal of frequency correction distributed among the VPP assets, and decrease the steady-state error of system frequency with a secondary frequency control type of coordination. This, in turn, provides flexible ancillary service to the frequency regulation of a power system.

Microgrids

VPPs are not MGs, as the latter usually have complete sensing and control of the grid among the controlled assets [33]. Yet, MGs have several aspects in common with VPPs, one of them being the ability to provide frequency and voltage support. The researchers in [34] study the impact of high shares of MGs on the load frequency control (LFC) of power systems. The paper develops a dynamic model of MGs to study its contribution to LFC. MGs are classified into two categories of different features with respect to the LFC. The work in [35] proposes a coordinated frequency control approach for the electric vehicles and renewable energy resources (RESs) included in a MG. The approach is based on an adaptive PI controller. The paper shows that the proposed controller is robust to the volatility of the RESs. In [36] an extended virtual synchronous generator (VSG) is presented for MGs based on the concept of virtual rotor to procure primary and secondary control. This virtual asset is inspired by the conventional synchronous generator. Its effectiveness is verified via of an MG test system.

In [37] the authors propose a distributed secondary frequency and voltage control approach for islanded MGs. The effectiveness of the proposed controller is evaluated by means of small-signal stability and time domain analyses. The paper shows that the controller is able to restore the frequency to its nominal value, restore bus voltages, as well as to ensure a proper reactive power sharing. Also, a power control strategy for doubly fed induction generators (DFIG) in an MG is introduced in [38]. The control strategy makes use of a 10% wind power margin to provide frequency support. Moreover, reference [38] proposes a variable coefficient strategy in order to utilize different control parameters at different wind speeds and, in this way, improve the DFIG frequency support.

Some works have presented hierarchical controllers in MGs and VPPs to provide frequency support. The work in [39] proposes a hierarchical controller for an MG comprising wind turbines and battery units. The aim is to provide primary and secondary frequency support to a weak grid by regulating the power flow over the tie-line. In [40], a control strategy based on the IEC/ISO 62264 standard for hierarchical control for MGs and VPPs with energy storage system is presented. The paper also reviews MG design principles, as well as the types and roles of VPPs. The authors in [41] propose a model-free based generalized droop control. The control uses an adaptive neuro-fuzzy inference system for voltage coexistent and frequency regulation in the islanded MGs. This control structure can help reduce the imbalance between generation and consumption.

Energy storage systems

ESSs and energy management systems (EMSs) are crucial elements of the frequency control of asynchronous devices and, in turn, of VPPs. If properly controlled, energy storage can also be utilized to provide virtual inertia and fast frequency control. The work in [42] summarizes the requirements of the capacity for the connection of DERs and ESSs to the grid. A comprehensive review of the technical aspects of the VPP, the active power control strategy applied in DERs, as well as the charging and discharging characteristics of ESS is also provided. In [43], an EMS strategy is presented to provide the load power demand in distinct operating status by means of achieving the active power regulation in a small-scale VPP. The paper also proposes a hybrid AC/DC connection scheme for VPPs. The devices are integrated in the VPP via a DC-bus to reduce the number of required interface power converters.

Two other aspects of VPPs and ESSs are very important; the actuation and allocation of energy storage systems. Research presented in [44] specifies that islanded and low-inertia systems are more sensitive to load–generation imbalances and suffer from larger frequency swings at these occurrences. Fast acting storage systems are employed in this proposal as dynamic frequency control support assets. The study is focused on islanded power systems with high shares of renewable generation that cannot contribute with inertia. Similarly, in [45], the authors study how to allocate ESSs efficiently with the aim to minimize storage system sizes and, thus, their costs. Hence, frequency reserves service is procured at reduced social costs. As such minimum size designs might underperform in cases of frequency overshooting, dissipating emergency resistors are also proposed.

A relevant question is whether the frequency response provided by VPPs has both technical and economic feasibility. For example, the work in [46] describes the authors' experience with a 1.6 MW/0.4 MWh lithium-ion battery ESS that provides primary frequency regulation in the framework of a 100% renewable Danish energy market by 2050. Results indicate that the investment in the lithium-ion battery ESS can be profitable in the Danish market if it is committed to at least 10 years.

Among existing ESSs, electric vehicles (EVs) play a special role as they are expected to dramatically increase their capacity in the near future. EVs can, in effect, be used as a VPP, as discussed in [47], in which EVs are utilized to support primary reserve in smart grids. By using online information of EVs like the initial state of charge, arriving time, the required state of charge for the next trip, and the temperature time, the primary frequency reserve not only averts reduction in frequency, but also improves the frequency response of the modeled Great Britain's power system. A fixed droop coefficient for each EV is individually considered to keep frequency nadir above 49.85 Hz, thereby improving the primary frequency response and avoiding overshoots caused by adaptive droop controllers.

The study in [48] notes the impact of three different EVs charging strategies, namely, proportional response, soft control, and aggressive control, on the frequency response of a system under a variety of wind generation scenarios. It is concluded that the participation of EVs in frequency control significantly improves the dynamic behavior of the system. The proposed EV control strategies have similar performance. In [49] EVs are used as storage for wind generators in order to enable their participation to the day-ahead electricity markets. This is achieved through VPPs that aggregate many EVs and wind generators. The paper shows through a realistic case study based on real data that this approach is profitable for both EVs and wind generators.

EVs can be also included in MGs. The authors of [50] present a frequency control strategy of an MG supported by VPP coordination. The MG consists of various distributed generation and storage resources and loads. By applying the VPP concept control method, the active power balance is sustained, thereby controlling the frequency of the MG. Particle swarm optimization (PSO) and firefly algorithm (FA) metaheuristic algorithms are used to tune the PID controllers of the MG for frequency control. The PID-FA has minimized the frequency deviation, as opposed to PID-PSO, PI-PSO, and PI-FA. The control of EVs and their communication are of vital importance for the effective integration of EVs into VPPs. The research in [51] focuses on the control architecture and communication requirements for EVs-VPP. The paper emphasizes the importance of a reliable, secure, and inexpensive communication infrastructure to effectively operate the distributed resources included in the EVs-VPP.

An efficient control of EVs considering charging rate issues is proposed in [52]. This work shows that an aggregator can effectively exploit the frequency regulation service that can be provided by EVs, while keeping in mind that these assets are also loads and seek to serve their primary purpose as mobility devices. The charging rates are thoroughly assessed to better perform the overall control. Similarly, in [53], it is aimed to balance the frequency regulation provision by EVs while ensuring that the EV battery is charged to the user's preferred level for mobility reasons. Adaptive frequency droop control and EV charging with frequency regulation are the two control strategies proposed in the context of this work. All control is proposed to be implemented in a decentralized manner abiding by aspirations for a more robust management strategy.

6.2 Taxonomy

In this section and in view of how they procure frequency control, VPPs are categorized according to different types of components, frameworks, control methods, frequency regulation stages, operators, and grid strength. For each category, a table with relevant references is given.

6.2.1 Components

A VPP is most typically composed of four parts, namely DERs, ESSs, ICTs, and controllable loads. Table 6.1 shows the VPP components and relevant references.

Table 6.1 VPP components.

VPP components	References
DERs	[10,17–25,27–32,34,35,40, 42,43,46–50]
ESSs	[10,18–22,24,26–28,31,32, 35,40,42,43,46]
ICT	[18–20,28,31,40,43,51]
Controllable loads	[21,22,24,35,47–51]

Distributed energy resources

DERs as VPP components can be categorized according to the type of the primary energy source, the capacity of DG, the ownership of DG, and the operational nature of DG.

(a) **Primary Energy Source Type.** Primary energy sources are divided into RES based like wind-based generators, photovoltaic arrays, solar-thermal systems, and small hydro-plants, and non-RES based like combined heat and power (CHP), biomass, biogas, diesel generators, gas turbines, and fuel cells (FC).

(b) **DG Capacity.** The DERs can be classified as per capacity of DG. The small-scale capacity DGs that must be connected to the VPP to increase access to the electricity market or they could be connected with controllable loads to form MGs. Also, the medium- and large-scale capacity DGs that can independently take part in the electricity market. However, they may opt for being connected to VPP to gain optimal steady revenue.

(c) **DG Ownership.** The ownership of DGs can be considered for the classification of DERs. Residential-, commercial-, and industrial-owned DGs, also known as domestic DGs (DDGs), are utilized for supplying all/part of its load. Utility-owned DGs, also known as public DGs (PDGs), are used to support the main grid supply shortage. Commercial company-owned DGs, also known as independent power producers DGs (IPPDGs), are to generate profits from selling power production to the network.

(d) **DG Operational Nature.** When it comes to wind or PV systems as DG units, the output power is uncontrollable because it highly relies on a variable input resource. To address this, such DGs must be equipped with battery storage to control the output power. This operational nature of DGs is called stochastic nature. Other DG technologies such as FCs and micro-turbines have an operational dispatchable nature. They are able to alter their operation rapidly. Thus, VPP should incorporate controllable loads, energy storage elements (ESE), and dispatchable DGs to compensate for the vulnerability of the stochastic nature-DG type.

Energy storage systems

To narrow the gap between the generation and demand, ESSs have an important role as they can arbitrate energy. ESSs can be categorized according to their applications as either energy or power supply, as below:

(a) **For energy supply:** hydraulic pumped energy storage (HPES) and compressed air energy storage (CAES).
(b) **For power supply:** flywheel energy storage (FES), superconductor magnetic energy storage (SMES), and supercapacitors.

Information and communication technologies

The heart of a VPP is an EMS that coordinates the power flows from the generators, controllable loads, and storages. So, EMS is the backbone of information and communication systems. Receiving information about the status of each element inside the VPP, forecasting RES primary sources and output power, loads forecasting and management, power flow coordination between the VPP elements, and also operation control of DGs, storage elements, and controllable loads are the most important duties of the EMS. The main aims and objectives of the EMS are generation cost minimization, energy losses minimization, greenhouse gases minimization, profit maximization, voltage profile improvement, and power quality enhancement.

Controllable loads

The controllable loads can be vehicle-to-grid (V2G) functionalities of electric vehicles, refrigerators, freezers, air conditioners, water heaters, heat pumps, battery storage, heat storage, etc. Below, different types of the controllable loads are provided:

(a) **Type a** includes residential loads such as fridges, washing machines, air conditioners, space cooling/heating, water heating, etc., which are interrupted or shifted by the load's utilities monitor. These loads cannot inject power to the network at any time.
(b) **Type b** includes battery storage, vehicle-to-grid (V2G), the combined cooling heating and power (CCHP), etc. As opposed to Type-a controllable loads, these loads are able to inject power into the network. They can be charged from or discharged to the network. They have also more considerable flexibility to be scheduled and, thus, they are tailored to network needs.
(c) **Type c** includes microgrid, VPP, etc. Although microgrid and VPP have distributed generators, battery storage, renewable energy, etc., the loads take a great proportion in these systems.

6.2.2 Frameworks

From the perspective of the nature of the entity and their topology, VPPs are classified into commercial VPP (CVPP) and technical VPP (TVPP). CVPPs operate just like traditional generators, bidding in the electricity markets without considering the effect of their operation on the local grid. By contrast, TVPPs employ DERs to handle the local grid in terms of thermal and voltage congestions. The TVPPs are also able to provide ancillary services to increase the security of the system. The operation and duties of CVPPs and TVPPs are explained below. Also, relevant references are indicated in Table 6.2.

Table 6.2 VPP frameworks.

VPP frameworks	References
TVPP	[10,18–21,24–26,28–32,35, 40,42,43,46–48,50,51]
CVPP	[17,21,22,25,27,28,30,40, 46,49]

Technical virtual power plant

The primary responsibility of a TVPP is to properly dispatch the DERs and the ESSs to manage the energy flow inside the VPP cluster, and offer the ancillary services accordingly. A TVPP receives information from the CVPP about the contractual DGs and the controllable loads. The most important data that TVPP must consider are the maximum capacity and commitment of each DER unit, the prediction of production and consumption, the location of DER units and loads, the capacity and the locations of the ESSs, the available control strategy of the controllable loads at all times during the day as per contractual obligations between the VPP and the loads. On the whole, some duties of TVPPs are as follows:

- Managing the local distribution network
- Providing balancing, management of the network, and execution of ancillary services
- Providing visibility of the DERs in the distribution grids to the TSO, thereby setting the stage for DG and demand to make contribution to the transmission system management activities
- Monitoring the DER operation based on the requirements obtained by the CVPP (system status information)
- Constantly monitoring the status for the retrieval of equipment historical loadings

Commercial virtual power plant

A CVPP carries out bilateral contracts with the DG units and the customers. The data of these contracts is sent out to the TVPP to consider the amount of the contracted power all through the performance of technical research. The most important responsibilities of CVPP are:

- Production scheduling as per the predicted needs of consumers
- Trading in the wholesale electricity market
- Providing services to the system operator

- Submitting characteristics, costs, and maintenance of DERs
- Predicting production and consumption as per weather forecasting and demand profiles
- Providing outage demand management
- Selling DER power in the electricity market

6.2.3 Control methods

Table 6.3 shows VPP control methods and relevant references. For optimal operation of VPPs in terms of power loss minimization, cost reduction, profit maximization, and environmental emission reduction, a wide variety of numerical and heuristic control methods are used.

Table 6.3 VPP control methods.

VPP control methods	References
Numerical methods (e.g., LP, NLP)	[21,25,28,49,51]
Metaheuristic methods (e.g., GA, PSO)	[17,28,41,50]
Hybrid optimization algorithm based methods (e.g., GA-PSO, FLC-ANN)	[28,50]

Numerical methods

The most widely used numerical methods are linear programming (LP) which is capable of addressing optimization problems in terms of optimal DER power and DG energy extraction; nonlinear programming (NLP) to determine the length of time of several DGs; gradient search, sequential quadratic programming (SQP), dynamic programming (DP), and exhaustive search for finding several purposes like optimal DG locations, optimal DG sizes, minimization of cost, and also loss minimization.

Heuristic methods

Due to the potential and proven capabilities of metaheuristic methods for solving optimization problems, a wide variety of metaheuristic algorithms have been presented. The most common ones are genetic algorithm (GA), particle swarm optimization (PSO), fuzzy logic controller (FLC), artificial neural network (ANN), tabu search (TS), ant colony optimization (ACO), artificial bee colony (ABC), harmony search (HS), cat swarm optimization, and firefly algorithm (FA). The most important purposes that have been considered in objective functions of the above algorithms were optimal placement, size, and type of DG, power loss minimization, energy loss minimization, profit maximization, voltage profile improvement, maximization of DG penetrations, power quality improvement, and reliability indices. It is worth pointing out that over the years, combination and hybrid metaheuristic algorithms have come to researchers' attention with the same purposes, yet better performances, GA-PSO, FLC-ANN, PSO-FA, to name but a handful.

6.2.4 Frequency regulation

This service brings back the frequency to the nominal operating level after any deviation occurrence due to the physical unbalance between generation and demand. This is attainable by adjusting the active power reserves of the system through automatic and rapid responses. The TSOs need to plan, in advance, to make sure that the correct levels of active power reserves are available in real-time, as well as that the TSOs must take remedial actions, when it comes to a shortage. Active power reserves embrace generator units, storage, and sometimes demand response. The main ancillary services for frequency regulation are shown in Table 6.4.

Table 6.4 VPP frequency regulation.

VPP frequency regulation	References
Inertia (and inertial emulation) and primary response (load following)	[10–13,16,18–21,26,28,32,35,36,38–41,44–48,50,53]
Secondary and tertiary	[10,12,13,24,25,28,34,36,37,39,40,43]
Procurement/concerns for frequency reserves	[17,21,27,28,52]

Frequency containment reserve (FCR)/primary frequency control

FCR is the first control action to be activated, usually within 30 s, in a decentralized fashion over the synchronous area.

Frequency restoration reserve (FRR)/secondary frequency control

FRR is the centralized automated control, enabled from the TSO in the time interval between 30 s and 15 min from the unbalance occurrence. FRR can be categorized according to reserves with automatic activation (automatic frequency restoration reserve, aFRR) and reserves with manual activation (manual frequency restoration reserve, mFRR).

Replacement reserves (RR)/tertiary frequency control

RR is a manual control. The typical activation time for the RR is from 15 min after the unbalance occurrence up to hours.

Procurement/concerns for frequency reserves

The procurement methods are compulsory provision, bilateral contracts, tendering, and spot markets. In the first method, a class of generators is involved to provide specific reserves of ancillary services. This engagement rises through the national regulations and network codes. In the second method, the TSO negotiates with each provider the quantity and price of the offered ancillary service. This permits the TSO to purchase only specific ancillary services and to deal with sellers to minimize the overall expense. The last two methods refer to an ancillary service exchange process characterized by increased competition. Although the tendering market is usually composed of long-duration services, the spot market involves shorter and less standardized products.

6.2.5 Operation

The owner or operator of the VPP may be controlling the VPP assets either as an actual entity, like system operator control room personnel, or as a software framework, similar to distribution or energy management systems (DMS or EMS). Regardless of this detail, the VPP operator will need to abide by the level of the electric grid to which the size of the VPP and its assets are operating. The classification of the VPP operation and relevant references are listed in Table 6.5.

Table 6.5 VPP operation.

VPP operation	References
Distribution level	[10,18,19,22,27–31,42, 43,50]
Transmission level	[10,17,19,20,22,24–26, 28–30,32,46,47,49,51]

Distribution level

At the distribution network level, VPPs are not expected to contribute directly to frequency regulation services as the aggregated power of their assets cannot easily justify the procurement of reserves and their release upon disturbances. Furthermore, for any single generating or storage asset, no standardized code/regulation explicitly required them to contribute to the frequency regulation until recently when a framework was discussed (IEEE 1547-2018). Nevertheless, aggregators of assets or collectives of VPP operators may coordinate and bid their reserves for services to the market operator, provided proper regulatory framework exists.

Transmission level

Most of the resources connected at the transmission network level are expected to contribute with frequency regulation services to support the grid stability. Except for renewables, all other generators have been typically designed with governors that respond to frequency deviations at either the primary or secondary level. In terms of market involvement, generating assets at the transmission level submit bids for up and down reserves (i.e., to increase their active power output or decrease it, according to the frequency deviations, respectively) and the operator clears and assigns them. In this sense, VPP operators are expected to handle both the market and technical aspect of frequency regulation and the procurement of reserves in cases the VPP assets are connected at the transmission network level.

6.2.6 Grid strength (inertia and/or sensitivity to load changes)

An important aspect of frequency regulation (of any service not just VPPs) is whether this service is offered in strong or weak grids. This distinction affects how frequently this service is activated and whether higher reserves are required to implement it. Relevant references on this topic can be found in Table 6.6.

Table 6.6 Grid strength (inertia and/or sensitivity to load changes).

Grid strength (inertia and/or sensitivity to load changes)	References
Strong/interconnected	[10–14,16,18–20,22,24–29,31,32, 34,36,38,40,43,45–47,49–53]
Weak/islanded or high inverter share	[17,21,29,37–41,44,48]

Strong grids

Strong grids, with considerably high number of rotating masses (conventional generators and motors) connected, tend to be less affected by typical load–generation imbalances, hence, the frequency in such systems changes less sharply and in tighter deviations intervals. This implies that a VPP connected to a strong grid will not be required to procure considerable reserves or activate them as much. As a by-product this VPP might not have to be involved in primary frequency control, as other rotating generators may be assigned this role.

Weak grids

Contrary to strong grids, weak grids have few rotating machines (generators and motors) and, in some cases, inverter-interfaced DGs and storage systems. In these grids, frequency deviations are more frequent and severe, and VPP assets might be expected to respond more frequently and in broader operating ranges, causing added wear and tear. This implies that the frequency regulation provided by VPPs connected to weak grids has a higher cost than that of VPPs included in strong grids. Additionally, the VPP assets need to be clearly and more thoroughly accounted for in stability studies of weak grids.

6.3 Examples

The section illustrates the dynamic behavior of VPPs when their resources are controlled in order to provide frequency support. Two case studies are discussed. Section 6.3.1 focuses on primary frequency control while Section 6.3.2 discusses secondary frequency control. The dynamic response of various regulation and coordination strategies applied to VPPs are compared.

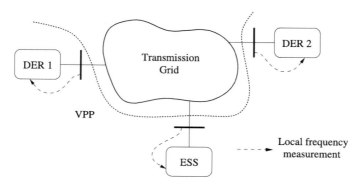

Figure 6.2 Illustration of the transmission-system VPP topology.

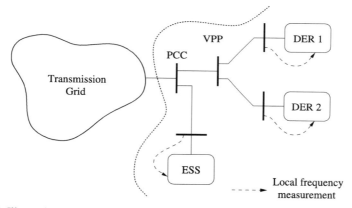

Figure 6.3 Illustration of the distribution-system VPP topology.

6.3.1 Primary frequency control

The aim of VPP primary frequency control is to improve the recovery of the frequency to its reference value in short-term transients following a power unbalance. The primary frequency control of the DERs and ESSs included in the VPP is commonly implemented via droop or proportional-integral (PI) controllers. Relevant references that utilize this controller are [32,40,41,47,50].

At the system level, two different VPP topologies are relevant: (i) transmission system VPP (TS-VPP), where DERs and ESSs are connected directly to the high-voltage transmission system; and (ii) distribution-system VPP (DS-VPP), for which the devices are connected with the transmission grid via a point of common coupling (PCC). These two topologies are illustrated in Figs. 6.2 and 6.3. The TS-VPP topology is widely adopted to combine geographically dispersed and/or high-capacity DERs, whereas the DS-VPP is suitable for the geographically close and medium/small capacity DERs.

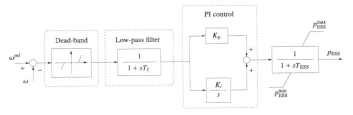

Figure 6.4 Typical frequency controller of an ESS [54].

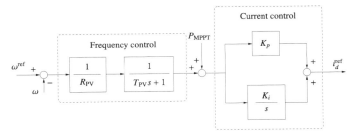

Figure 6.5 Typical frequency controller of a solar photovoltaic power plant [55].

Frequency controllers

This section presents typical schemes of the frequency controllers utilized for ESSs [54], solar photovoltaic power plants [55], and wind power plants [11].

Fig. 6.4 depicts a simplified ESS scheme for active power control, which aims at regulating the deviation of the measured frequency ω, with a reference frequency ω^{ref}. The controller is composed of a PI regulator coupled with the dead-band and low-pass filter (LPF), which reduces the sensitivity of the input signal and filters out noises, respectively. The scheme in Fig. 6.4 also represents a very simplified model of the storage device, modeled as a first-order low-pass transfer function with capacity limits. More detailed ESS models are given, for example, in [16] and [56].

Fig. 6.5 shows a simple scheme of the frequency controller of solar photovoltaic power plants. In this case, a droop control for frequency regulation composed of droop gain and LPF is introduced. Then, the output signal of the frequency controller is fed to a PI controller with respect to the maximum power point tracking (MPPT) reference power signal. Finally, the controller determines the reference d-axis current (i_d^{ref}).

The frequency control approach of wind turbine couples the output of the MPPT with the deviation of the measured frequency via two parallel channels to regulate the frequency deviation (droop control) and/or the rate of change of frequency (RoCoF) control. Such a control is depicted in Fig. 6.6. The outputs of the droop and RoCoF controllers go through a deal-band and are finally added to the output of the MPPT and define the reference active power that the controller of the converter of the wind turbine will track.

The controllers described so far are local and decentralized. For completeness, a "coordinated" primary frequency control is also presented below. In [19] a coordi-

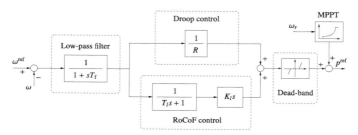

Figure 6.6 Typical frequency controller of a wind turbine [11].

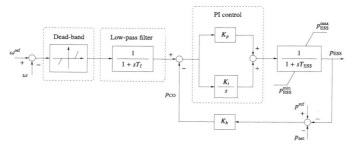

Figure 6.7 Primary frequency controller of an ESS belonging to a VPP with inclusion of a coordinated control signal [19].

nated control approach of the ESS which combines the DERs active power output in VPP is proposed to allow extra frequency support from the VPP following a contingency. With this aim, a feedback signal, p_{CO}, with the information of active power outputs from wind turbines and SPVGs (p_{net}) with respect to a reference, p^{ref}, is added between the LPF and PI controller of the ESS control loop. Fig. 6.7 illustrates the control scheme of the ESS with a coordinated signal from the DERs of the VPP.

Comprehensive discussions on the advantages and drawbacks of the controllers introduced in this section are given in [18] and [19].

Test system

The case study presented in this section considers the TS-VPP and DS-VPP topologies, both based on a modified version of the well-known WSCC 9-bus, 3-machine system (see [57]).

The setup of the grid for the TS-VPP is shown in Fig. 6.8 and consists of the following modifications with respect to the original network:

- The active and reactive powers of the original load at bus 6 are reduced to 0.578 MW and 0.117 MVAr, respectively.
- One solar PV plant, two wind power plants, and one ESS are connected at buses D8, D5, D6, and D9, respectively. The initial active power generation of the wind

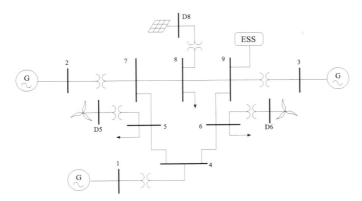

Figure 6.8 Modified WSCC 9-bus, 3-machine system with a TS-VPP topology.

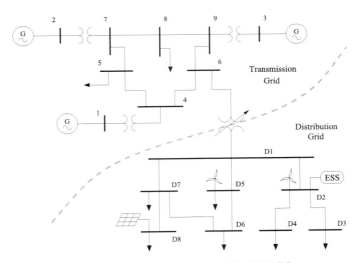

Figure 6.9 Modified WSCC 9-bus, 3-machine system with a DS-VPP.

power plants and the solar PV plant are 15 MW each, whereas the power rate of the ESS is 10 MW.

The setup of the grid for the DS-VPP is shown in Fig. 6.9. In this scenario, the load at bus 6 in the original network is replaced with a distribution system that includes a VPP. The distribution system consists of 8-buses at 38 kV and is connected to the transmission system through an automatic under-load tap changer (ULTC) type step-down transformer [58]. The total active and reactive power of the loads in the distribution system is 0.578 MW and 0.117 MVar, respectively. Finally, the VPP includes one solar PV plant, two wind power plants, one ESS, with the same parameters as the TS-VPP, that are connected at buses D8, D5, D2, and D2, respectively.

Case study

This case study studies the impact of both TS-VPP and DS-VPP on power system transient response considering five scenarios, corresponding to five frequency control setups of the VPP, as follows:

- Without ESS and DERs frequency control
- With ESS but without DERs frequency control
- Without ESS but with DERs frequency control
- With uncoordinated ESS and with DERs frequency control
- With coordinated ESS and with DERs frequency control

Note that, in this context, "uncoordinated ESS" means that the ESS regulates the frequency using the scheme shown in Fig. 6.4, whereas "coordinated ESS" indicates the ESS with the frequency control and additional power set-point signal as shown in Fig. 6.7.

The contingency considered in this scenario is a 3% instantaneous load demand increase at $t = 1$ s. In this case study, all simulations are solved using the Python-based software tool DOME [59].

The performance of the system for different control schemes is studied by comparing the frequency of the center of inertia (CoI). Figs. 6.10 and 6.11 show the frequency responses following the disturbance of all five VPP control structures for both TS-VPP and DS-VPP. These frequency trajectories are compared to the frequency CoI of the original 9-bus system with the same disturbance, where the load at bus 6 in the original system has been also reduced to 0.578 MW and 0.117 MVar, respectively.

Simulation results show that the ESS can significantly improve the frequency response of the system even without DERs frequency control. Moreover, the coordinated ESS is able not only to reduce the frequency deviation but also to almost eliminate the frequency steady-state error (see Figs. 6.10b and 6.11b). In fact, the coordination acts similarly to a secondary frequency control by "rescheduling" the ESS power output following a contingency.

Figs. 6.10 and 6.11 also show that the effect of the TS-VPP and DS-VPP for frequency support are similar. The DS-VPP has a slightly better performance than TS-VPP for the control scenarios where the VPP does not include ESS and DERs frequency control, with ESS but without DERs frequency control, as well as with coordinated ESS and with DERs frequency control. This is due to the fact that the DERs and ESS use the same frequency signal measured at the PCC in the DS-VPP, while in the TS-VPP, the frequency signal for the devices is measured at their local buses. Local frequencies are conditioned by the dynamic behavior of the closest synchronous generators and, thus, might not always be the best signals for asynchronous devices [18,60,61].

Overall, and as expected, a better response is achieved through a coordinated control of the ESS in terms of maximum deviation and post-disturbance equilibrium.

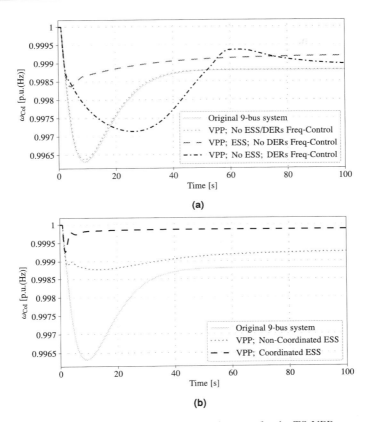

Figure 6.10 Frequency of the CoI following the contingency for the TS-VPP.

6.3.2 *Secondary frequency control*

The provision of the secondary frequency control by VPP is of utmost importance for managing the real-time balance between generation and demand. For example, in recent years there are many DERs units that operate as a VPP and participate in the ancillary service market to provide frequency regulation to the TSOs [62]. The TSOs have specific rules for generators that provide these services. In particular, they require that the power output of a VPP increases linearly during the ramp-up time [62]. If that is not the case and the VPP rate of change of power is higher than the agreed linear ramp, then the TSO does not pay for the excess power, whereas if the VPP provides less power than the scheduled linear ramp, then the TSO fines the VPP. Linear ramping requirements are common in the European power system [63]. On the other hand, providing a linear ramping response is a challenge for the VPP as different generators have different characteristics, e.g., different capacity, response, and ramping time. This problem is illustrated in Figs. 6.12 and 6.13 and duly discussed in [62] and [25].

The points in Fig. 6.12 have the following meaning:

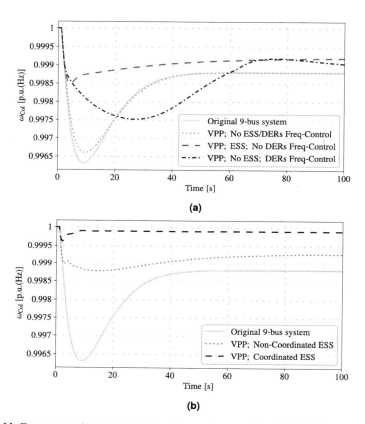

Figure 6.11 Frequency of the CoI following the contingency for the DS-VPP.

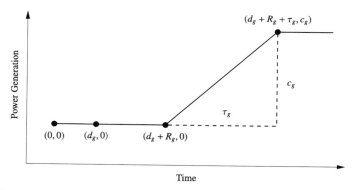

Figure 6.12 Power production of a single small generator [62].

Frequency control and regulating reserves by VPPs

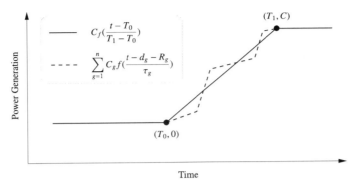

Figure 6.13 Power production of a single large power plant (continuous line) and that of a collection of small generators (dashed line) [62].

- $(0, 0)$ represents the time when the TSO tells the VPP to go to the maximum production.
- $(d_g, 0)$ represents the delay time of the VPP, i.e., the time to send a signal to the gth generator to start the production.
- $(d_g + R_g, 0)$ represents the time at which the gth generator transitions from the minimum to the ramping production, where R_g is the response time of the gth generator of the VPP.
- $(d_g + R_g + \tau_g, c_g)$ represents the time at which the gth generator of the VPP transitions from the ramping to the maximum production, where τ_g is the ramping time and c_g is the maximum capacity of the gth generator.

The points in Fig. 6.13 have the following meaning:

- $(T_0, 0)$ represents the start of the ramping of the VPP.
- (T_1, C) represents the stop of the ramping of the VPP, with C being the total capacity of the small generators of the VPP.

Furthermore, the functions

$$C f \left(\frac{t - T_0}{T_1 - T_0} \right), \tag{6.2}$$

$$\sum_{g=1}^{n} C_g f \left(\frac{t - d_g - R_g}{\tau_g} \right) \tag{6.3}$$

represent the active power generated of a single large power plant (linear), and the total active power output of small generators of the VPP (piecewise linear), respectively.

Two approaches for the provision of secondary frequency control through VPPs are discussed below. The first is based on an MILP problem that optimally schedules the small generators of the VPP. The second is based on an AGC approach.

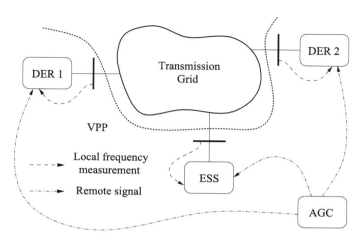

Figure 6.14 Illustration of the AGC coordination scheme.

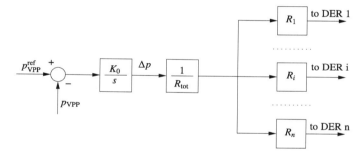

Figure 6.15 Basic AGC control scheme for active power regulation of VPPs.

AGC-based VPP

Secondary frequency regulation based on AGC is widely used by TSOs to eliminate the steady-state frequency error remained after the primary frequency control and keep the interchange between different control areas at the scheduled values. Conventional AGCs operate in the time scales ranging from minutes to tens of minutes.

As an example, this section considers a VPP where all its units are connected to an AGC, as illustrated in Fig. 6.14. The AGC considered here regulates the total active power produced by the VPP rather than the frequency. This setup allows for a direct comparison with the MILP-based scheduling approach described in the previous section.

The control scheme of the AGC is shown in Fig. 6.15. The signal $p_{\text{VPP}}^{\text{ref}}$ represents the reference power signal sent by the TSO to the VPP, whereas p_{VPP} represents the sum of the measured active power of the DERs included in the VPP. Then the AGC includes an integrator with gain K_0 in order to reduce the active power steady-state error to zero. Finally, the output of the integrator is sent to the turbine governors of the machines proportionally to their droops R.

MILP-based VPP

The MILP problem proposed in [62] is considered in this section as an example of optimal scheduling approach to coordinate the units that compose the VPP and obtain a ramping rate that is as close to linear as possible. The mathematical formulation of the MILP problem utilized in the case study presented in this section is recalled below:

$$\min \sum_t (p_{a,t} + K p_{b,t}) \tag{6.4}$$

such that

$$p_{g,t} \leq C_g, \quad \forall g, t, \tag{6.5}$$

$$p_{g,t} = p_{g,t-1} + R_g (b_{g,t} - \overline{b_{g,t}}), \quad \forall g, t, \tag{6.6}$$

$$b_{g,t} \geq b_{g,t-1}, \quad \forall g, t, \tag{6.7}$$

$$\overline{b_{g,t}} \geq \overline{b_{g,t-1}}, \quad \forall g, t, \tag{6.8}$$

$$\sum_t (b_{g,t} - \overline{b_{g,t}}) = \tau_g, \quad \forall g, \tag{6.9}$$

$$\sum_g p_{g,t} + p_{b,t} - p_{a,t} = \frac{t \sum_g C_g}{|\tilde{T}|}, \quad \forall t, \tag{6.10}$$

$$b_{g,t}, \overline{b_{g,t}} \in \{0, 1\}, \quad \forall g, t, \tag{6.11}$$

$$p_{g,t}, p_{a,t}, p_{b,t} \geq 0, \quad \forall g, t, \tag{6.12}$$

where $p_{a,t}$ and $p_{b,t}$ represent continuous variables that model the distances above and below the target linear characteristic at time t, respectively (see Fig. 6.13); K is a penalty multiplier for when the actual ramping rate is below the target line. This is needed as the VPP is penalized if it provides less power but is not if it provides more than the scheduled power. In this work, a value of $K = 10$ is considered. Eq. (6.5) models the capacity limits of single small generators, where $p_{g,t}$ represents the active power generation of the gth generator at time period t. Equalities in (6.6) model the ramping limits of generating units, where the binary variables $b_{g,t}$ model the status of generating units when they are generating (1 if producing and 0 otherwise), while the binary variables $\overline{b_{g,t}}$ model the status of generating units when they are generating at maximum capacity (1 if true and 0 otherwise). Eqs. (6.7) and (6.8) model the logic of the binary variables. Eq. (6.9) models the generators ramp time (τ_g), i.e., the sum of the differences $b_{g,t} - \overline{b_{g,t}}$ must equal τ_g. Eq. (6.10) models the target ramping line, i.e., $\frac{t \sum_g C_g}{|\tilde{T}|}$, with $|\tilde{T}|$ representing the total number of time periods. Finally, Eqs. (6.11) and (6.12) represent variable declarations.

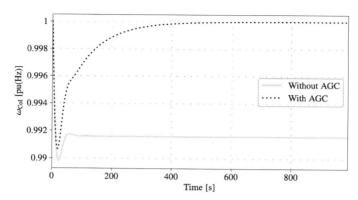

Figure 6.16 Frequency response of the system with and without a conventional AGC.

Case study

The effectiveness of both MILP- and AGC-based approaches is evaluated by means of a modified IEEE 39-bus system [64]. The VPP is assumed to be composed of 10 small gas power plants, which are connected to buses 10–19. These DERs are equipped with conventional synchronous generators and primary frequency and voltage regulators. The data of the VPP can be found in [25]. In particular, a total capacity of VPP equal to 20% of the total load is considered. In order to simulate the performance of the VPP ramping response, it is assumed that at $t = 1$ s there is a 20% instantaneous load increase. Because of this contingency, it is assumed that the TSO sends a signal to the VPP to start the production and cover the increase of the load within 15 minutes.

It is relevant to discuss first the performance of the conventional AGC, namely, the AGC installed in the control center of TSOs is discussed. Fig. 6.16 shows the transient behavior of the system for two cases, namely, with and without a conventional AGC. The AGC improves the frequency nadir and, as expected, removes the steady-state error of the frequency.

Next, the performance of the VPP approaches from the dynamic point of view of the system is compared below. In all cases, it is assumed that there is a conventional AGC installed in the system.

Fig. 6.17 shows the results for three scenarios, namely, MILP-based VPP, AGC-based VPP and VPP (i.e., without any control). In the long term, the AGC-based VPP leads to a better dynamic behavior than the MILP-base approach and achieves a null steady-state frequency error. On the other hand, the MILP-based approach has a slightly faster response than the AGC-based one. Since the MILP-based control starts-up the generators one by one (e.g., starting from the cheapest one) until they reach the set-point sent by the TSO, it is not able to track perfectly the reference frequency. Moreover, the MILP-based approach causes several "ripples" in the frequency that correspond to the start-up of the units included in the VPP. The AGC-based VPP, on the other hand, starts-up the DERs all at the same time and then smoothly increases their power output. These results suggest that the continuous control provided by the AGC-based approach is to be preferred from a dynamic performance point of view.

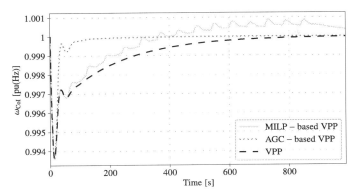

Figure 6.17 Comparison of the frequency response between the AGC-based VPP, MILP-based VPP and VPP without secondary frequency control.

However, since all units are started at the same time, it is possible that the AGC-based solution is more expensive than the MILP-based approach.

6.4 Conclusions

VPPs are expected to participate actively in the procurement and offer of ancillary services. Most typical and crucial of the ancillary services is frequency control, as it expresses the effort to retain generation–demand equilibrium at any given moment. This chapter reviews the roles that researchers and policy makers have proposed for the involvement of VPPs in frequency control. The chapter introduces first a background on conventional frequency control of power systems and why VPP is a paradigm that can and should contribute substantially to the frequency regulation. Then the chapter provides a comprehensive review with respect to the VPP frequency regulation. Next, the chapter defines various classifications for VPPs in the context of frequency control procurement and realization. For each category, relevant works are reviewed in the chapter. The classification of these works shows that most of the literature focuses on how DERs are integrated into a VPP so as to offer primary frequency control. This is not surprising. As large conventional power plants are decommissioned, in fact, the services they provided have to be covered by DERs.

Then, the chapter discusses the performance of some common VPP primary and secondary frequency control approaches. Specifically, the primary frequency control from DERs and ESSs within a VPP is implemented via droop and PI controllers. The performance of these controllers is tested through a case study based on a modified version of the WSCC 9-bus system. It is shown that ESS can significantly improve the dynamic behavior of the system even without DERs frequency control. This case study also shows that a better dynamic behavior is achieved through a coordinated control of ESS with respect to maximum deviation and post-disturbance equilibrium. The second case study based on a modified IEEE 39-bus system evaluates the performance of two common VPP secondary frequency control approaches, namely, one based on MILP scheduling and another on AGC (the problem of VPP linear ramping response is considered). It is shown that the AGC-based VPP leads to a better long-term dynamic behavior of the system compared to the MILP-based method.

Given the results of the two test cases and the extensive review on the existing research and policy literature, we may safely conclude that there needs to be a shift of focus on the subject of frequency control by VPPs. Much effort has been exerted on the individual actions of DER and ESSs and less attention has been paid on VPP-level frequency control in light, also, of broader system dynamics. There are numerous indications that as the inertia from conventional resources is displaced by inverter-interfaced DERs and ESSs, the matter of frequency control of asynchronous resources as a potential cause for system instability and a source of low-frequency oscillations will become a growing concern [65]. VPPs are uniquely positioned to actively aggregate, control, and manage the effects of DERs and ESSs at the highest level of coordination that ensures grid stability. VPP control has to handle the matter of system stability from a disadvantaged position as it lacks grid visibility (e.g., does not have information about the topology of the grid). In this sense, data-heavy models, as well as sensing and novel state estimation techniques are needed to enable VPPs in this complicated role in modern power systems. Information and communication infrastructure and innovations are also needed to support VPP operation. VPPs, in fact, need to respond to frequency variations and fast acting inverter-driven controls that may lead to negative and undesired grid dynamics.

References

[1] Australian Energy Market Operator, Energy explained frequency, https://aemo.com.au/en/newsroom/energy-live/energy-explained-frequency.
[2] IEEE/IEC International Standard – Measuring relays and protection equipment – Part 118-1: Synchrophasor for power systems – Measurements, IEC/IEEE Std., 2018.
[3] F. Milano, Á. Ortega, Frequency Variations in Power Systems: Modeling, State Estimation, and Control, John Wiley & Sons, 2020.
[4] F. Milano, F. Dörfler, G. Hug, D.J. Hill, G. Verbič, Foundations and challenges of low-inertia systems (invited paper), in: Power Systems Computation Conference (PSCC), June 2018, pp. 1–25.
[5] A. Berizzi, The Italian 2003 blackout, in: IEEE PES General Meeting, vol. 2, 2004, pp. 1673–1679.
[6] ENTSO-E, Interim report system disturbance on 4 November 2006, https://eepublicdownloads.azureedge.net/clean-documents/pre2015/publications/ce/otherreports/IC-Interim-Report-20061130.pdf.
[7] H. Haes Alhelou, M.E. Hamedani-Golshan, T.C. Njenda, P. Siano, A survey on power system blackout and cascading events: research motivations and challenges, Energies 12 (4) (2019) 682.
[8] F. Arrigo, C. Mosca, E. Bompard, P. Cuccia, Frequency models and control in normal operation: the Sardinia case study, in: 55th International Universities Power Engineering Conference (UPEC), 2020, pp. 1–6.

[9] M. Liu, J. Chen, F. Milano, On-line inertia estimation for synchronous and non-synchronous devices, IEEE Transactions on Power Systems (2020).

[10] R.A. Ahangar, A. Sheykholeslami, Bulk virtual power plant, a novel concept for improving frequency control and stability in presence of large scale RES, International Journal of Mechatronics, Electrical and Computer Technology 4 (10) (2014) 1017–1044.

[11] J. Morren, S.W. De Haan, W.L. Kling, J. Ferreira, Wind turbines emulating inertia and supporting primary frequency control, IEEE Transactions on Power Systems 21 (1) (2006) 433–434.

[12] G. Ramtharan, N. Jenkins, J. Ekanayake, Frequency support from doubly fed induction generator wind turbines, IET Renewable Power Generation 1 (1) (2007) 3–9.

[13] K. Vidyanandan, N. Senroy, Primary frequency regulation by deloaded wind turbines using variable droop, IEEE Transactions on Power Systems 28 (2) (2012) 837–846.

[14] A. Yazdani, P.P. Dash, A control methodology and characterization of dynamics for a photovoltaic (PV) system interfaced with a distribution network, IEEE Transactions on Power Delivery 24 (3) (2009) 1538–1551.

[15] P. Moutis, A. Vassilakis, A. Sampani, N. Hatziargyriou, DC switch driven active power output control of photovoltaic inverters for the provision of frequency regulation, IEEE Transactions on Sustainable Energy 6 (4) (2015) 1485–1493.

[16] Á. Ortega, F. Milano, Generalized model of VSC-based energy storage systems for transient stability analysis, IEEE Transactions on Power Systems 31 (5) (2015) 3369–3380.

[17] P. Moutis, N.D. Hatziargyriou, Decision trees-aided active power reduction of a virtual power plant for power system over-frequency mitigation, IEEE Transactions on Industrial Informatics 11 (1) (2014) 251–261.

[18] Á. Ortega, F. Milano, Frequency control of distributed energy resources in distribution networks, IFAC-PapersOnLine 51 (28) (2018) 37–42.

[19] W. Zhong, M.A.A. Murad, M. Liu, F. Milano, Impact of virtual power plants on power system short-term transient response, Electric Power Systems Research 189 (2020) 106609.

[20] K.S. Nielsen, Test system for determining a frequency response of a virtual power plant, Sept. 1 2015, US Patent 9,122,274.

[21] Y. Liu, H. Xin, Z. Wang, D. Gan, Control of virtual power plant in microgrids: a coordinated approach based on photovoltaic systems and controllable loads, IET Generation, Transmission & Distribution 9 (10) (2015) 921–928.

[22] N. Ruiz, I. Cobelo, J. Oyarzabal, A direct load control model for virtual power plant management, IEEE Transactions on Power Systems 24 (2) (2009) 959–966.

[23] E.A. Setiawan, Concept and Controllability of Virtual Power Plant, Kassel University Press GmbH, 2007.

[24] J. Yang, Q. Zheng, J. Zhao, X. Guo, C. Gao, Control strategy of virtual power plant participating in the system frequency regulation service, in: 4th International Conference on Systems and Informatics (ICSAI), IEEE, 2017, pp. 324–328.

[25] T. Kërçi, M.T. Devine, M.A.A. Murad, F. Milano, Impact of the aggregate response of distributed energy resources on power system dynamics, in: IEEE PES General Meeting, IEEE, 2020, pp. 1–5.

[26] M. Swierczynski, D.I. Stroe, A.I. Stan, R. Teodorescu, Primary frequency regulation with Li-ion battery energy storage system: a case study for Denmark, in: IEEE ECCE Asia Downunder, IEEE, 2013, pp. 487–492.

[27] N. Etherden, M.H. Bollen, J. Lundkvist, Quantification of ancillary services from a virtual power plant in an existing subtransmision network, in: IEEE PES ISGT Europe 2013, IEEE, 2013, pp. 1–5.

[28] L. Yavuz, A. Önen, S. Muyeen, I. Kamwa, Transformation of microgrid to virtual power plant – a comprehensive review, IET Generation, Transmission & Distribution 13 (11) (2019) 1994–2005.
[29] H. Khan, S. Dasouki, V. Sreeram, H.H. Iu, Y. Mishra, Universal active and reactive power control of electronically interfaced distributed generation sources in virtual power plants operating in grid-connected and islanding modes, IET Generation, Transmission & Distribution 7 (8) (2013) 885–897.
[30] D. Pudjianto, C. Ramsay, G. Strbac, Virtual power plant and system integration of distributed energy resources, IET Renewable Power Generation 1 (1) (2007) 10–16.
[31] N. Etherden, V. Vyatkin, M.H. Bollen, Virtual power plant for grid services using IEC 61850, IEEE Transactions on Industrial Informatics 12 (1) (2015) 437–447.
[32] J. Kim, E. Muljadi, V. Gevorgian, M. Mohanpurkar, Y. Luo, R. Hovsapian, V. Koritarov, Capability-coordinated frequency control scheme of a virtual power plant with renewable energy sources, IET Generation, Transmission & Distribution 13 (16) (2019) 3642–3648.
[33] C. Schwaegerl, L. Tao, The microgrids concept, Microgrids (2013) 1–24.
[34] A. Ghafouri, J. Milimonfared, G.B. Gharehpetian, Classification of microgrids for effective contribution to primary frequency control of power system, IEEE Systems Journal 11 (3) (2015) 1897–1906.
[35] P. Jampeethong, S. Khomfoi, Coordinated control of electric vehicles and renewable energy sources for frequency regulation in microgrids, IEEE Access (2020).
[36] A. Fathi, Q. Shafiee, H. Bevrani, Robust frequency control of microgrids using an extended virtual synchronous generator, IEEE Transactions on Power Systems 33 (6) (2018) 6289–6297.
[37] X. Wu, C. Shen, R. Iravani, A distributed, cooperative frequency and voltage control for microgrids, IEEE Transactions on Smart Grid 9 (4) (2016) 2764–2776.
[38] J. Zhao, X. Lyu, Y. Fu, X. Hu, F. Li, Coordinated microgrid frequency regulation based on DFIG variable coefficient using virtual inertia and primary frequency control, IEEE Transactions on Energy Conversion 31 (3) (2016) 833–845.
[39] X. Zhao-Xia, Z. Mingke, H. Yu, J.M. Guerrero, J.C. Vasquez, Coordinated primary and secondary frequency support between microgrid and weak grid, IEEE Transactions on Sustainable Energy 10 (4) (2018) 1718–1730.
[40] O. Palizban, K. Kauhaniemi, J.M. Guerrero, Microgrids in active network management—part I: hierarchical control, energy storage, virtual power plants, and market participation, Renewable & Sustainable Energy Reviews 36 (2014) 428–439.
[41] H. Bevrani, S. Shokoohi, An intelligent droop control for simultaneous voltage and frequency regulation in islanded microgrids, IEEE Transactions on Smart Grid 4 (3) (2013) 1505–1513.
[42] T. Sikorski, et al., A case study on distributed energy resources and energy-storage systems in a virtual power plant concept: technical aspects, Energies 13 (12) (2020) 3086.
[43] X. Dominguez, M. Pozo, C. Gallardo, L. Ortega, Active power control of a virtual power plant, in: IEEE Ecuador Technical Chapters Meeting (ETCM), IEEE, 2016, pp. 1–6.
[44] G. Delille, B. Francois, G. Malarange, Dynamic frequency control support by energy storage to reduce the impact of wind and solar generation on isolated power system's inertia, IEEE Transactions on Sustainable Energy 3 (4) (2012) 931–939.
[45] A. Oudalov, D. Chartouni, C. Ohler, Optimizing a battery energy storage system for primary frequency control, IEEE Transactions on Power Systems 22 (3) (2007) 1259–1266.
[46] M. Świerczyński, D.I. Stroe, R. Lærke, A.I. Stan, P.C. Kjær, R. Teodorescu, S.K. Kær, Field experience from Li-ion BESS delivering primary frequency regulation in the Danish energy market, ECS Transactions 61 (37) (2014) 1.

[47] H.H. Alhelou, P. Siano, M. Tipaldi, R. Iervolino, F. Mahfoud, Primary frequency response improvement in interconnected power systems using electric vehicle virtual power plants, World Electric Vehicle Journal 11 (2) (2020) 40.

[48] L. Casasola-Aignesberger, S. Martinez, Electric vehicle recharge strategies for frequency control in electrical power systems with high wind power generation, in: IEEE International Conference on Environment and Electrical Engineering and IEEE Industrial and Commercial Power Systems Europe (EEEIC / I CPS Europe), 2020, pp. 1–5.

[49] M. Vasirani, R. Kota, R.L. Cavalcante, S. Ossowski, N.R. Jennings, An agent-based approach to virtual power plants of wind power generators and electric vehicles, IEEE Transactions on Smart Grid 4 (3) (2013) 1314–1322.

[50] R. Khan, N. Gogoi, J. Barman, A. Latif, D.C. Das, Virtual power plant enabled co-ordinated frequency control of a grid connected independent hybrid microgrid using firefly algorithm, in: IEEE Region 10 Symposium (TENSYMP), IEEE, 2019, pp. 795–800.

[51] B. Jansen, C. Binding, O. Sundstrom, D. Gantenbein, Architecture and communication of an electric vehicle virtual power plant, in: First IEEE International Conference on Smart Grid Communications, IEEE, 2010, pp. 149–154.

[52] S. Han, S. Han, K. Sezaki, Development of an optimal vehicle-to-grid aggregator for frequency regulation, IEEE Transactions on Smart Grid 1 (1) (2010) 65–72.

[53] H. Liu, Z. Hu, Y. Song, J. Lin, Decentralized vehicle-to-grid control for primary frequency regulation considering charging demands, IEEE Transactions on Power Systems 28 (3) (2013) 3480–3489.

[54] B.C. Pal, A.H. Coonick, I.M. Jaimoukha, H. El-Zobaidi, A linear matrix inequality approach to robust damping control design in power systems with superconducting magnetic energy storage device, IEEE Transactions on Power Systems 15 (1) (2000) 356–362.

[55] B. Tamimi, C. Cañizares, K. Bhattacharya, Modeling and performance analysis of large solar photo-voltaic generation on voltage stability and inter-area oscillations, in: IEEE PES General Meeting, IEEE, 2011, pp. 1–6.

[56] F. Milano, Á. Ortega, Converter-Interfaced Energy Storage Systems – Context, Modelling and Dynamic Analysis, 1st ed., Cambridge University Press, 2019.

[57] P.W. Sauer, M.A. Pai, Power System Dynamics and Stability, vol. 101, Wiley Online Library, 1998.

[58] C. Murphy, A. Keane, Local and remote estimations using fitted polynomials in distribution systems, IEEE Transactions on Power Systems 32 (4) (2016) 3185–3194.

[59] F. Milano, A Python-based software tool for power system analysis, in: IEEE PES General Meeting, IEEE, 2013, pp. 1–5.

[60] F. Milano, Á. Ortega, Frequency divider, IEEE Transactions on Power Systems 32 (2) (March 2017) 1493–1501.

[61] Á. Ortega, F. Milano, Impact of frequency estimation for VSC-based devices with primary frequency control, in: IEEE International Conference on Innovative Smart Grid Technologies (IEEE ISGT Europe), Turin, Italy, Sept. 2017.

[62] P. Beagon, M.D. Bustamante, M.T. Devine, S. Fennell, J. Grant-Peters, C. Hall, R. Hill, T. Kërçi, G. O'Keefe, Optimal scheduling of distributed generation to achieve linear aggregate response, in: Proceedings from the 141st European Study Group with Industry, 2018, available at: http://www.maths-in-industry.org/miis/758/.

[63] INTERRFACE Consortium, TSO-DSO-Consumer INTERFACE architecture to provide innovative grid services for an efficient power system, [Online]. Available: http://www.interrface.eu/.

[64] Illinois Center for a Smarter Electric Grid (ICSEG), IEEE 39-Bus System, http://publish.illinois.edu/smartergrid/ieee-39-bus-system/.
[65] U. Markovic, O. Stanojev, P. Aristidou, E. Vrettos, D.S. Callaway, G. Hug, Understanding small-signal stability of low-inertia systems, IEEE Transactions on Power Systems (2021).

VPP's participation in demand response aggregation market

Ali Shayegan-Rad[a] and Ali Zangeneh[b]
[a]MAPNA Electric and Control, Engineering & Manufacturing Co. (MECO), MAPNA Group, Karaj, Iran, [b]Electrical Engineering Department, Shahid Rajaee Teacher Training University, Tehran, Iran

7.1 Introduction

The virtual power plant has a potential to integrate the capacity of various distributed energy resources (renewable or fossil based) like gas turbines, wind turbines, battery energy storage systems, and so on, into a coordinated uniform power plant to supply the entire load demands [1]. Since the behavior of renewable energy resources depends on stochastic nature of primary sources, reliable supply of load demands have obtained particular importance in the VPP operation. Besides, the rapid variation of electricity consumptions may increase the complexity of the equalization between power generation and load demands. One of the available methods to overcome these drawbacks is to schedule controllable loads as a unit agent in the distribution system. The idea is to aggregate individual DR programs in a controlled and intelligent way to act like a small power plant as a part of the VPP. They are able to modify their behavior under the received commands of the VPP [2].

Electricity consumption of different types of customers, including residential, commercial, and industrial, varies based on new price tariffs and incentives using demand response programs [3]. According to the US Federal Energy Regulatory Commission, the demand response programs have been presented to encourage responsive loads to vary their electric consumption due to the changes in the electricity billing prices, incentive payments during peak electricity market prices, or emergency conditions when distribution network reliability is jeopardized [4]. Consequently, VPP-DRA implements DR programs via signing incentive contracts, as well as setting different price tariffs during peak and off-peak intervals of a day to encourage responsive loads to decrease electricity consumption in a certain period and shift a portion of electricity demand from peak periods to off-peak periods of the distribution network [5]. Customers perform load curtailment (LC) via switching-off interruptible appliances such as heating, ventilation system, lightings, and refrigerators. In such DR programs, electricity consumption is decreased during particular times without shifting to another time periods. On the other hand, in the load shifting (LS) programs, shiftable appliances like dishwasher and washer–dryer can shift their working periods to other parts of the day [6].

From the VPP point of view, maximum contribution of the customers are achieved in various kinds of DR applications. Furthermore, DR programs significantly improve distribution system reliability, reduce peak demand, power losses, market price volatilities, fuel consumption, and carbon emission, as well as they can enhance economic efficiency [7–10].

Despite of the mentioned benefits, employing DR applications would be also a challenging duty in reality. Since individual customers offer only a little portion of their entire consumption, they do not have noticeable negotiation power to affect DR prices. Furthermore, DSO lacks technical skills in planning and implementing DR mechanisms at such a large scale [11]. Thus, to facilitate involvement of individual customer in DR programs, a demand response aggregator (DRA) is established as a sub-unit of VPP to aggregate DR applications [12]. In other words, the VPP-DRA is introduced to play as a small power plant beside other stochastic generations, e.g., wind turbines and photovoltaics, in distribution systems. To avoid complexity, it is assumed that the VPP under study has just a DR aggregator and its scheduled generation units are not considered in this chapter.

7.2 Single-level model of DSO without DR programs

In this model, the DSO has no choice except for importing the required energy from the wholesale market. Consequently, the DSO supplies its customers by purchasing the entire power demand from the power electricity market. Fig. 7.1 indicates the scheduling model of the DSO without applying DR programs.

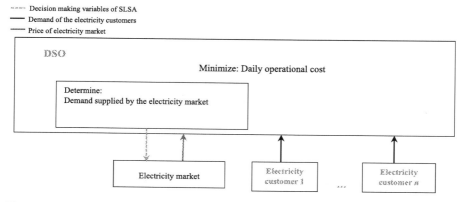

Figure 7.1 The SLSA model of the DSO without DR programs.

Eq. (7.1) calculates the net imported energy cost from the wholesale market. This function is minimized subject to the technical constraints (Eqs. (7.2)–(7.6)) regarding the distribution system limitations. The power transaction with the upstream network is limited in (7.2) due to the technical constraint of the substation transformer. An approximate power flow analysis is carried out in the distribution system to guarantee convexity as well as reduce calculation burden of the proposed scheduling approaches

[14]. It is worth noting that the validity of the proposed approximate power flow analysis is verified for radial distribution systems with high power factor and R/X rates [15]. The power flow between nodes n and m is calculated using (7.3), and Eq. (7.4) ensures the power balance at all distribution system nodes. Power flows in the branches of the distribution system are limited by (7.5). Finally, minimum up and down limits of voltage magnitude for node n are represented by (7.6):

$$\underset{P_{sub}^{EM}, V_n}{Min} \sum_h \left[\rho^{EM}(h) \times P_{sub}^{EM}(h) \right]; \tag{7.1}$$

$$P_{sub}^{EM}(h) \leq P_{sub}^{EM,\max} \quad \forall sub \in SUB; \quad \forall h \in H; \tag{7.2}$$

$$P_{nm}(h) \cong \frac{|V_n(h)|.(|V_n(h)| - |V_m(h)|)}{|Z_{nm}|} \quad \forall n \in N; \quad \forall m \in M; \quad \forall h \in H; \tag{7.3}$$

$$P_n^{gen}(h) - P_n^{dem}(h) = \sum_{m \in M} \frac{|V_n(h)|.(|V_n(h)| - |V_m(h)|)}{|Z_{nm}|} \quad \forall n \in N; \quad \forall h \in H; \tag{7.4}$$

$$\left| \frac{|V_n(h)|.(|V_n(h)| - |V_m(h)|)}{|Z_{nm}|} \right| \leq P_{nm}^{\max} \quad \forall n \in N; \quad \forall m \in M; \quad \forall h \in H; \tag{7.5}$$

$$V_n^{\min} \leq V_n(h) \leq V_n^{\max} \quad \forall n \in N; \quad \forall h \in H. \tag{7.6}$$

7.3 Single-level scheduling model of DSO with DR programs

The main objective function of this model is to simultaneously minimize the operation costs of DSO and VPP-DRA taking into account the procurement of all electrical demand. To this end, DSO has two supply options, namely power from the wholesale electricity market and available responsive loads for each operation time interval. In other words, DSO purchases a portion of its electricity demand from VPP-DRA by executing the responsive loads and purchases all the remaining demand from the electricity market or procures the entire electricity from the upstream electricity market. Fig. 7.2 indicates the structure of SLSA of a DSO with DR programs.

Responsive loads typically have two different strategies: first, curtailing load during peak time intervals, and second, shifting load from peak-time to off-peak intervals. The first and second terms of the objective function in (7.7) calculate the payments to VPP-DRA for performing load curtailment and shifting applications, respectively, and the last term calculates incurred cost of importing electricity from the wholesale market. Eqs. (7.8)–(7.10) represent the minimum and maximum of responsive loads to participate in the load curtailment, shifting, and recovery, respectively. Eq. (7.11) indicates that the electricity shifted during requested time intervals has to be recovered during another times of the day. Note that the distribution system technical constrains

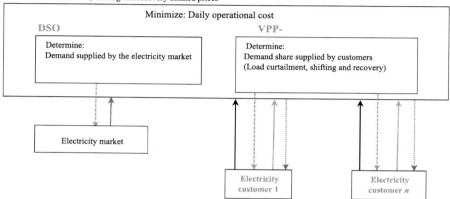

Figure 7.2 The SLSA model of DSO with DR programs.

(7.12)–(7.16) have to be satisfied as mentioned in Section 7.2:

$$\underset{P_i^{LC}, P_i^{LS}, P_{sub}^{EM}, V_n}{Min} \sum_h \left[\lambda_{DRP}^{LC} \times \sum_i P_i^{LC}(h) + \lambda_{DRP}^{LS} \times \sum_i P_i^{LS}(h) \\ + \rho^{EM}(h) \times P_{sub}^{EM}(h) \right]; \quad (7.7)$$

$$P_i^{LC,\min}(h) \leq P_i^{LC}(h) \leq P_i^{LC,\max}(h) \quad \forall i \in I; \quad \forall h \in H; \quad (7.8)$$

$$P_i^{LS,\min}(h) \leq P_i^{LS}(h) \leq P_i^{LS,\max}(h) \quad \forall i \in I; \quad \forall h \in H; \quad (7.9)$$

$$P_i^{LR,\min}(h) \leq P_i^{LR}(h) \leq P_i^{LR,\max}(h) \quad \forall i \in I; \quad \forall h \in H; \quad (7.10)$$

$$\sum_h [P_i^{LS}(h) - P_i^{LR}(h)] = 0 \quad \forall i \in I; \quad (7.11)$$

$$P_{sub}^{EM}(h) \leq P_{sub}^{EM,\max} \quad \forall sub \in SUB; \quad \forall h \in H; \quad (7.12)$$

$$P_{nm}(h) \cong \frac{|V_n(h)| \cdot (|V_n(h)| - |V_m(h)|)}{|Z_{nm}|} \quad \forall n \in N; \quad \forall m \in M; \quad \forall h \in H; \quad (7.13)$$

$$P_n^{gen}(h) - P_n^{dem}(h) = \sum_{m \in M} \frac{|V_n(h)| \cdot (|V_n(h)| - |V_m(h)|)}{|Z_{nm}|} \quad \forall n \in N; \quad \forall h \in H; \quad (7.14)$$

$$\left| \frac{|V_n(h)| \cdot (|V_n(h)| - |V_m(h)|)}{|Z_{nm}|} \right| \leq P_{nm}^{\max} \quad \forall n \in N; \quad \forall m \in M; \quad \forall h \in H; \quad (7.15)$$

$$V_n^{\min} \leq V_n(h) \leq V_n^{\max} \quad \forall n \in N; \quad \forall h \in H. \quad (7.16)$$

7.4 Bi-level scheduling model between DSO and VPP-DRA

Section 7.3 addressed a single-level scheduling framework of the DSO that considered DR programs. However, the previous model executed DR programs by just minimizing the operation cost of DSO and VPP, and did not model the accurate interaction between them. Therefore, imposed costs and profits would not be allocated fairly and correctly between DSO and VPP-DRA. To overcome this weakness, a BLSA between DSO and VPP-DRA is implemented in this section (Fig. 7.3) [16]. The VPP-DRA, the leader, is regarded in the upper level of the model while the DSO, the follower, is set in the lower level. The VPP-DRA in the upper level determines the most suitable LC and LS prices that would maximize its obtained profit. Consequently, the DSO in the lower-level problem minimizes its payments, reacting to the received DR prices of the VPP-DRA in an interactive competition [17,18]. The lower-level problem is transformed into the upper-level problem using Karush–Kuhn–Tucker (KKT) optimality conditions. Hence, the bi-level problem is converted into a single level nonlinear problem [19,20].

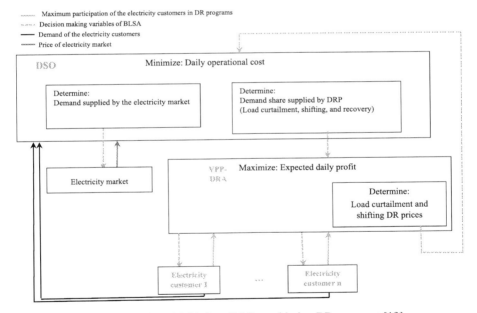

Figure 7.3 The structure of the BLSA for a DSO considering DR programs [13].

7.4.1 Upper level: profit function of the VPP demand response aggregator (VPP-DRA)

The goal of the VPP-DRA is to optimize the proposed LS and LC bidding prices so as to maximize its profit. If it bids a high price, DSO might prefer to provide its required demand from the wholesale market and not call DR programs; conversely, if it bids a low price, enough profit might not be guaranteed. Therefore, the VPP-DRA should consider the conditions and behaviors of DSO to propose its optimal bidding prices. The objective function of VPP-DRA in the upper level is described in (7.17). The first and second terms indicate the income by executing load curtailment and shifting, respectively. The third and fourth terms are payments to customers for performing the load curtailment and shifting, respectively:

$$\underset{\lambda_{DRP}^{LC}, \lambda_{DRP}^{LS}}{Max} \sum_h \left[\begin{array}{c} \lambda_{DRP}^{LC} \times \sum_i P_i^{LC}(h) + \lambda_{DRP}^{LS} \times \sum_i P_i^{LS}(h) \\ - \sum_i \pi_i^{LC}(h) \times P_i^{LC}(h) - \sum_i \pi_i^{LS}(h) \times P_i^{LS}(h) \end{array} \right]. \quad (7.17)$$

7.4.2 Lower level: cost function of the distribution system operator (DSO)

The DSO always tends to minimize its overall payment in the lower level. The DSO decides on the power supply share from either VPP-DRA or the wholesale electricity market, or both, to provide the customers' electricity demand. The objective function (7.8) is minimized subject to DR constraints (7.19)–(7.22), as well as technical constraints of distribution system (7.23)–(7.27):

$$\underset{P_i^{LC}, P_i^{LS}, P_{sub}^{EM}, V_n}{Min} \sum_h \left[\begin{array}{c} \lambda_{DRP}^{LC} \times \sum_i P_i^{LC}(h) + \lambda_{DRP}^{LS} \times \sum_i P_i^{LS}(h) \\ + \rho^{EM}(h) \times P_{sub}^{EM}(h) \end{array} \right]; \quad (7.18)$$

$$P_i^{LC,\min}(h) \le P_i^{LC}(h) \le P_i^{LC,\max}(h) \quad \forall i \in I; \quad \forall h \in H; \quad (7.19)$$

$$P_i^{LS,\min}(h) \le P_i^{LS}(h) \le P_i^{LS,\max}(h) \quad \forall i \in I; \quad \forall h \in H; \quad (7.20)$$

$$P_i^{LR,\min}(h) \le P_i^{LR}(h) \le P_i^{LR,\max}(h) \quad \forall i \in I; \quad \forall h \in H; \quad (7.21)$$

$$\sum_h [P_i^{LS}(h) - P_i^{LR}(h)] = 0 \quad \forall i \in I; \quad (7.22)$$

$$P_{sub}^{EM}(h) \le P_{sub}^{EM,\max} \quad \forall sub \in SUB; \quad \forall h \in H; \quad (7.23)$$

$$P_{nm}(h) \cong \frac{|V_n(h)| \cdot (|V_n(h)| - |V_m(h)|)}{|Z_{nm}|} \quad \forall n \in N; \quad \forall m \in M; \quad \forall h \in H; \quad (7.24)$$

$$P_n^{gen}(h) - P_n^{dem}(h) = \sum_{m \in M} \frac{|V_n(h)| \cdot (|V_n(h)| - |V_m(h)|)}{|Z_{nm}|} \quad \forall n \in N; \quad \forall h \in H; \quad (7.25)$$

$$\left| \frac{|V_n(h)| \cdot (|V_n(h)| - |V_m(h)|)}{|Z_{nm}|} \right| \le P_{nm}^{\max} \quad \forall n \in N; \quad \forall m \in M; \forall h \in H; \quad (7.26)$$

$$V_n^{\min} \le V_n(h) \le V_n^{\max} \quad \forall n \in N; \quad \forall h \in H. \quad (7.27)$$

7.4.3 Equivalent single-level programming problem

The equivalent single-level model is achieved by applying KKT optimality conditions associated with the lower level of the BLSA. Eq. (7.28) states the objective function of the equivalent single-level model. DR and technical constraints of the distribution system are represented in (7.29)–(7.32) and (7.33)–(7.37), respectively. Finally, the equations obtained based on the KKT optimality conditions are listed in (7.38)–(7.54), which are taken from [13].

$$\underset{\substack{\lambda_{DRP}^{LC}, \lambda_{DRP}^{LS}, P_i^{LC}, \\ P_i^{LS}, P_i^{LR}, P_{sub}^{EM}, V_n}}{Max} \sum_h \left[\begin{array}{c} \lambda_{DRP}^{LC} \times \sum_i P_i^{LC}(h) + \lambda_{DRP}^{LS} \times \sum_i P_i^{LS}(h) \\ - \sum_i \pi_i^{LC}(h) \times P_i^{LC}(h) - \sum_i \pi_i^{LS}(h) \times P_i^{LS}(h) \end{array} \right]; \tag{7.28}$$

$$P_i^{LC,\min}(h) \le P_i^{LC}(h) \le P_i^{LC,\max}(h) \quad \forall i \in I; \quad \forall h \in H; \tag{7.29}$$

$$P_i^{LS,\min}(h) \le P_i^{LS}(h) \le P_i^{LS,\max}(h) \quad \forall i \in I; \quad \forall h \in H; \tag{7.30}$$

$$P_i^{LR,\min}(h) \le P_i^{LR}(h) \le P_i^{LR,\max}(h) \quad \forall i \in I; \quad \forall h \in H; \tag{7.31}$$

$$\sum_h [P_i^{LS}(h) - P_i^{LR}(h)] = 0 \quad \forall i \in I; \tag{7.32}$$

$$P_{sub}^{EM}(h) \le P_{sub}^{EM,\max} \quad \forall sub \in SUB; \quad \forall h \in H; \tag{7.33}$$

$$P_{nm}(h) \cong \frac{|V_n(h)| \cdot (|V_n(h)| - |V_m(h)|)}{|Z_{nm}|} \quad \forall n \in N; \quad \forall m \in M; \quad \forall h \in H; \tag{7.34}$$

$$P_n^{gen}(h) - P_n^{dem}(h) = \sum_{m \in M} \frac{|V_n(h)| \cdot (|V_n(h)| - |V_m(h)|)}{|Z_{nm}|} \quad \forall n \in N; \quad \forall h \in H; \tag{7.35}$$

$$\left| \frac{|V_n(h)| \cdot (|V_n(h)| - |V_m(h)|)}{|Z_{nm}|} \right| \le P_{nm}^{\max} \quad \forall n \in N; \quad \forall m \in M; \quad \forall h \in H; \tag{7.36}$$

$$V_n^{\min} \le V_n(h) \le V_n^{\max} \quad \forall n \in N; \quad \forall h \in H; \tag{7.37}$$

$$\lambda_{DRP}^{LC} - \theta_i(h) + \overline{\varsigma}_i(h) - \underline{\varsigma}_i(h) = 0; \quad \forall i \in I; \quad \forall h \in H; \tag{7.38}$$

$$\lambda_{DRP}^{LS} - \theta_i(h) - \omega_i + \overline{\mu}_i(h) - \underline{\mu}_i(h) = 0; \quad \forall i \in I; \quad \forall h \in H; \tag{7.39}$$

$$\theta_i(h) + \omega_i + \overline{\gamma}_i(h) - \underline{\gamma}_i(h) = 0; \quad \forall i \in I; \quad \forall h \in H; \tag{7.40}$$

$$\lambda^{EM}(h) - \theta_{sub}(h) + \overline{\alpha}_{sub}(h) - \underline{\alpha}_{sub}(h) = 0; \quad \forall sub \in SUB; \quad \forall h \in H; \tag{7.41}$$

$$\theta_n(h) \times \sum_{m \in M} \frac{(2 \times |V_n(h)| - |V_m(h)|)}{|Z_{nm}|} - \sum_{m \in M} \theta_m(h) \times \frac{|V_m(h)|}{|Z_{nm}|}$$

$$+ \sum_{m \in M} (\overline{\varphi}_{nm}(h) - \underline{\varphi}_{nm}(h)) \times (\frac{2 \times |V_n(h)| - |V_m(h)|}{|Z_{nm}|})$$

$$+ \sum_{m \in M} (\underline{\varphi}_{mn}(h) - \overline{\varphi}_{mn}(h)) \times (\frac{|V_m(h)|}{|Z_{nm}|})$$

$$+ \overline{\beta}_n(h) - \underline{\beta}_n(h) = 0; \quad \forall n \in N; \quad \forall h \in H; \quad (7.42)$$

$$\overline{\varsigma}_i(h) \times (P_i^{LC}(h) - P_i^{LC,\max}(h)) = 0; \quad \overline{\delta}_i(h) \geq 0; \quad \forall i \in I; \quad \forall h \in H; \quad (7.43)$$

$$\underline{\varsigma}_i(h) \times (P_i^{LC,\min}(h) - P_i^{LC}(h)) = 0; \quad \underline{\varsigma}_i(h) \geq 0; \quad \forall i \in I; \quad \forall h \in H; \quad (7.44)$$

$$\overline{\mu}_i(h) \times (P_i^{LS}(h) - P_i^{LS,\max}(h)) = 0; \quad \overline{\mu}_i(h) \geq 0; \quad \forall i \in I; \quad \forall h \in H; \quad (7.45)$$

$$\underline{\mu}_i(h) \times (P_i^{LS,\min}(h) - P_i^{LS}(h)) = 0; \quad \underline{\mu}_i(h) \geq 0; \quad \forall i \in I; \quad \forall h \in H; \quad (7.46)$$

$$\overline{\gamma}_i(h) \times (P_i^{LR}(h) - P_i^{LR,\max}(h)) = 0; \quad \overline{\gamma}_i(h) \geq 0; \quad \forall i \in I; \quad \forall h \in H; \quad (7.47)$$

$$\underline{\gamma}_i(h) \times (P_i^{LR,\min}(h) - P_i^{LR}(h)) = 0; \quad \underline{\gamma}_i(h) \geq 0; \quad \forall i \in I; \quad \forall h \in H; \quad (7.48)$$

$$\overline{\alpha}_{sub}(h) \times (P_{sub}^{EM}(h) - P_{sub}^{EM,\max}) = 0; \quad \overline{\alpha}_{sub}(h) \geq 0;$$
$$\forall sub \in SUB; \quad \forall h \in H; \quad (7.49)$$

$$\underline{\alpha}_{sub}(h) \times P_{sub}^{EM}(h) = 0; \quad \underline{\alpha}_{sub}(h) \geq 0; \quad \forall sub \in SUB; \quad \forall h \in H; \quad (7.50)$$

$$\overline{\varphi}_{nm}(h) \times (\frac{|V_n(h)|.(|V_n(h)| - |V_m(h)|)}{|Z_{nm}|} - P_{nm}^{\max}) = 0; \quad \overline{\varphi}_{nm}(h) \geq 0;$$
$$\forall n \in N; \quad \forall m \in M; \quad \forall h \in H; \quad (7.51)$$

$$\underline{\varphi}_{nm}(h) \times (-\frac{|V_n(h)|.(|V_n(h)| - |V_m(h)|)}{|Z_{nm}|} - P_{nm}^{\max}) = 0; \quad \underline{\varphi}_{nm}(h) \geq 0;$$
$$\forall n \in N; \quad \forall m \in M; \quad \forall h \in H; \quad (7.52)$$

$$\overline{\beta}_n(h) \times (V_n(h) - V_n^{\max}) = 0; \quad \overline{\beta}_n(h) \geq 0; \quad \forall n \in N; \quad \forall h \in H; \quad (7.53)$$

$$\underline{\beta}_n(h) \times (V_n^{\min} - V_n(h)) = 0; \quad \underline{\beta}_n(h) \geq 0; \quad \forall n \in N; \quad \forall h \in H. \quad (7.54)$$

7.5 Numerical studies and discussions

The methodology presented in this chapter is carried out on a 34-bus modified distribution test system. This distribution system is connected to the main grid via node 1.

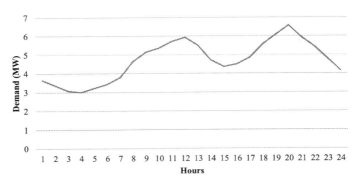

Figure 7.4 Total demand in the IEEE 34-bus distribution system [13].

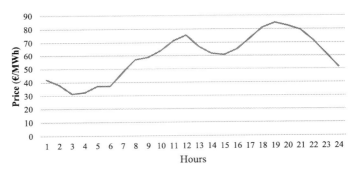

Figure 7.5 Hourly day-ahead electricity market price [13].

Figs. 7.4 and 7.5 depict hourly demands and electricity market prices, respectively. The presented daily time is divided into three discrete intervals including valley time interval (1 to 6 a.m.), shoulder time interval (7 to 10 a.m., 2 to 6 p.m. and 10 to 12 p.m.) and finally peak time interval (11 a.m. to 1 p.m. and 7 to 9 p.m.). The VPP-DRA refund responsive loads for participating in DR programs based on Table 7.1 [13]. It is assumed that electricity customers allocate up to 15%, 20%, and 30% of their hourly demand profile to participate in load curtailment, shifting, and recovery, respectively [13].

In this chapter, five different scenarios have been considered to show effectiveness of performing DR programs in both SLSA and BLSA as follows:

- Scenario A – the DSO purchases electricity from the wholesale market to meet its obligation to fulfill electricity customers without using DR programs.
- Scenario B – the DSO's concept in this scenario is developed to motivate individual customers for participating in LC model of DR program in the SLSA. The bid price of LC program is assumed to be 53.7 (€/MWh) for this scenario.
- Scenario C – this scenario is similar to scenario B, but the LS program is also considered. The bid prices of LC and LS programs are assumed to be 53.7 and 39.3 (€/MWh) for this scenario.

Table 7.1 Demand response cost [13].

	DR cost (€/MWh)
Valley	28
Shoulder	30
Peak	35

Table 7.2 Optimal bidding prices of VPP-DRA [13].

Case	Load curtailment price (€/MWh)	Load shifting price (€/MWh)
Scenario D	62.70	
Scenario E	63.21	45.94

- Scenario D – VPP-DRA aggregates individual customers and negotiates with the DSO to determine optimal contract prices. A BLSA for LC model of DR program is applied.
- Scenario E – this scenario is similar to scenario D, but the LS program is also considered.

Table 7.2 indicates the obtained optimal contract prices of the VPP-DRA for scenarios D and E. Since the shifted load will be recovered in another interval of day time horizon, LS price is considered lower than LC price.

Figs. 7.6 and 7.7 indicate hourly participation of electricity customers in DR programs in scenarios B and C, respectively. The DSO is obligated to purchase electricity from the electricity market to meet its customers and also cover total losses of the distribution system. Hence, the DSO calculates two different quantities: first, the incurred payment for importing power from electricity market to meet its demand and covering distribution power losses; second, incurred payment for executing DR programs to reduce a portion of electricity demands. Finally, the DSO compares two different choices to decide on a strategy that will yield the minimum payment. In scenario B, the DSO prefers to perform LC during hours 7–24. According to DSO point of view, shifting a portion of electricity demand from peak intervals to off-peak intervals is profitable, if difference between peak and off-peak prices be more than LS contract price. Hence, DSO decides to curtail electricity during hours 8–24 and also shift a portion of demand from hours 18–21 to hours 1–6 of daily time horizon.

Hourly participation of electricity customers in DR applications in scenarios D and E is shown in Figs. 7.8 and 7.9, respectively. Since higher DR prices are determined than in scenarios B and C, DSO decides to perform DR applications in fewer hours. However, DSO implements LC program during hours 8–23 in scenario D. Also, in scenario E, LC program is applied during hours 8–23, and LS program is considered to shift a portion of demand from hours 18–21 to hours 2–6.

Fig. 7.10 depicts demand patterns for five considered scenarios. Since DSO does not use DR applications in scenario A, its demand profile is the same as the initial

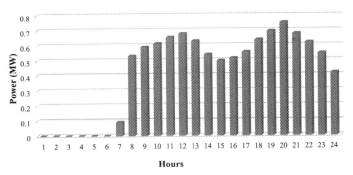

Figure 7.6 Hourly participation of electricity customers in DR applications in scenario B.

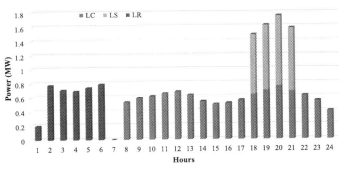

Figure 7.7 Hourly participation of electricity customers in DR applications in scenario C [13].

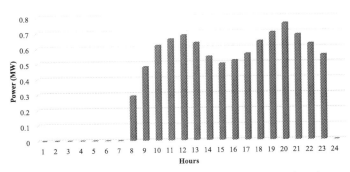

Figure 7.8 Hourly participation of electricity customers in DR applications in scenario D [13].

demand of the DSO. The demand profiles of scenarios B and D are more flat than for scenario A because the LC program is performed. Finally, more flat profiles in scenarios C and E demonstrate the effectiveness of joint LC and LS of DR applications. It should be noted that importing electricity from the electricity market to meet the modified demand seems inevitable. Thus, the power purchased from the electricity market is the same as in modified demand profiles in Fig. 7.10.

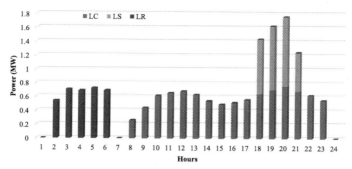

Figure 7.9 Hourly participation of electricity customers in DR applications in scenario E [13].

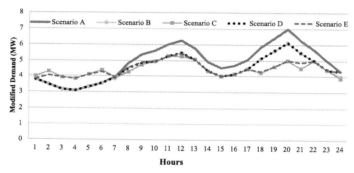

Figure 7.10 Demand patterns for five considered scenarios [13].

Fig. 7.11 indicates the electricity losses of the network in the five different scenarios. As it can be seen, the hourly pattern of electricity losses is the same as for the modified demand profile. Consequently, in scenarios B and D, DSO decreased its electricity losses using LC application in comparison with scenario A. Furthermore, in off-peak intervals of the scenarios C and E, DSO has more demand and hereby more electricity losses than scenarios B and D. Conversely, in peak intervals, DSO has less demand, therefore, less electricity losses than in scenarios B and D.

Fig. 7.12 indicates the hourly incurred cost of the DSO in the five different scenarios. In scenarios B and D, DSO mainly reduces its hourly costs using LC application in comparison to scenario A. During hours 2–6 of scenarios C and E, the cost of DSO is increased in comparison to other scenarios. This is due to the fact that DSO has more demand within these intervals. Conversely, in peak intervals, DSO pays lower cost due to supplying less demand.

Table 7.3 lists the overall cost and network electricity losses of the DSO for the five different scenarios. In scenarios B and D, performing DR programs decreases the total electricity losses of the network and also reduces the overall operation cost of the DSO. Furthermore, performing joint LC and LS applications is more effective and valuable in loss and also cost reduction.

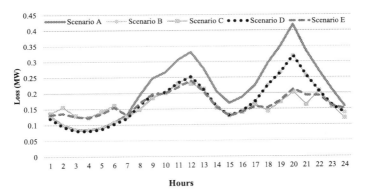

Figure 7.11 Electricity losses of the network for five different scenarios [13].

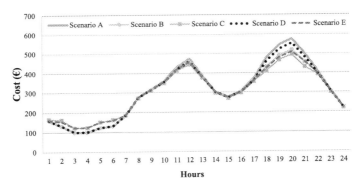

Figure 7.12 Hourly incurred cost of the DSO in the five different scenarios [13].

Table 7.3 System power losses and cost of DSO in the five case studies [13].

	Power losses (MW)	Cost of DSO (€)
Scenario A	5.15	7458.32
Scenario B	4.12	7258.18
Scenario C	3.90	7168.80
Scenario D	4.14	7310.88
Scenario E	4.06	7295.87

The amount of VPP-DRA net profit is computed as obtained revenue from providing DR applications minus incurred cost of performing DR applications. Table 7.4 indicates the net profit of VPP-DRA for scenarios D and E. As shown in Table 7.4, executing both LC and LS applications causes more benefit for the VPP-DRA.

Table 7.4 Income, cost and net profit of VPP-DRA in scenarios D and E [13].

	Scenario D	Scenario E
Income (€)	594.18	753.39
Cost (€)	304.28	419.15
Profit (€)	289.90	334.24

7.6 Conclusion

This chapter develops DR application in both the single and bi-level scheduling approaches. The SLSA is considered based on the DSO offering prices to those responsive loads that have preferences to participate in LC and LS demand response applications. Subsequently, sharing amounts of responsive loads in the LC and LS applications of the DSO scheduling problem are determined. In BLSA, the fair transaction between DSO and VPP-DRA is obtained to determine the optimal contract prices of LC and LS applications. The KKT optimality conditions are implemented to convert the BLSA into a single-level problem. The developed approaches result in cost savings, demand profile improvement, and loss decrease for the DSO. Hence, the allocated contract prices not only benefit DSO but can also increase VPP-DRA obtained profit. Numerical results and discussions verify the efficiency and validity of the presented programming approach to guarantee the maximum profit for the DSO.

Nomenclature

Acronyms
BLSA Bi-level scheduling approach
DR Demand response
DSO Distribution system operator
VPP-DRA Virtual power plant – demand response aggregator
KKT Karush–Kuhn–Tucker
LC Load curtailment
LR Load recovery
LS Load shifting
SLSA Single-level scheduling approach

Indices
H Index of hourly intervals
I Index of customers participating in DR applications
n, m Index of nodes
Sub Index of substations

Parameters
$P_n^{dem}(h)$ Active power demand in node n at hour h (MW)
$P_i^{LC,\min}(h), P_i^{LC,\max}(h)$ Minimum and maximum limits of the load curtailment for the ith aggregated customer (MW)

$P_i^{LR,\min}(h), P_i^{LR,\max}(h)$ Minimum and maximum limits of the load recovery for the ith aggregated customer (MW)

$P_i^{LS,\min}(h), P_i^{LS,\max}(h)$ Minimum and maximum limits of the load shifting for the ith aggregated customer (MW)

$P^{EM,\max}$ Maximum capacity of the substation (MVA)

$P_{nm}^{sub,\max}$ Maximum limit for active power flow from node n to node m (MVA)

$V_n^{\min}, V_n^{\max}$ Minimum and maximum limits on node voltage n (KV)

Z_{nm} Impedance of the line connecting nodes n and m (Ω)

$\pi_i^{LC}(h)$ Price of the ith electricity customer participating in load curtailment at hour h (€/MWh)

$\pi_i^{LS}(h)$ Price of the ith electricity customer participating in load shifting at hour h (€/MWh)

$\rho^{EM}(h)$ Electricity market price at hour h (€/MWh)

Variables

$P_i^{LC}(h)$ Load curtailment of the ith electricity customer at hour h (MW)

$P_i^{LR}(h)$ Load recovery of the ith electricity customer at hour h (MW)

$P_i^{LS}(h)$ Load shifting of the ith electricity customer at hour h (MW)

$P_{sub}^{EM}(h)$ Active power purchased from electricity market at hour h (MW)

$P_n^{gen}(h)$ Active power generation of node n at hour h (MW)

$P_{nm}(h)$ Active power flow from node n to node m at hour h (MVA)

$V_n(h)$ Voltage of node n at hour h (kV)

λ_{DRP}^{LC} Load curtailment contract price of DRP (€/MWh)

λ_{DRP}^{LS} Load shifting contract price of DRP (€/MWh)

$\overline{\alpha}_{sub}(h), \underline{\alpha}_{sub}(h)$ Dual variable associated with the constraint of upper and lower limits for purchasing power from the main grid at hour h

$\overline{\beta}_n(h), \underline{\beta}_n(h)$ Dual variable associated with the constraint of minimum and maximum limits on node voltage at hour h

$\overline{\gamma}_i(h), \underline{\gamma}_i(h)$ Dual variable associated with the constraint of minimum and maximum limits for recovering demand of the ith electricity customer at hour h

$\overline{\varsigma}_i(h), \underline{\varsigma}_i(h)$ Dual variable associated with the constraint of minimum and maximum limits for curtailing demand of the ith electricity customer at hour h

$\overline{\mu}_i(h), \underline{\mu}_i(h)$ Dual variable associated with the constraint of minimum and maximum limits for shifting demand of the ith electricity customer at hour h

$\overline{\varphi}_{nm}(h), \underline{\varphi}_{nm}(h)$ Dual variable associated with the constraint of maximum limit for active power flow from node n to node m at hour h

ω_i Dual variable associated with the constraint of load shifting and load recovery power balance for the ith electricity customer

$\theta_n(h)$ Dual variable associated with the constraint of active power balance in node n at hour h

References

[1] S. Yin, Q. Ai, Z. Li, Y. Zhang, T. Lu, Energy management for aggregate presumes in a virtual power plant: a robust Stackelberg game approach, Int. Trans. Electr. Energy Syst. 117 (2020) 105605.

[2] M. Peik-Herfeh, H. Seifi, M.K. Sheikh-El-Eslami, Decision making of a virtual power plant under uncertainties for bidding in a day-ahead market using point estimate method, Int. J. Electr. Power Energy Syst. 44 (1) (2013) 88–98.

[3] M.H. Albadi, E.F. El-Saadany, A summary of demand response in electricity markets, Electr. Power Syst. Res. 78 (11) (2008) 1989–1996.

[4] U.S. Federal Energy Regulatory Commission, Assessment of demand response and advanced metering staff report 2012, Retrieved January 15, 2013, from http://www.ferc.gov/legal/staff-reports/12-20-12-demand-response.pdf, 2012.

[5] H. Farham, L. Mohammadian, H. Alipour, J. Pouladi, Robust performance of photovoltaic/wind/grid based large electricity consumer, Sol. Energy 174 (2018) 923–932.

[6] A. Zakariazadeh, S. Jadid, P. Siano, Smart microgrid energy and reserve scheduling with demand response using stochastic optimization, Int. J. Electr. Power Energy Syst. 63 (2014) 523–533.

[7] R.N. Boisvert, P.A. Cappers, B. Neenan, The benefits of consumer participation in wholesale electricity markets, Electr. J. 15 (3) (Apr. 2002) 41–51.

[8] H.J. Wellinghoff, D.L. Morenoff, Recognizing the importance of demand response: the second half of the wholesale electric market equation, Energy Law J. 28 (2) (2007) 389–419.

[9] S.S. Reddy, Optimizing energy and demand response programs using multi-objective optimization, Electr. Eng. 99 (2017) 397–406, https://doi.org/10.1007/s00202-016-0438-6.

[10] D.S. Kirschen, Demand-side view of electricity markets, IEEE Trans. Power Syst. 18 (2) (May 2003) 520–527.

[11] L. Gkatzikis, I. Koutsopoulos, T. Salonidis, The role of aggregators in smart grid demand response markets, IEEE J. Sel. Areas Commun. 31 (7) (July 2013).

[12] A. Zakariazadeh, S. Jadid, P. Siano, Economic-environmental energy and reserve scheduling of smart distribution system: a multiobjective mathematical programming approach, Energy Convers. Manag. 78 (2014) 151–164.

[13] A. Shayegan-Rad, A. Zangeneh, A stochastic bilevel scheduling model for determining load shifting and curtailment in demand response programs, J. Electr. Eng. Technol. 13 (3) (2018) 1069–1078.

[14] N. Mahmoudi, T.K. Saha, M. Eghbal, Modelling demand response aggregator behavior in wind power offering strategies, Appl. Energy 133 (2014) 347–355.

[15] W. El-Khattan, K. Bhattacharya, Y. Hegazy, M.M.A. Salama, Optimal investment planning for distributed generation in a competitive electricity market, IEEE Trans. Power Syst. 20 (4) (2005) 1718–1727.

[16] S. Dempe, Foundations of Bilevel Programming, Kluwer, Dordrecht, The Netherlands, 2002.

[17] A. Zangeneh, A. Shayegan-Rad, F. Nazari, A multi leader-follower game theory for optimal contract pricing of virtual power plants in smart distribution networks, IET Gener. Transm. Distrib. 12 (2018) 5747–5752.

[18] A. Shayegan-Rad, A. Zangeneh, Optimal contract pricing of load aggregators for direct load control in smart distribution systems, J. Electr. Eng. Technol. 13 (3) (2018) 1069–1078.

[19] E.G. Kardakos, C.K. Simoglou, A.G. Bakirtzis, Optimal bidding strategy in transmission-constrained electricity markets, Electr. Power Syst. Res. 109 (4) (2014) 141–149.

[20] X. Fang, Q. Hu, F. Li, B. Wang, Coupon-based demand response considering wind power uncertainty: a strategic bidding model for load serving entities, IEEE Trans. Power Syst. 31 (2) (2015) 1–13.

VPP's participation in demand response exchange market

Ali Shayegan-Rad[a] and Ali Zangeneh[b]
[a]MAPNA Electric and Control, Engineering & Manufacturing Co. (MECO), MAPNA Group, Karaj, Iran, [b]Electrical Engineering Department, Shahid Rajaee Teacher Training University, Tehran, Iran

8.1 Introduction

Distributed energy resources (DERs) are growing rapidly in the distribution systems due to their low environmental impact, variety of energy resources, advancement of technology, and improved energy efficiency. Consequently, some portion of large conventional fuel-based power stations are going to be replaced by the small DERs [1,2]. However, low capacity and uncertain behavior of renewable sources are significant obstacles for playing a significant role in the electricity and regulation markets as a large conventional power plant. Therefore, considering a proper infrastructure that is able to control the generation of DERs, providing required regulation and improving the system capability are crucial for taking part in the electricity markets. A promising solution to cover these drawbacks is to establish the concept of virtual power plants (VPPs) in the distribution systems [3].

Using VPP infrastructure, individual DERs will have the same controllability, visibility, and market functionality as the traditional power stations. Furthermore, aggregated DERs will benefit from VPP intelligence to gain optimal revenue opportunities. In other words, DERs are scheduled so that the total profit of VPP can be maximized [4–8]. Besides, responsible loads can be optimally aggregated in the VPP framework as well [9,10]. DR programs benefit VPPs in reducing peak power demand, saving costly generation, decreasing electricity losses, reducing fuel consumption and carbon emission, restoring quality of services, ensuring reliability of distribution system, and providing regulation services [11–15].

Among the advantages, the particular importance of regulation service provision is widely recognized, which would not only benefit the independent system operator (ISO), but also increases the VPP's profit for participating in the electricity market. The potential capacity of responsive loads can play a key role in providing RR under VPP's participation in the electricity market. The VPP maximizes its expected profit by integrating and coordinating the energy and regulation of responsive loads with other available DERs.

8.2 VPP scheduling framework

VPP clusters various DERs like renewable and fuel based units to participate in the exchange market. In this chapter, some energy resources including wind turbines, gas turbines, and demand response of customers are available to be scheduled by a VPP. Three different customers are considered in the VPP as residential, commercial, and industrial with various load patterns. Fig. 8.1 schematically depicts the VPP framework. The VPP aggregates the generation of DERs to participate in the joint energy and RR market. To this end, dispatchable units like gas turbines participate in both the energy and regulation reserve modes, while the wind turbines are not able to provide RR service due to their intermittent nature. Besides, a portion of customer demands is allocated for providing energy and RR based on the signed contracts between VPP and customers.

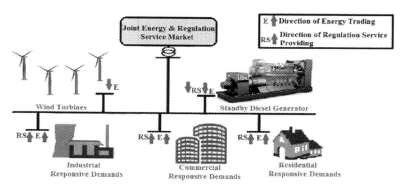

Figure 8.1 Typical scheme of the VPP framework [16].

8.3 VPP scheduling model

The VPP aims to optimize its overall profit by participating in the joint energy and RR market. Eq. (8.1) calculates the profit of the VPP as

$$profit(h) = \sum_h \text{Re}_E^{DA}(h) + \text{Re}_{RR}^{DA}(h) - Co^{GT}(h) - Co^{DR}(h). \tag{8.1}$$

The first and second terms indicate revenue for participating in the energy and RR market, respectively. The operational cost of the gas turbine is represented in the third term, and the last term shows the cost of implementing DR applications. Eq. (8.2) states that the gas turbine operation cost consists of two terms: the first denotes standby and generation cost (8.3), and the second is the penalty cost of CO_2 emissions (8.4). The ramp up and down constraints of the gas turbine are presented in (8.5). The upper and lower generation limits of the gas turbine are considered in (8.6). Eqs. (8.7) and

(8.8) prevent the gas turbine from conflicting operation. Minimum up and down time constraints are guaranteed with (8.9) and (8.10), respectively. Finally, (8.11) and (8.12) indicate the upper and lower constraints of the RR service provision:

$$Co^{GT}(h) = Co^{GT}_{P\&S}(h) + Co^{GT}_{Em}(h), \tag{8.2}$$

$$Co^{GT}_{P\&S}(h) = a.P^{GT}(h)^2 + b.P^{GT}(h) + c.U^{GT}(h) + SUC \times U^{GT}_{on}(h), \tag{8.3}$$

$$Co^{GT}_{Em}(h) = P^{GT}(h).CO2^{GT}.\lambda^{co2}, \tag{8.4}$$

$$RDN \le P^{GT}(h) - P^{GT}(h-1) \le RUP, \tag{8.5}$$

$$P^{GT}_{min} \times U^{GT}(h) \le P^{GT}(h) \le P^{GT}_{max} \times U^{GT}(h), \tag{8.6}$$

$$U^{GT}_{on}(h) - U^{GT}_{off}(h) \le U^{GT}(h) - U^{GT}(h-1), \tag{8.7}$$

$$U^{GT}_{on}(h) + U^{GT}_{off}(h) \le 1, \tag{8.8}$$

$$[X^{on}(h-1) - H^{on}] \times [U^{GT}(h-1) - U^{GT}(h)] \ge 0, \tag{8.9}$$

$$[X^{off}(h-1) - H^{off}] \times [U^{GT}(h) - U^{GT}(h-1)] \ge 0, \tag{8.10}$$

$$RR^{GT}(h) + P^{GT}(h) \le P^{GT}_{max}.U^{GT}_{on}(h), \tag{8.11}$$

$$RR^{GT}(h) \le P^{GT}(h). \tag{8.12}$$

The uncertain nature of wind turbine generation is modeled through Weibull distribution function [17]. Eq. (8.13) relates the output power of the wind turbine ($P^{WT}(h)$) to its wind speed ($V^{WT}(h)$) as

$$P^{WT}(h) = \begin{cases} 0, & 0 \le V^{WT} \le V^{WT}_{ci}, \\ (\dfrac{V^{WT}(h) - V^{WT}_{ci}}{V^{WT}_r - V^{WT}_{ci}}) \times P^{WT}_r, & V^{WT}_{ci} \le V^{WT} \le V^{WT}_r, \\ P^{WT}_r, & V^{WT}_r \le V^{WT} \le V^{WT}_{co}, \\ 0, & V^{WT} \ge V^{WT}_{co}. \end{cases} \tag{8.13}$$

The VPP defines incentive payments to share the benefits of participation in DR programs with customers. To this end, it offers price–quantity package and, subsequently, manages existing responsive loads in the DR program. Fig. 8.2 depicts the proposed price–quantity curve. Based on the proposed curve, higher DR participation is associated with higher prices. Responsive loads receive a payment for two different market services, namely power energy and regulation reserve. The DR payments are calculated in (8.14) and the amounts of DR provision at each block of price–quantity curve (Fig. 8.2) are defined in (8.15) and (8.16). Eq. (8.17) determines the overall DR provided by the customers. Participation of electricity customers in RR provision should be within their upper and lower bounds as considered by Eqs. (8.18) and (8.19):

$$Co^{DR}(h) = \sum_j [\sum_{l=1}^{Nl} \lambda^{EDR}_l(j) \times p_l(j) + \lambda^{RRDR}(j,h) \times RR^{DR}(j,h)], \tag{8.14}$$

$$0 \le p_l(j) \le P_l(j), \quad l = 1, \tag{8.15}$$

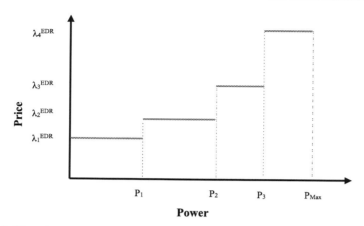

Figure 8.2 The price–quantity curve related to responsive loads [16].

$$0 \leq p_l(j) \leq P_{l+1}(j) - P_l(j), \quad l = 2, 3, \ldots, \tag{8.16}$$

$$P^{DR}(j,h) = \sum_{l=1}^{Nl} p_l(j), \tag{8.17}$$

$$P^{DR}(j,h) + RR^{DR}(j,h) \leq P_{Max}(j), \tag{8.18}$$

$$RR^{DR}(j,h) \leq P^{DR}(j,h). \tag{8.19}$$

The VPP participates in the wholesale energy market to sell the aggregated energy. Eq. (8.20) calculates the revenue of the VPP from the energy market. Additionally, it takes part in the RR market to help reduce the mismatch between committed day-ahead generation and real-time electricity generation. Therefore, regulation up and down services are defined as follows [18]:

- When the generation is lower than the commitment, i.e., regulation up service, VPP will be paid according to the regulation and spot market prices for its readiness declaration and generation enhancement.
- When the generation is more than the commitment, i.e., regulation down service, VPP will be paid according to the regulation market price for its readiness declaration and generation reduction.

Consequently, Eq. (8.21) represents the revenue of RR having two terms: the first is the income from readiness declaration to procure regulation and the second is the income/cost of up/down RR delivery based on the spot prices. The spot prices are determined using random parameters p_1 and p_2 according to (8.22):

$$\text{Re}_E^{DA}(h) = P^{DA}(h) \times \lambda^E(h), \tag{8.20}$$

$$\text{Re}_{SR}^{DA}(h) = RR^{DA}(h) \times \lambda^{RR}(h) + (\Pr_{up}^{reg}(h) + \Pr_{down}^{reg}(h)) \times RS^{DA}(h) \times \lambda^{spot}(h), \tag{8.21}$$

$$\lambda^{spot}(h) = \begin{cases} (1+p_1) \times \lambda^E(h), & 0 \leq p_1 \leq 0.2, \ h \in [12, 23], \\ (1+p_2) \times \lambda^E(h), & -0.1 \leq p_2 \leq 0.1, \ \text{otherwise.} \end{cases} \quad (8.22)$$

Eq. (8.23) is implemented to ensure a balance between the energy produced and sold in the electricity market. Similarly, the balance between regulation provided and sold is kept using (8.24):

$$P^{GT}(h) + P^{WT}(h) + \sum_j P^{DR}(j,h)$$
$$= P^{DA}(h) + (\Pr_{up}^{reg}(h) - \Pr_{down}^{reg}(h)) \times RR^{DA}(h), \quad (8.23)$$

$$RR^{GT}(h) + \sum_j RR^{DR}(j,h) = RR^{DA}(h). \quad (8.24)$$

8.4 Uncertainties arising from VPP scheduling

Due to uncertain parameters, decision-making will be a challenging task for the VPP to schedule its resources appropriately. The uncertainties are the energy and regulation market prices, wind turbine production, different customer behaviors, and probabilities of up/down regulation calling requests. To model such stochastic behaviors, the well-known scenario tree method [19] is applied to generate enough scenarios shown in Fig. 8.3.

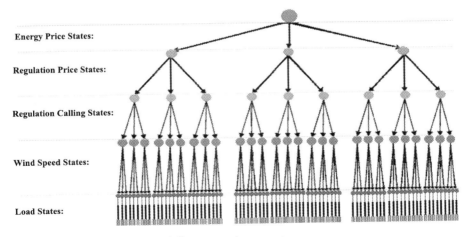

Figure 8.3 Scenario tree for modeling uncertain parameters.

8.5 Numerical studies and discussions

The presented scheduling approach is applied over a daily time horizon. Fig. 8.4 depicts the day-ahead energy and regulation market prices. As described, the VPP comprises a gas turbine, four similar wind turbines, and three kinds of responsible loads. Table 8.1 lists the characteristics of the gas turbine, including operational constraints, fuel costs, CO_2 emission rate, and related penalty cost [20]. VPP pays 40 $/kW for providing regulation via the considered gas turbine [16]. Table 8.2 lists the required characteristics of mounted wind turbines and Fig. 8.5 depicts different amounts of calling requests from the electricity market for delivering RR [16].

Figure 8.4 Energy and regulation market prices.

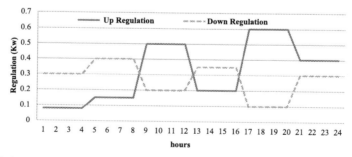

Figure 8.5 Amounts of up/down regulation calling requests [16].

Three kinds of electricity customers, including 100 residential customers, 2 medium industrial customers, and 2 commercial customers, are located in the VPP. Fig. 8.6 depicts hourly demands of customers with various demand patterns [21]. It is considered that electricity customers prefer to reduce their consumptions up to 30% of their hourly loads. Based on agreed DR contract with customers, different prices equal to 0.027, 0.036, and 0.05 $/kWh for valley (hours 1 to 6), shoulder (hours 7 to 14 and 20 to 24), and peak (remaining hours) intervals are allocated, respectively.

The block prices in Fig. 8.2 are allocated according to the agreed DR contract. Price steps are regarded as 1.1, 1.15, 1.25, and 1.4 of the corresponding energy prices, and

Table 8.1 Gas turbine characteristics.

P_{min} (kW)	P_{max} (kW)	RUP (kW)	RDN (kW)	RU (h)	RD (h)	b ($/kW)	c ($)	SUC ($)	$CO2SDG$ (kg/kWh)	$CO2\lambda$ ($/kg)
30	300	100	100	1	1	0.053	1.5	0.15	0.89	0.006

Table 8.2 Wind turbine characteristics.

V_{ci}^{WT} (m/s)	V_r^{WT} (m/s)	V_{co}^{WT} (m/s)	P_r^{WT} (kW)
3	12	25	30

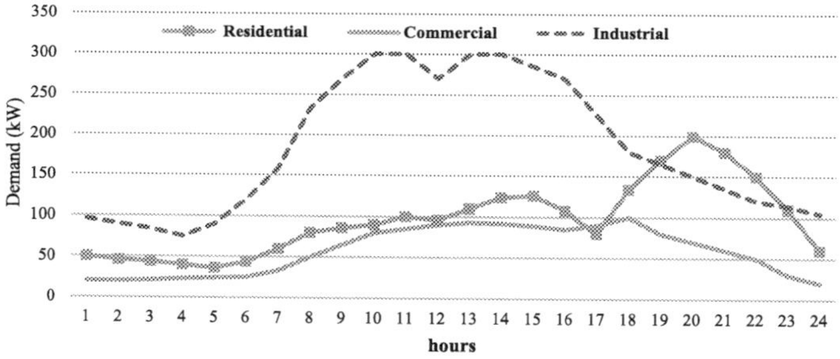

Figure 8.6 Hourly demand pattern for different customers.

quantity steps are 0.3, 0.6, 0.8, and 1 of the corresponding upper permissible participation in the DR applications. Moreover, VPP offers to pay electricity customers for providing RR equal to 1.05 of the hourly energy prices.

Four different scenarios are considered to discuss the proposed model as follows:

- Scenario A. VPP participates in the energy market without considering DR programs
- Scenario B. VPP participates in the energy market considering DR programs
- Scenario C. VPP participates in the both energy and RR markets without considering DR programs
- Scenario D. VPP participates in the both energy and RR markets considering DR programs

The optimal operation of the gas turbine to provide energy and regulation is indicated in Figs. 8.7 and 8.8. In scenarios A and B, the VPP operates the gas turbine to generate energy during hours 16–18, in which electricity market prices are high. In scenarios C and D, VPP increases its gas turbine generation and allows it to provide regulation in comparison to scenarios A and B. During hours 16–18 VPP provides regulation, in which regulation prices are appropriate. Due to noticeable regulation price at hour 16, VPP prefers to provide more regulation than during hours 17 and 18. Consequently, VPP operates the gas turbine during hour 15 to be ready for high power production at hour 16 considering its ramp up limitation. However, the gas turbine is not operated at the remaining hours because the energy market prices are not high enough. It should be noticed that the effect of emission penalty cost on the gas turbine operation is considerable.

VPP clusters various kinds of electricity customers, including residential, commercial, and industrial customers, with different load patterns. The energy provided using DR applications in scenarios B and D is depicted in Figs. 8.9 and 8.10, respectively. VPP compares the electricity market prices with the DR energy cost at each hour to decide whether to execute DR programs or not. Therefore, during hours 17, 18, and 21

VPP's participation in demand response exchange market

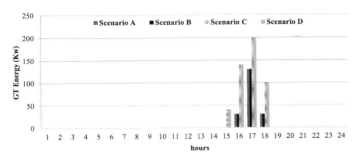

Figure 8.7 Hourly energy procured by the gas turbine [16].

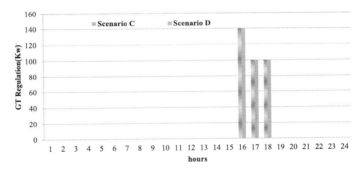

Figure 8.8 Hourly regulation provided by the gas turbine [16].

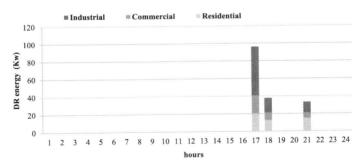

Figure 8.9 Hourly energy provided by the electricity customers in scenario B [16].

executing DR programs seems reasonable in scenario B. In scenario D, VPP has more tendency to execute DR applications because also a potential of regulation provision will be available. Hence, VPP prefers to provide regulation within mentioned hours, which is shown by Fig. 8.11. Since hour 16 has the most expensive regulation price, VPP executes the maximum possible DR application to provide desirable regulation.

VPP aggregates production of individual DERs to act as a unit power player. Figs. 8.12 and 8.13 indicate the overall energy and regulation sold to the electricity market, respectively, in which energy is provided via the wind turbine, gas turbine,

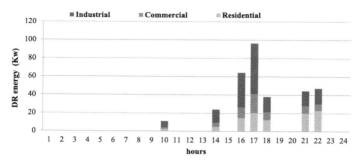

Figure 8.10 Hourly energy provided by the electricity customers in scenario D [16].

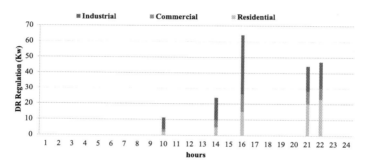

Figure 8.11 Hourly regulation provided by the electricity customers in scenario D [16].

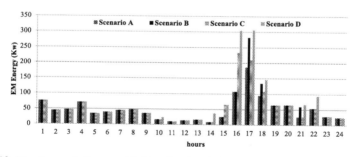

Figure 8.12 Hourly regulation provided by the electricity customers [16].

and responsive loads, and regulation is provided via the gas turbine and responsive loads.

Hourly profits of VPP in the four different scenarios are shown in Fig. 8.14. During hours 1–9, 11–13, 19, 20, 23, and 24, different approaches have the same effect on the VPP obtained profits because of the same strategy. During hours 10 and 14 of scenario D, VPP uses responsive loads and consequently achieves more profit. It should be noted that at hour 15 of scenarios C and D, VPP increases its gas turbine generation. Consequently, more emission penalty cost and less profit are provided in comparison to scenarios A and B. At hour 16, provision of regulation has a specific

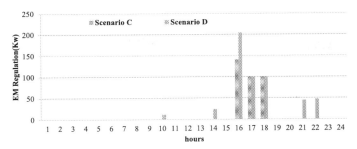

Figure 8.13 Hourly regulation provided by the electricity customers [16].

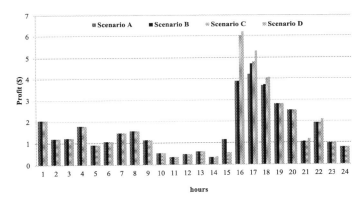

Figure 8.14 Hourly obtained profit of the VPP [16].

Table 8.3 Overall obtained profit in the four scenarios [16].

	Profit ($)
Scenario A	37.632
Scenario B	38.183
Scenario C	40.097
Scenario D	41.164

effect on obtained profit because the regulation price is so attractive. During hours 17, 18, and 21, scenario D has significant advantages over other scenarios because both the regulation reserve and DR programs are considered simultaneously.

Finally, the overall obtained profits of VPP in the four different scenarios are listed in Table 8.3. As shown, the largest profit is achieved in scenario D. Therefore, performing DR applications in the joint energy and regulation markets is more effective and valuable in profit optimization.

8.6 Conclusion

In this chapter, a stochastic programming approach is applied to present a VPP optimal scheduling in both the electricity and regulation reserve markets. Besides energy procurement, the VPP provides RR to maximize its expected profit. The VPP aggregates wind turbines, gas turbine, and responsive loads as available DERs to play a role as a unit market player. Furthermore, a portion of provided RR is delivered to the electricity market based on the market decision. The VPP motivates its electricity customers to execute DR programs based on incentive price–quantity curves. Based on the proposed model, responsive loads are able to provide energy and regulation in the implemented DR programs. Moreover, the stochastic nature of the energy and regulation market prices, wind turbine generation, regulation up/down calling probabilities, and customer behaviors are implemented through different scenario trees. The obtained results demonstrate that executing DR applications is effective in not only enhancing VPP revenues but also in reducing imbalance errors of renewable generations. Moreover, it is concluded that scenario D has significant advantages over other scenarios because both the regulation reserve and DR programs are considered simultaneously. Numerical results and discussions verify the robustness and validity of the developed scheduling approach to guarantee the maximum profit of the VPP.

Nomenclature

Indices
H Index of time interval
J Index of customer participating in demand response programs
L Index of steps in energy type of demand response programs, $l = 2, 3, ..., N_l$

Parameters
a, b, c   Production cost coefficients of gas turbine ($)
$CO2^{GT}$ CO_2 emission rate of gas turbine (ton/kW)
H^{on}, H^{off} Minimum up and down times of gas turbine (h)
$P_{Max}(j)$ Maximum demand response participation that customer J can carry out
$P_{min}^{GT}, P_{max}^{GT}$ Minimum and maximum limits of gas turbine output power (kW)
$P^{WT}(h)$ Power generation of wind turbines at hour h (kW)
P_r^{WT} Rated power of the wind turbines (kW)
$Del_{up}^{reg}(h), Del_{down}^{reg}(h)$ Amount of up/down regulation calling requests at hour h
RUP, RDN Ramp-up and ramp-down rate limits of gas turbine (kW/h)
SUC Gas turbine start-up cost ($)
$V_{ci}^{WT}, V_r^{WT}, V_{co}^{WT}$ Cut in, rated and cut-out speeds of the wind turbine (m/s)
$V^{WT}(h)$ Hourly wind speed (m/s)
$\lambda^{EDR}(h)$ Price associated to energy type of demand response programs at hour h ($/kWh)
$\lambda^{RRDR}(h)$ Price associated to regulation type of demand response programs at hour h ($/kWh)
λ^{co2} CO_2 penalty price of emissions ($/ton)
$\lambda^{Dem}(h)$ Price of demand procurement at hour h ($/kWh)
$\lambda^{E}(h)$ Price of day-ahead energy market at hour h ($/kWh)
$\lambda^{RR}(h)$ Price of day-ahead regulation reserve market at hour h ($/kWh)
$\lambda^{spot}(h)$ Price of spot market at hour h ($/kWh)

Continuous variables

$p_l(j)$ Amount of accepted load reduction of consumer j in step l of price–quantity offer package (kW)
$P^{DA}(h)$ Total energy sold in day-ahead market at hour h (kW)
$P^{DR}(j,h)$ Total energy provided by consumer j in demand response programs at hour h (kW)
$P^{GT}(h)$ Gas turbine power generation at hour h (kW)
$RR^{DA}(h)$ Total provided regulation reserve for day-ahead market at hour h (kW)
$RR^{DR}(j,h)$ Total regulation reserve provided by consumer j in demand response programs at hour h (kW)
$RR^{GT}(h)$ Gas turbine regulation provision at hour h (kW)
$X^{on}(h), X^{off}(h)$ Time duration that gas turbine has been on and off at hour h (h)

Binary variables

$U^{GT}(h)$ Binary decision variable for on/off state of gas turbine at hour h (on=1, off=0)
$U^{GT}_{on}(h), U^{GT}_{off}(h)$ Binary decision variables for start-up/shut-down state of gas turbine at hour h

References

[1] ENIRDGnet: Concepts and opportunities of Distributed Generation: The Driving European Forces and Trend, ENIRDGnet Project deliverable D3, 2003.
[2] C. Ziogou, D. Ipsakis, P. Seferlis, S. Bezergianni, S. Papadopoulou, S. Voutetakis, Optimal production of renewable hydrogen based on an efficient energy management strategy, Energy 55 (2013) 58–67.
[3] G. Koeppel, Distributed generation literature review and outline of the Swiss situation, in: EEH Power Systems Laboratory, Internal Report, Zurich, Switzerland, Nov 2003.
[4] R. Caldon, A. Rossi, R. Turri, Optimal control of a distribution system with a virtual power plant, in: Bulk Power System Dynamics and Control-VI, Italy, Aug 2004.
[5] I. Kamphuis, J. Kok, C. Wamer, M. Hommelberg, Massive coordination of residential embedded electricity generation and demand response using the power matcher approach, in: The 4th International Conference on Energy Efficiency in Domestic Appliances and Lighting–EEDAL06, 2007.
[6] R. Skovmark, J.H. Jacobsen, Analysis, Design and Development of a Generic Framework for Power Trading, MSc thesis, available from Technical University of Denmark, 2007.
[7] F.F. Wu, K. Moslehi, A. Bose, Power system control centers: past, present, and future, Proc. IEEE 93 (Nov 2005) 1890–1907.
[8] F. Bignucolo, R. Caldon, V. Prandoni, The voltage control on MV distribution networks with aggregated DG units (VPP), in: Proceedings of the 41st International, vol. 1, Sep 2006, pp. 187–192.
[9] E. Ghorbankhani, A. Badri, A bi-level stochastic framework for VPP decision making in a joint market considering a novel demand response scheme, Int. Trans Electr. Energy Syst. 28 (2017) 1–18.
[10] J-F. Toubeau, Z. De Grève, F. Vallée, Medium-term multi-market optimization for virtual power plants: a stochastic-based decision environment, IEEE Trans. Power Syst. 33 (2) (2018) 1399–1410.
[11] Sh. Nolan, M. O'Malley, Challenges and barriers to demand response deployment and evaluation, Appl. Energy 152 (2015) 1–10.
[12] C. Lang, E. Okwelum, The mitigating effect of strategic behavior on the net benefits of a direct load control program, Energy Econ. 49 (2015) 141–148.

[13] N. Ruiz, I. Cobelo, J. Oyarzabal, A direct load control model for virtual power plant management, IEEE Trans. Power Syst. 24 (2) (2009) 959–966.
[14] H. Mohsenian-Rad, Optimal demand bidding for time-shiftable loads, IEEE Trans. Power Syst. 30 (2) (2015) 939–951.
[15] S. Nojavan, H.A. Aalami, Stochastic energy procurement of large electricity consumer considering photovoltaic, wind turbine, micro-turbines, energy storage system in the presence of demand response program, Energy Convers. Manag. 103 (2015) 1008–1018.
[16] A. Shayegan-Rad, A. Badri, A. Zangeneh, M. Kaltschmitt, Risk-based optimal energy management of virtual power plant with uncertainties considering responsive loads, Int. J. Energy Res. (2019) 1–16.
[17] J. Qiu, K. Meng, Y. Zheng, Z.Y. Dong, Optimal scheduling of distributed energy resources as a virtual power plant in a transactive energy framework, IET Gener. Transm. Distrib. 11 (13) (2017) 3417–3427.
[18] S.J. Kazempour, M. Parsa Moghaddam, M.R. Haghifam, G.R. Yousefi, Risk constrained dynamic self-scheduling of a pumped storage plant in the energy and ancillary service markets, Energy Convers. Manag. 50 (5) (2009) 1368–1375.
[19] A.J. Conejo, M. Carrion, J.M. Morales, Decision Making Under Uncertainty in Electricity Markets, Springer, New York, NY, USA, 2010.
[20] A. Zakariazadeh, S. Jadid, P. Siano, Smart microgrid energy and reserve scheduling with demand response using stochastic optimization, Electr. Power Syst. Res. 63 (2014) 523–533.
[21] S. Papathanassiou, N. Hatziargyriou, K. Strunz, A benchmark low voltage microgrid network, in: Proc. CIGRE Symp Power Systems with Dispersed Generation, Athens, Greece, April 17–20, 2005.

Uncertainty modeling of renewable energy sources

Davood Fateh, Mojtaba Eldoromi, and Ali Akbar Moti Birjandi
Electrical Engineering Department, Shahid Rajaee Teacher Training University, Tehran, Iran

9.1 Introduction

High penetration of renewable energy sources (RESs) has led current distribution systems to the next generation of intelligent systems [1]. The condition of uncertainty comes in the form of weather conditions for the RES [2]. The three different RES identified for the integration into the grid system are a wind generation system, solar PV system, and a micro-hydro system.

Due to its geographically wide distribution and low capacity, distributed RES is difficult to be managed at a grid-scale. Thus, the concept of a virtual power plant (VPP) that integrates multi-regional distributed energy resources (DERs) into a coordinated uniform power utility could improve the competitiveness of RES to participate in all aspects of the electricity market [3]. Also, a detailed inspection of the modeling, sizing, and cost analysis of a PV/wind/battery hybrid system considering resource and load uncertainty for electrification of a remote area is rarely found. However, high variability and uncertainty associated with renewable energy sources pose major challenges in designing isolated power systems.

Meanwhile, DERs in the USA have grown almost three times faster than the net total generation capacity (168 GW vs. 57 GW) on a 5-year basis (2015–2019) [4]. This increase brings huge challenges to practical power system and electricity market operations. The traditional "passive" distribution systems are transforming into an active network, which requires operators to revise their operation strategies to accommodate the transition from the traditional top-down power flow to a bottom-up paradigm [5] and update almost all of their control applications.

As a typical tool for distribution automation, distribution system state estimation (DSSE) enables continuous and reliable monitoring of DERs, which is critically important in practice as it allows distribution system operators (DSOs) to perceive the operating status of the system. However, the intermittency and poor predictability of DER outputs introduce great uncertainty into DSSE [6]. This chapter provides a comprehensive review of the state-of-the-art methodology, techniques, and application issues in the uncertainty modeling of RESs in distribution systems.

VPP is a combination of renewable sources, energy storage system (ESS), small conventional power plants, and interruptible loads that can supply market actions as a single power plant [7]. Because of the uncertainty and intermittent nature of PV and WT powers, it is usually possible that the actual values of these variables are different from the corresponding schemed values. Therefore, grid operators attempt to have a definite level of reserve in the grid. Thus, they would be capable of compensating for the uncertainty in the output power of these sources to protect the system security [8,9].

A VPP is a network of decentralized, medium-scale power generating units such as wind farms, solar parks, and combined heat and power (CHP) units, as well as flexible power consumers and storage systems. The interconnected units are dispatched through the central control room of the virtual power plant but nonetheless remain independent in their operation and ownership. The objective of a virtual power plant is to relieve the load on the grid by smartly distributing the power generated by the individual units during the periods of peak load. Additionally, the combined power generation and consumption of the networked units in the virtual power plant is traded on the energy exchange. The concept of virtual power plants turns on its head the more traditional idea of relying on centralized power plants for predictable power. Generally, the USA has depended on controllable power from big, centralized plants, often coal or natural gas plants. Power flow has been in one direction, from the utility to the business or consumer. But in recent years, small and large independent power producers have entered the scene, generating solar, wind, and other renewable resources from all corners of the USA. Suddenly, power flow has become bi-directional. This clean power has disrupted the energy grid and created the need for new models.

Another important impact that arises with the integration of RESs into the electricity market is that as RES units provide more and more generation offerings on the demand side, all individual customers and DER owners will get the opportunity to take part in the energy market [10]. Consequently, the traditional centralized market structure is gradually transforming into a decentralized one, which enables better interactions between the distribution system operator and customers.

In order to compensate for the uncertainty of wind and PV powers, pumped hydro energy storage (PHES), compressed air energy storage (CAES), flywheel energy storage, superconducting magnetic energy storage (SMES), supercapacitor energy storage, and battery energy storage are used [11]. The storage units have two major benefits: first, these are loads with limited abilities, and second, they can turn into generators. Some papers have concentrated on VPP and different associated aspects and concepts of their interactions with the power system. In [12,13] the VPP is considered as a centralized entity, including some micro combined heat and power (CHP) units connected to a low voltage distribution system.

An optimal operation approach of a VPP composed of CHP units is offered based on a decentralized control strategy [14]. The optimal utilization of CHP units has been specified as the main aim, but the important role of electrical storages and demand response resources was not considered. In [15], a control method of VPP, which includes photovoltaic panels and controllable loads, is solved by using mixed-integer programming so that the power output of the VPP can be set in a wide range flexibly. In [16], the VPP attempts to maximize its expected turnover via participating in both the day-ahead and balancing markets using a two-stage stochastic mixed-integer linear programming model. A heuristic game theory-based virtual power market model for security restricted unit commitment strategy was proposed in [17]. A price-based unit commitment method allowing a VPP to exchange energy with an upstream network for the day-ahead market sale/purchase bids was employed in [18]. In [19] optimal scheduling of a microgrid, including PV, WT, fuel cell (FC), in an isolated load area has been formulated for a 24-hour period and 1 h time intervals. Likewise, in [20] the scheduling of DERs in an isolated grid has been investigated that the optimization problem has been solved by the branch-and-bound technique and then used by an artificial neural network (ANN) to better manage the DER. In [21], and energy management method is investigated for VPPs, and the cost and emission impacts of VPPs formation and electric vehicle penetration are analyzed.

9.2 Modeling of RESs

Fig. 9.1 shows a hypothetical power grid with different kinds of generation. Wind turbines and PV plants are the main sources of renewable energy that are widely used in power grids. Along with renewable sources, there are many conventional generation units, such as fossil fuel, hydro, cogeneration, and steam units. Each of them injects a different amount of uncertainty into the grid with different characteristics in terms of magnitude and frequency of occurrence. Despite recent developments in load forecasting techniques, there is still uncertainty coming from load forecasting error.

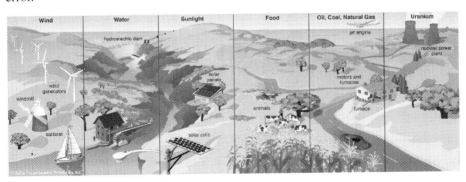

Figure 9.1 Renewable energy sources [22].

In the proposed model, the output power for generators and demand power for loads is modeled with two terms:

$$P(t) = \mu(t) + \epsilon(t), \tag{9.1}$$

where $\mu(t)$ is the average of power signal (MW) at each time instant or, in other words, it is what we expect to have for each component ahead of time. For example, we can consider the forecasted value for the load or renewable generation or scheduled output of a conventional generation unit for the next day as $\mu(t)$. In this model $\epsilon(t)$ represents the uncertainty (MW) which can come from the forecast error or mismatch in the output power for conventional units. In this study, the load and wind output power forecast errors are considered as uncertainty. Fig. 9.2 shows the placement of different components of the proposed model.

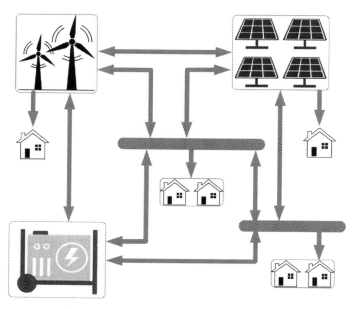

Figure 9.2 Placement of components of the proposed model.

9.3 Modeling of VPP

VPP, whose structure is shown in Fig. 9.3, is just a cloud-based distributed power plant that combines the capabilities of heterogeneous DER to increase power generation, as well as to trade or sell electricity on an open market.

VPP integrates various kinds of DER in different areas and achieves coordinated dispatching through the connection of DERs and energy management system (EMS). In the power industry chain, VPP interacts with market participants [23], assists in network management, and provides ancillary services [24].

9.4 Classification and description of uncertainties in VPP

With the upgrading of large-scale grid-connected power generation and the reform of the renewable energy market, the uncertainty in the grid is gradually increasing, which makes safe and stable operation of the grid more challenging. Uncertainty limits the

Figure 9.3 A virtual power plant [25].

accurate estimation of variables during VPP operation and optimal scheduling. Therefore, it is necessary to identify the sources of uncertainty and choose the appropriate description method [26].

9.4.1 Renewable energy uncertainty

Renewable energy uncertainty is mainly reflected in the alternating and temporally variable properties of wind and solar. In order for a VPP to work continuously, its various specifications must be fully considered in the design process.

9.4.2 Market price uncertainty

VPP uncertainty is another economic issue. Over the life of an energy system, market prices, such as of electricity, natural gas and oil, are unlikely to remain stable and will vary with changes in the energy, climate, economy, and global and local markets.

9.4.3 Load demand uncertainty

In addition to cost, load demand uncertainty is also a common uncertainty in VPP optimization problems. Load power is highly time dependent, including deterministic and random components. The deterministic part is reproducible and is influenced by factors such as time and weather. The random part is derived from prediction and measurement errors. Increasing the ratio of controllable loads, such as EV, increases the uncertainty of load demand in VPP. At the same time, VPPs allow customers to change their power consumption and actively participate in the smart power distribution. However, cargo demand fluctuates not only with seasonal changes, but also with consumer psychology, current economic performance, productive activity, and emergencies. These uncertain features complicate the demand for cargo.

9.5 Optimization approaches of VPP with uncertainties

The highest goal in VPP optimization is to maximize revenue or minimize costs [28]. In other words, optimal planning is mainly based on maximizing revenue or minimizing costs subject to the constraints of the power system. In fact, the virtual power generated by the VPP is a force that is saved by reducing the load with flexible adjustment. Therefore, in order to achieve the goal of adjusting the demand burden economically, a comprehensive review of the objectives of the various VPP components is necessary and significant. The mathematical expression of these goals mainly depends on various factors such as the capacity of distributed producers, ESS, market prices, customer preferences, policy, and risk [29].

9.5.1 Optimization approaches considering renewable power uncertainty

Uncertainty about renewable energy has a significant impact on the overall performance of VPPs. Therefore, finding a targeted optimization approach is essential.

9.5.1.1 Monte Carlo simulation (MCS)

The most popular and accurate method of random programming. This is a common tool for quantifying uncertainty with a variable process. In this method, it is assumed that PDF input parameters are known and each input parameter is modeled with a statistical distribution to generate random samples. The required output parameters are selected by statistical analysis.

9.5.1.2 Robust optimization (RO)

This is an effective way to solve the problem of data uncertainty, especially when lacking complete information. The nature of RO is to find solutions that are not sensitive to data disturbances. In other words, the main idea of RO is to find the optimal solution in the worst case. Only the uncertainty set is needed to describe the uncertainty of the input parameters, while the probability distribution is not. Therefore, the RO method has a high computational efficiency so that it can be widely used in power system planning decision problems.

9.5.1.3 Rolling horizon approach (RHA)

RHA has attracted widespread attention as an effective way to deal with computational complexity and dynamic uncertainty of large-scale planning problems. The nature of RHA [28] refers to finding the optimal value in the current limited time frame. And until the next sampling, this optimization starts again, so it spins again and again.

Through the use of RHA in the planning phase, uncertainty in using the forecast of renewable energy production as input data is reduced.

9.5.2 Optimization approaches considering market price uncertainty

The price of electricity in the market is determined by many factors, such as the price of electricity generation, the local climate pattern, government subsidies, transmission and distribution infrastructure, industry regulation, mechanisms, and market rules. Price uncertainty puts VPP members at a great risk of losses. VPP optimization with price uncertainty has become an important research topic in recent years. Here, two typical optimization algorithms are introduced in detail as follows:

9.5.2.1 First-order stochastic dominance constraints (FSD)

The concept of random dominance constraints is a common principle in rational decision making. No need to adjust the function. Based on the partial preferential information obtained by the decision maker, a relatively sub-optimal solution can be eliminated and a set of investment possibilities is provided for the decision maker.

9.5.2.2 Information gap decision theory (IGDT)

This does not require any assumption of probability distribution from unknown factors and only addresses the gap between known and unknown information. This theory effectively solves the problems of describing uncertainty and evaluation criteria in severely uncertain environments. Therefore, IGDT is widely used in electricity market risk management with serious uncertainty.

9.5.3 Optimization approaches considering load demand uncertainty

The energy matching between power generation and user demand for VPPs is affected not only by the uncertainty of renewable energy but also by uncertainty in load demand. Therefore, the optimization method for uncertainty load is also necessary.

9.5.3.1 Fuzzy chance constraint programming (FCCP)

A type of mathematical programming uncertainty based on probability and fuzzy set theories. FCCP means that the chance limit is directly equivalent to the determinant given the level of assurance. FCCP can be used to solve the problem of optimal balance between economy and reliability of VPP performance [30].

9.5.3.2 Interval analysis (IA)

In this method, all indeterminate factors are represented by intervals. In other words, the upper limit, the lower limit, and the confidence level are the main components of the interval algorithm. Based on distance analysis, a range of values is used instead of a single point to perform interval calculations. By calculating the optimization of each interval, all the optimal sets of problems can be obtained with a certain accuracy.

9.5.3.3 Point estimation method (PEM)

A statistical method of probability that can calculate the moments of the objective function by distributing the probability of input variables. PEM reduces computation time and also ensures high computational accuracy, so PEM can be adapted to the application of uncertainty in the engineering system.

9.6 Problem formulation

The objective function and the corresponding constraints of the optimal operation management are described in this section [27]. VPPs can combine distributed photovoltaics, wind energy, energy storage equipment, regenerative boilers, and controllable loads to maximize the benefits of overall power supply. In urban buildings, VPP can control flexible and controllable loads such as central air conditioning and EV in real time, so that VPP automatically adjusts and optimizes response quality according to supply and demand. In addition, VPPs can also participate in multiple markets for optimal planning and bidding. Depending on the electricity market such as the day-ahead market (DAM), as well as real-time and ancillary services markets, where VPP participates, different market models can be created, such as DAM, balanced market (BM), bilateral contract (BC) market, and so on.

9.6.1 Objective function

This section presents an energy management scheme in a VPP including renewable and conventional DGs. The objective function is to minimize the VPP operational cost by the control of the local production, as well as the interactive relationship with the grid. In fact, the VPP will try to supply its consumers locally. However, if the total power produced by DGs is not enough or too expensive to cover the supplying loads, then energy is bought from the grid (as the upstream network) and will be sold to the consumers or stored in the storage devices. Consequently, it can be concluded that the objective cost consists of the power that is exchanged between the VPP and the grid, the fuel cost for DGs, and the start-up or shut-down cost of the power sources used in the VPP. The objective function is similar to [31], but the principal difference is the cost of the energy loss reduction occurred into distribution system lines.

This function is stated as follows:

$$\text{Min } f = \sum_{t=1}^{T} Cost = \sum_{t=1}^{T} \begin{pmatrix} P_{Grid}(t) \times G_{Grid}(t) \\ + U_{WT}(t) \times P_{WT}(t) \times C_{WT}(t) \\ + U_{PV}(t) \times P_{PV}(t) \times C_{PV}(t) \\ + U_{FC}(t) \times P_{FC}(t) \times C_{FC}(t) \\ + U_{MT}(t) \times P_{MT}(t) \times C_{MT}(t) \\ + \sum_{j=1}^{N_s} U_j(t) \times P_{Sj}(t) \times C_j(t) \\ + \sum_{i=1}^{N_g} S_{Gi} \times |U_i(t) - U_i(t-1)| \\ + \sum_{j=1}^{N_s} S_{Gj} \times |U_j(t) - U_j(t-1)| \\ - \Delta P(t) \times C_{\Delta P}(t) \end{pmatrix}, \quad (9.2)$$

$$\sum_{t=1}^{T} \Delta P(t) = \sum_{t=1}^{T} (P_{original\ losses}(t) - P_{new\ losses}(t)), \quad (9.3)$$

$$P_{final\ losses}(t) = \sum_{t=1}^{T} \sum_{i=1}^{N_{br}} R_i \times |I_i(t)|^2, \quad (9.4)$$

where $C_{WT}(t)$, $C_{PV}(t)$, $C_{FC}(t)$, $C_{MT}(t)$, and $C_j(t)$ are the bids of the wind turbine, photovoltaic, fuel cell, micro turbine, and storage devices at hour t, respectively; S_{Gi} and S_{Gj} are the start-up or shut-down costs for the ith DG and jth storage, respectively; $P_{Grid}(t)$ is the active power which is bought (sold) from (to) the utility at time t and $G_{Grid}(t)$ is the bid of utility at time t; N_g and N_s represent the total number of generation and storage units, respectively; $U_{WT}(t)$, $U_{PV}(t)$, $U_{FC}(t)$, $U_{MT}(t)$, and $U_j(t)$ represent the ON or OFF states of all units at hour t of the day. Also, $\Delta P(t)$, $C_{\Delta P}(t)$, R_i, and $I_i(t)$ are the difference between the original losses and the new losses of feeders and its cost at hour t, as well as the resistance and the actual current of the ith branch, respectively [32].

9.6.2 Constraints

The power balance in each time period t is

$$\sum_{i=1}^{T} \left(P_{Grid}(t) + P_{WT}(t) + P_{PV}(t) + P_{FC}(t) + P_{Battery\ discharge}(t) \right)$$

$$= \sum_{i=1}^{T} \left(Load(t) + P_{Battery\ discharge}(t) + P_{Loss}(t) \right). \quad (9.5)$$

Wind turbine generation limits in each time period t are given by

$$P_{WTmin}(t) \leq P_{WT}(t) \leq P_{WTmax}(t); \quad t = 1, \ldots, T. \tag{9.6}$$

Photovoltaic generation limits in each time period t are

$$P_{PVmin}(t) \leq P_{PV}(t) \leq P_{PVmax}(t); \quad t = 1, \ldots, T. \tag{9.7}$$

Fuel cell limits in each time period t are

$$P_{FCmin}(t) \leq P_{FC}(t) \leq P_{FCmax}(t); \quad t = 1, \ldots, T. \tag{9.8}$$

Microturbine limits in each time period t are

$$P_{MTmin}(t) \leq P_{MT}(t) \leq P_{MTmax}(t); \quad t = 1, \ldots, T. \tag{9.9}$$

Utility bounds in each time period t are given by

$$P_{Gridmin}(t) \leq P_{Grid}(t) \leq P_{Gridmax}(t); \quad t = 1, \ldots, T. \tag{9.10}$$

Storage battery limits in each time period t are

$$P_{Sjmin}(t) \leq P_{Sj}(t) \leq P_{Sjmax}(t); \quad t = 1, \ldots, T. \tag{9.11}$$

Due to the restriction on charge and discharge rate of storage devices during each time period, the following equation and constraints can be considered:

$$W_{ess}(t) = W_{ess}(t-1) + \eta_{charge} P_{charge} \Delta t - \frac{1}{\eta_{discharge}} P_{discharge} \Delta t, \tag{9.12}$$

$$W_{essmin} \leq W_{ess}(t) \leq W_{essmax}; \quad t = 1, \ldots, T, \tag{9.13}$$

$$P_{charge}(t) \leq P_{chargemax} \cdot X(t); \quad t = 1, \ldots, T; \; X \in \{0, 1\}, \tag{9.14}$$

$$P_{discharge}(t) \leq P_{dischargemax} \cdot Y(t); \quad t = 1, \ldots, T; \; Y \in \{0, 1\}. \tag{9.15}$$

Storage battery cannot charge and discharge at the same time, thus

$$X(t) + Y(t) \leq 1; \quad t = 1, \ldots, 24; \; X \text{ and } Y \in \{0, 1\}, \tag{9.16}$$

where $W_{ess}(t)$ and $W_{ess}(t-1)$ are the quantities of energy stored inside the battery at hour t and $t-1$, respectively, P_{charge} and $P_{discharge}$ are the allowed rates of charge and discharge through a period of time $\Delta t = 1$ h, η_{charge} and $\eta_{discharge}$ are the charge and discharge efficiency.

9.7 Tools used to solve optimization problems of VPP with uncertainties

Many solvers can be used directly in the VPP optimization model with uncertainty. Researchers should only focus on building and transforming models. The solution process is done by specialized tools. The most widely used tools are:

9.7.1 General algebraic modeling system (GAMS)

GAMS is the most common application software to deal with optimal dispatching problem with uncertainties of VPP. It is suitable for complex and large-scale modeling applications, and is very good at numerical analysis [33].

9.7.2 Matrix laboratory (MATLAB®)

MATLAB integrates many powerful functions, such as numerical analysis, matrix calculation, modeling and simulation of non-linear dynamic systems, into a visual window environment. In order to solve the scheduling optimization problem in VPP with uncertainty, Optimal Toolbox can be chosen. Optimize Toolbox provides a function to find the optimal solution and satisfy constraints at the same time. The toolbox provides functions *linprog*, *quadprog*, and *fmincon* to solve LP, QP, and NLP problems, respectively.

9.7.3 LINGO

LINGO is a mathematical modeling language designed to formulate and solve optimization problems, including linear, quadratic, and nonlinear programming problems. LINGO modeling language is simple and easy to understand. It can express a set of similar constraints with a concise expression. For example, a constraint is applied to a member of a group of objects. If this restriction also applies to other members of the group, there is no need to apply a single restriction to all members of the group. LINGO can express the limitations of the whole group in a brief statement. In addition, LINGO has built-in solutions, which can automatically select the appropriate solver.

9.8 Case study

The proposed method is tested on a 24-bus system typical network to evaluate the impact of uncertainty on flow of lines. To see the impacts of renewable energy penetration into the grid, 6 conventional generators of the original system are replaced with photovoltaic (PV), wind turbine (WT), CHP, fuel cell (FC), and microturbine (MT). Using a power flow approximation, the power flowing from bus 1 to 24 is calculated. It should be noted that bus 4 is considered as a reserve bus (11,17,18, and 22) as shown in Fig. 9.4. All contribution units are fed by a substation transformer. The distribution of line flow uncertainty was implemented in this case study. As the renewable energy plants are fed by free sources (solar, wind, and fuel cell) and in support of clean energies, CHP and microturbine charge is considered for these resources.

This system consists of several types of RESs, such as WT, PV, FC, MT, and also CHP units. The analysis is implemented for a 24 h time interval to see the performance of each power unit better. It is assumed that all the DGs generate active power at a unit

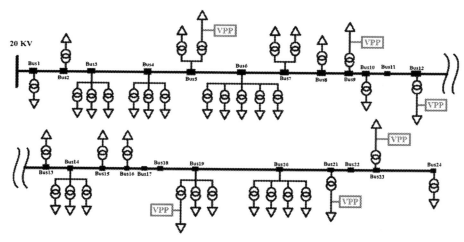

Figure 9.4 Local network of VPP.

Table 9.1 The limitations and bids of RESs and the utility.

Type	Min power (kW)	Max power (kW)	Bid (€ct/kW)	Start-up/shut-down cost (€ct)
PV	0	500	2.584	0
WT	0	600	1.073	0
MT	50	300	0.457	0.96
FC	60	600	0.294	1.65
CHP	80	400	1.4	1.84

power factor. In addition, the heat load demand is not considered in the proposed system.

The maximum and minimum generation limits and bids' information for each of the DGs are shown in Table 9.1. As it can be seen from this table, although the WT and PV do not require any fuel for power generation, their bid is higher than for the other DGs. This is because of their high capital cost. The higher bid of these RESs offers to return the extra capital cost, as well as to pay for renewal and maintenance costs. Also, the cost of start-up/shut-down for each unit in euro cents (€ct) per kilowatt hour (kWh) is given in Table 9.1 [27].

In order to better understand our study, three different scenarios are considered. In the first scenario, all the DGs are dispatched according to their real constraints and worked with their ideal power. In the second scenario, all DGs are allowed to start up or shut down according to the uncertainty condition. In fact, the effect of shadows and wind speed and the level of participation of the employees of each unit is considered. The utility is assumed as an unconstrained unit that can exchange energy with VPP without any restriction. In the third scenario, it's assumed that all the DGs act within their power limits but the utility behaves as an unconstrained unit

Table 9.2 System configuration method and scenario conditions.

Bus number	Total power (kW)	Components	Percentage of participation of each department in the scenario according to the operating in 24 hours		
			1st scenario	2nd scenario	3rd scenario
5	900	PV & CHP	100	40	0
9	500	PV	100	40	0
12	400	CHP	100	60	100
19	600	WT	100	50	100
21	300	MT	100	50	100
23	400	FC	100	50	100

and exchanges energy with the VPP without any restriction. The system configuration method and scenario conditions are presented in Table 9.2.

To further investigate this issue, the proposed system is divided into three sections in 24 hours. Each section is presented in the form of a scenario. In the first scenario, the presence of all units is ideal and with 100% nominal power. It is assumed that the performance of these conditions is related to direct sunlight and the operation of wind turbines at nominal speed. In the second scenario, lack of solar radiation, reduced wind speed, and the absence of all employees have caused the system to not operate under nominal conditions. In the third scenario, the two parts are in operation and the amount of power provided by them is zero. Other units are trying to compensate for this absence. The voltage of the buses is shown in Fig. 9.5. In the third scenario, due to the non-performance of the two system sections completely and the absence of staff, due to the reduction of existing costs, it has also provided good performance. Fewer units are used, and the resulting voltage drop is acceptable. The average voltages are shown in Fig. 9.6.

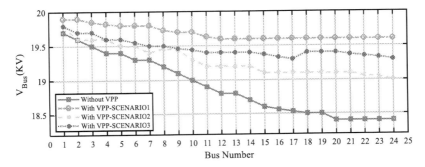

Figure 9.5 Comparison of bus voltages in scenarios.

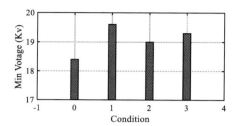

Figure 9.6 Average voltage under the studied conditions.

9.9 Conclusion

In this chapter, we discuss the impacts of RESs on VPP and then present a review of the uncertainty modeling. Upon this review, we can conclude on the precise and effective uncertainty modeling. Uncertainty in VPP optimization is an inevitable and significant issue. In this paper, an attempt has been made to classify VPP uncertainties into three categories, including renewable power generation uncertainties, market price uncertainties, and load demand uncertainties. In order to quantify uncertainty, probability distributions are employed, i.e., probabilistic and distance descriptions are introduced. In addition, the components of objective functions and constraints, including uncertainty, are specified separately. The advantages and disadvantages of optimization algorithms, which are used to solve VPP optimization problems, are summarized and compared. Then, the tools and operating systems that support the implementation of optimal algorithms, including GAMS, MATLAB, and LINGO, are summarized. To investigate the performance of uncertainties, three different scenarios have been proposed, and numerical results showed that uncertainty can be compensated by proper planning in different VPP units.

References

[1] Mir Hadi Athari, Zhifang Wang, Modeling the uncertainties in renewable generation and smart grid loads for the study of the grid vulnerability, in: 2016 IEEE Power & Energy Society Innovative Smart Grid Technologies Conference (ISGT), IEEE, 2016, pp. 1–5.
[2] Manzar Ahmed, Uzma Amin, Suhail Aftab, Zaki Ahmed, Integration of renewable energy resources in microgrid, Energy and Power Engineering 7 (01) (2015) 12.
[3] Elaheh Mashhour, Seyed Masoud Moghaddas-Tafreshi, Bidding strategy of virtual power plant for participating in energy and spinning reserve markets—part I: problem formulation, IEEE Transactions on Power Systems 26 (2) (2010) 949–956.
[4] Federal Energy Regulatory Commission, Distributed Energy Resources Technical Considerations for the Bulk Power System, Energy Transformation Network of Ontario, 2018.
[5] Walid El-Khattam, Magdy M.A. Salama, Distributed generation technologies, definitions and benefits, Electric Power Systems Research 71 (2) (2004) 119–128.
[6] Anggoro Primadianto, Chan-Nan Lu, A review on distribution system state estimation, IEEE Transactions on Power Systems 32 (5) (2016) 3875–3883.

[7] Ozan Erdinc, Mehmet Uzunoglu, The importance of detailed data utilization on the performance evaluation of a grid-independent hybrid renewable energy system, International Journal of Hydrogen Energy 36 (20) (2011) 12664–12677.
[8] Alireza Soroudi, Raphaël Caire, Nouredine Hadjsaid, Mehdi Ehsan, Probabilistic dynamic multi-objective model for renewable and non-renewable distributed generation planning, IET Generation, Transmission & Distribution 5 (11) (2011) 1173–1182.
[9] Miloš Pantoš, Stochastic optimal charging of electric-drive vehicles with renewable energy, Energy 36 (11) (2011) 6567–6576.
[10] Abdullah S. Emhemed, Graeme Burt, Olimpo Anaya-Lara, Impact of high penetration of single-phase distributed energy resources on the protection of LV distribution networks, in: 2007 42nd International Universities Power Engineering Conference, IEEE, 2007, pp. 223–227.
[11] Francisco Díaz-González, Andreas Sumper, Oriol Gomis-Bellmunt, Roberto Villafáfila-Robles, A review of energy storage technologies for wind power applications, Renewable & Sustainable Energy Reviews 16 (4) (2012) 2154–2171.
[12] Christian Schulz, Gerold Roder, Michael Kurrat, Virtual power plants with combined heat and power micro-units, in: 2005 International Conference on Future Power Systems, IEEE, 2005, p. 5.
[13] Christian Schulz, Business models for distribution power generation with combined heat and power micro-unit, in: 3rd International conference on the European Electricity Market, 2006, p. 8.
[14] Bernhard Wille-Haussmann, Thomas Erge, Christof Wittwer, Decentralised optimisation of cogeneration in virtual power plants, Solar Energy 84 (4) (2010) 604–611.
[15] Hrvoje Pandžić, Juan M. Morales, Antonio J. Conejo, Igor Kuzle, Offering model for a virtual power plant based on stochastic programming, Applied Energy 105 (2013) 282–292.
[16] Miadreza Shafie-Khah, Mohsen Parsa Moghaddam, Mohamad Kazem Sheikh-El-Eslami, Development of a virtual power market model to investigate strategic and collusive behavior of market players, Energy Policy 61 (2013) 717–728.
[17] Malahat Peik-Herfeh, H. Seifi, M.K. Sheikh-El-Eslami, Decision making of a virtual power plant under uncertainties for bidding in a day-ahead market using point estimate method, International Journal of Electrical Power & Energy Systems 44 (1) (2013) 88–98.
[18] Hugo Morais, Péter Kádár, Pedro Faria, Zita A. Vale, H.M. Khodr, Optimal scheduling of a renewable micro-grid in an isolated load area using mixed-integer linear programming, Renewable Energy 35 (1) (2010) 151–156.
[19] Z.A. Vale, Pedro Faria, H. Morais, H.M. Khodr, Marco Silva, Peter Kadar, Scheduling distributed energy resources in an isolated grid—an artificial neural network approach, in: IEEE PES General Meeting, IEEE, 2010, pp. 1–7.
[20] Pamela MacDougall, Anna Magdalena Kosek, Hendrik Bindner, Geert Deconinck, Applying machine learning techniques for forecasting flexibility of virtual power plants, in: 2016 IEEE electrical power and energy conference (EPEC), IEEE, 2016, pp. 1–6.
[21] Ali Ghahgharaee Zamani, Alireza Zakariazadeh, Shahram Jadid, Day-ahead resource scheduling of a renewable energy based virtual power plant, Applied Energy 169 (2016) 324–340.
[22] https://www.britannica.com/science/renewable-energy.
[23] Wang Han, Shariq Riaz, Pierluigi Mancarella, Integrated techno-economic modeling, flexibility analysis, and business case assessment of an urban virtual power plant with multi-market co-optimization, Applied Energy 259 (2020).
[24] Mouna Rekik, Zied Chtourou, Nathalie Mitton, Ahmad Atieh, Geographic routing protocol for the deployment of virtual power plant within the smart grid, Sustainable Cities and Society 25 (2016) 39–48.

[25] Songyuan Yu, Fang Fang, Yajuan Liu, Jizhen Liu, Uncertainties of virtual power plant: problems and countermeasures, Applied Energy 239 (2019) 454–470.
[26] Seyyed Mostafa Nosratabadi, Rahmat-Allah Hooshmand, Eskandar Gholipour, A comprehensive review on microgrid and virtual power plant concepts employed for distributed energy resources scheduling in power systems, Renewable & Sustainable Energy Reviews 67 (2017) 341–363.
[27] Mohammad Javad Kasaei, Majid Gandomkar, Javad Nikoukar, Optimal management of renewable energy sources by virtual power plant, Renewable Energy 114 (2017) 1180–1188.
[28] Shang-fei Xiong, Xiao-yan Zou, Value at risk and price volatility forecasting in electricity market: a literature review, Power System Protection and Control 42 (2014) 146–153.
[29] Songli Fan, Qian Ai, Longjian Piao, Fuzzy day-ahead scheduling of virtual power plant with optimal confidence level, IET Generation, Transmission & Distribution 10 (1) (2016) 205–212.
[30] Amjad Anvari Moghaddam, Alireza Seifi, Taher Niknam, Mohammad Reza Alizadeh Pahlavani, Multi-objective operation management of a renewable MG (micro-grid) with back-up micro-turbine/fuel cell/battery hybrid power source, Energy 36 (11) (2011) 6490–6507.
[31] S.M. Tabatabaei, B. Vahidi, Bacterial foraging solution based fuzzy logic decision for optimal capacitor allocation in radial distribution system, Electric Power Systems Research 81 (4) (2011) 1045–1050.
[32] Neil J. Vickers, Animal communication: when I'm calling you, will you answer too?, Current Biology 27 (14) (2017) R713–R715.
[33] Ji Chen, Haiyun Shi, Bellie Sivakumar, Mervyn R. Peart, Population, water, food, energy and dams, Renewable & Sustainable Energy Reviews 56 (2016) 18–28.

Frameworks of considering RESs and loads uncertainties in VPP decision-making

Zeal Shah[a], Ali Moghassemi[b], and Panayiotis Moutis[c]
[a]Electrical & Computer Engineering Department, University of Massachusetts Amherst, Amherst, MA, United States, [b]Department of Electrical Engineering, South Tehran Branch, Islamic Azad University, Tehran, Iran, [c]Wilton E. Scott Institute for Energy Innovation, Carnegie Mellon University, Pittsburgh, PA, United States

10.1 Introduction

Power systems have experienced significant changes in the last 30 years, one of which is the large-scale penetration of distributed energy resources (DERs). DERs are relatively small generators, dispersed over the grid – albeit, typically, closer to the distribution system (load customers), and usually powered by resources like diesel, gas, and renewables (RES). Among the most typical RES-based DERs are solar photovoltaics (PVs) and wind turbine generators. The recent increasing number of electric vehicle (EV) sales has also sparked the vision to account for their potential as DERs, too, provided bi-directional charging/discharging can be offered. Three major benefits of DERs are their relatively low-cost investments, increased flexibility, and improved resilience. But the penetration of DERs in the distribution system introduces numerous complexities in the planning, operation, and management of both the grid and DERs.

One major challenge regarding DERs is the highly volatile nature of most of the RESs that power them, which comes to aggravate the on-going concerns by the stochastic behavior of load customers. The reason why volatility is posed as a concern in power system engineering is because it prohibits a confident way of operating the grid and planning for it. If this kind of uncertainty is left unchecked, the grid may deteriorate all the way to a collapse. On the other hand, uncertainty mitigation with traditional means can lead to costly decisions and, thus, prohibitive increase in energy costs. As from the perspective of DERs specifically (meaning their value potential), uncertainty limits them (in most cases) to trade electricity in the energy markets on an individual basis. To handle the above issues, the concept of virtual power plants (VPPs) has been recently introduced. Leveraging the advances in networking technologies, internet-of-things, and cloud computing infrastructure, a VPP can aggregate, operate, and manage DERs as a single "fictitious" power plant. A VPP can also comprise controllable loads, energy storage systems, and conventional units to diversify the energy mix. In Fig. 10.1 a simple example of a VPP and its interactions with a local aggregator (could be a utility, an energy trader, etc.) is presented.

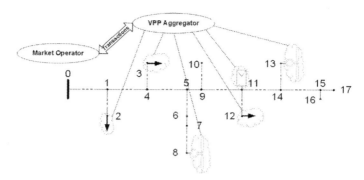

Figure 10.1 Example of a VPP aggregation of resources over a feeder and transactive exchanges with the market operator.

As VPPs are the concept that mainly aims to handle the uncertainty of its integrated assets, some additional context is offered. "Uncertainty" applies to a myriad of applications and fields of interest such as physics, finance, engineering, and industry applications. The uncertainty is usually referred to when we want to handle upcoming events that are of a volatile/unpredictable nature. In the context of power systems, the uncertainty can be linked to the scheduling and real-time control problems in power system operating and planning. The factors of uncertainty in power systems are, most typically, volatility of wind power, solar power, load, market price, marginal costs, outages of generation units and grid infrastructure, and, more recently, driving and recharging patterns of EVs. Among these factors, wind power, solar power, load, and market price are the most important elements of electrical grid uncertainty.

Once we quantify the effect of the uncertainty of each factor, it is paramount to find the modeling approaches which can best describe them and are fitting to the paradigm of aggregation of resources considered; in this case the VPP. Monte Carlo simulation (MCS), roulette wheel mechanism (RWM), Latin hypercube sampling (LHS), Weibull distribution, beta distribution, normal (or Gaussian) distribution, and scenario trees are the most commonly used uncertainty modeling approaches and are also consistently preferred in other fields, as well.

As the uncertainty in the power system sector is linked to the scheduling and planning of its operation, its inherent complexity needs to be pointed out. It is here reminded that the mathematical formulation of scheduling with real AC optimal power flow constraints is non-linear and non-convex, i.e., by algorithmic definition, there is no software that can efficiently optimize power system operation. Hence, adding uncertainty parameters and scenarios on top of the typical power system scheduling problem only make things worse for the electrical engineer researcher and practitioner. Understanding the merits and weaknesses of each method of uncertainty modeling and management is critical in handling the overall problem.

Different types and degrees of uncertainties are introduced by RESs, loads, and market dynamics. VPPs' main task is to address these uncertainties by developing accurate forecasting models for generation, load, and market prices, as well as optimization and control frameworks to facilitate VPPs' profitable participation in different energy markets. Further, VPP operators need to figure out strategies to deal with deviations from forecasted values to ensure that the grid balance is maintained and that a minimum level of profit is achieved. This is usually governed by the inclusion of energy storage systems, conventional generators, and demand response programs within a VPP.

Two major benefits may already be noted here as from the preliminary consideration of uncertainty mitigation through the VPP paradigm. Firstly, VPPs, by enabling RESs to participate in the market, support wider and meaningful penetration of cleaner sources of energy, thereby reducing pollution. Secondly, VPPs allow for active participation of end-users in the market, which has been otherwise not possible.

10.1.1 Sources of uncertainty in VPPs

The very concept of a VPP is to integrate and control heterogeneous generation systems in such a manner that they can trade electricity in the market, which are not allowed to on an individual basis, due to high degree of uncertainties associated with them, as also their typically small sizes. High penetrations of DERs increase uncertainties in power systems and make it much more difficult to maintain power system stability. VPP control and planning models are developed keeping this concern in mind and are capable of handling various uncertainties, which can be classified into three main categories: RES generation uncertainty, market price uncertainty, and load uncertainty.

RES generation uncertainties are mainly introduced by randomness associated with natural factors affecting the output of solar and wind generators like solar irradiation, wind speed, temperature, clouds, seasonal cycles, abrupt weather variations, etc. These uncertainties are further amplified by factors like degradation of PV modules, soiling of PV panels, freezing of wind turbines, and others. Typically, VPPs deal with RES energy uncertainties by relying on short-term PV and wind power output forecasting models.

The second set of uncertainties, energy market price uncertainties, are due to demand and supply availability, and the cost and constraints related to generation and transmission of power. These factors are also heavily influenced by climate and seasonal cycles and so is the energy market pricing. Furthermore, regulatory and institutional frameworks related to energy markets, coupled with the region's energy policies add to the stochasticity of energy market pricing.

The last set of uncertainties, those related to the demand, depend heavily on seasonal changes and customer behavior. Seasonal fluctuations in load are relatively easier to predict because the demand follows a typical pattern over time as seasons change. But it gets difficult to predict changes in load associated with customer behavior; customers can decide to switch their loads on or off depending on a variety of factors like market prices, emergencies, or even simple personal preferences. Additionally, a similar explanation applies to electric vehicles where it becomes difficult to predict the charging schedule of electric vehicles in a VPP. Therefore, if a VPP contains a high proportion of controllable loads, the uncertainties associated with demand increase significantly.

10.1.2 The cost of uncertainty in VPPs

As with any energy integration paradigm, VPPs as well seek to supply power at minimum cost or maximum profit. As VPP may comprise resources that lack flexibility or cannot be controlled (wind power, PV, etc.), aggregation with several other energy resources and flexible loads is what enables optimal dispatching of the VPP assets. This means that several types of costs can be considered for minimization in the objective functions, namely operation cost, market cost, ramping costs, no-load costs, fuel costs, imbalance costs imposed from the balancing market, production costs, the cost of unserved energy, and reserve curtailment, start-up costs, storage investment cost, operation and maintenance cost, reliability cost, production cost, wind power production cost, total emission cost, and the cost of providing upward uncertainty reserve capacity. The cost minimization can also be complemented with profit maximization factors such as revenue from electricity sales and bidding in the market, revenue from spinning and non-spinning reserve sales, payments for energy flexibility, time of use of service, and reimbursement for other types of ancillary services.

Objectives of VPP optimization models coupled with the constraints discussed before govern the bidding strategies and real-time operation of a VPP. In the majority of the cases, VPPs use a bi-level optimization model where the first level objective is to obtain a unit commitment schedule of VPP minimizing its expected costs or maximizing its expected profit, which is then used to submit a VPP's bid to the day-ahead market. Committed generation and served load along with bids submitted to the day-ahead market rely heavily on VPP's predictions of RES generation, market prices, and load profile on the next day. But due to the uncertainties associated with predicted variables, deviations from committed values are expected and are handled in real-time using the real-time objectives of a VPP optimization model. Real-time objectives include maintaining grid balance in case of deviation from committed values and ensuring profitable trading of electricity for the VPP while taking into consideration the spot values of RES output, market prices, and load. VPP submits bids in real-time based on the output of real-time optimization formulation, which assists in balancing the grid by providing reserves in the balancing market, leverages demand response strategies to increase the overall profits, and reduces the real-time operating costs. So, in most cases, the objective of a VPP's optimization model is to maximize VPP's expected profit or minimize VPP's expected costs in the longer term and the same for the operating costs in real-time control/dispatching. These objectives can be framed in different ways, e.g., a VPP can prioritize maximizing revenue from generation, taking into account the capital and operating costs, or through demand response strategies. It can also prioritize maximizing customer profits by aiming for net positive retail energy income. Furthermore, VPP objectives also include risk measures like conditional-value-at-risk (CVaR) that is aimed at ensuring a minimum level of profit for the VPP. Additionally, cost concerns include optimal battery operations by considering battery cycling and also costs associated with their degradation.

Demand-side management also plays an important role in the planning and operations of a VPP in terms of cost considerations. Demand-side management essentially entails managing demand by shifting or switching customer loads with changing market prices and generation volatility. Switching/shifting of loads can be planned beforehand by the operator looking at the forecasted generation and market prices, but the same can also be done in real-time to deal with deviations from forecasted values to ensure grid balance and profitable trading. A VPP can employ different kinds of demand response strategies and we can classify them under two large categories: incentive-based and price-based. In an incentive-based demand response system, the end-users voluntarily provide the VPP operator with control over some of their loads like air-conditioners, and in return, the end-users receive discounts on their electricity bill. Control over customer loads allows the VPP to efficiently manage demand on-the-fly to mitigate the impacts of deviation from committed generation, minimize the amount of abundant power and increase VPP revenue. A drawback of incentive-based demand response is that by handing operators control over some of the customers' loads, the comfort of a customer can be compromised to an unacceptable extent. In a price-based demand response strategy, the customer gets more flexibility because, instead of handing over the control to the VPP operator, the operator can send recommendations to a customer based on market prices to switch/shift their loads. This strategy allows the customers to obtain profits through retail energy income. Both strategies are aimed at benefiting the VPP as well as the end-user. At a higher level, the overall system reaps the benefits of efficient energy arbitrage not only in form of financial profits but also in terms of maintaining grid balance.

Even after taking all the measures into account, the unpredictability and volatile nature of RESs, load, and market prices are such that at times a VPP can struggle to make a profit or even cover up for the operating costs. This means a VPP is at "risk" of low or no profit scenarios, due to the high degree of uncertainties involved in the system. To overcome such risks, it becomes necessary to factor in the degree of risk as a measure in a VPP's planning and operations decision-making process. Risk is modeled and incorporated as a separate element into a VPP's objective function and it allows the VPP to achieve optimal risk-averse bidding strategies. Risk-aversion strategies protect the VPP operators from low-profit scenarios or suffering huge losses. Two popular risk assessment measures used in electricity markets are VaR and CVaR. Other risk metrics include, but are not limited to, mean-variance portfolio (MVP), shortfall probability or functions, expected shortage, mean standard deviation, and uncertainty budget.

10.1.3 The value of load forecasting in VPPs

Load forecasting has a prominent role in the next generation power systems, enabling smart grid applications, efficient energy management, and better power system planning. Due to the profound impact of load forecasting on the reliable operation of power systems and the economy, accurate load forecasting techniques are of vital importance. It ensures the reliable operation of the power system, thereby uninterruptable power supply to the consumer. Load forecasting is usually categorized into three classifications: (i) long-term load forecasting at a horizon of 1–10 years ahead, which is applied for power system planning, (ii) medium-term load forecasting in times of 1 month to 1 year ahead, which is normally related to power system operation and maintenance efficiency, and (iii) short-term load forecasting for horizons of 1 hour to 1 day or 1 week ahead, which happens to be the most typical kind of load forecasting. This is because of the impact of short-term load forecasting on unit commitment and spinning reserve. As a result, high forecast accuracy is required for multiple time horizons that are associated with regulation, dispatching, scheduling, and unit commitment of the power grid. In other words, to meet the needs and wants of demand response, an efficient, yet precise, load forecasting technique plays a leading role. Such a load forecasting technique can be parametric or non-parametric. Parametric techniques are called statistical techniques, namely time series technique, linear regression, autoregressive moving average, and stochastic time series. Non-parametric techniques which are called artificial intelligence (AI) techniques have absorbed many researchers. Over the past years, AI techniques and also metaheuristic optimization algorithms have been tilted towards as they are more than capable of addressing complex problems. Among non-parametric techniques, artificial neural network (ANN) and its hybrid with optimization algorithms have been mostly presented. This is because of the superb performance of ANN in solving complex problems, especially when it comes to decision making under uncertainty and prediction patterns.

10.2   Proposals for handling uncertainty within a VPP

A few works have attempted to understand, analyze, model, and manage the effects of uncertainty within VPPs. In the following, we discuss the main characteristics and the crucial outcomes of the most representative publications on the subject. We offer a brief description of the methodologies uses, the frameworks assumed, and the aims served by each work.

The review in [1] comments on three sources of uncertainty for VPPs, namely, renewable power uncertainty, market price uncertainty, and load demand uncertainty. It also presents three major categories of descriptions implemented in recent studies to measure the influence of uncertainties on VPP, probabilistic distribution description, possibilistic description, and interval description. The authors have summarized and discussed different optimization approaches and corresponding constraints for each uncertainty type and have also presented their pros and cons.

In [2] a probabilistic price based unit commitment (PBUC) technique using point estimate method (PEM) to model uncertainties associated with generation sources and market pricing is proposed. The aim is to determine optimal bidding strategies for a VPP in a day-ahead electricity market, by maximizing the expected VPP profit. PEM accounts for the stochastic trends in the uncertainty so that the formulations could be solved deterministically. Mixed-integer nonlinear programming is then preferred to solve the deterministic PBUC problem.

In [3] the authors present a stochastic scheduling model for VPP taking into consideration the uncertainties, price-driven demand response, and incentives-based demand response. VPP components consist of a wind power plant, PV, gas turbines, energy storage, and demand resource providers. A two-level stochastic scheduling model is proposed. The objective of the first level of scheduling is to maximize VPP's operating income using day-ahead predictions of wind and PV power. The second level (lower) is tasked to revise the scheduling scheme considering the actual output from PV and wind to minimize the system's net load and operating costs.

The main aim of [4] is to assess and evaluate the effect of risk-aversion on total and surplus profits of VPP in four different scenarios: VPPs do or do not participate in the reserve market, and single or dual pricing is used in the balancing (real-time) market. The proposal comprises a two-stage stochastic programming approach for risk-averse participation of VPP in the joint energy and spinning reserve markets. The uncertainties considered are those of renewables, load profile, reserve service requirements, day-ahead market prices, and also the prices in the reserve and balancing markets. The overall evaluation of the impact of risk-aversion on total and surplus profits is done by incorporating CVaR into the model and solving it as a mixed-integer linear program.

A day-ahead scheduling framework for VPP in the joint energy and regulation reserve markets is discussed in [5]. The model considers three major types of energy sources in day-ahead energy and regulation reserve markets: renewable sources, conventional small-scale power plants, and storage. A regulation reserve is provided based on stochastic price in the regulation market along with up/down regulation calling requests. Vehicle-to-grid option is also considered, about which the authors propose setting incentive contracts taking into account the CO_2 emission penalty, as well. The point estimate method is used to model uncertainties related to day-ahead energy and regulation reserve market prices, regulation up/down calling probabilities, renewable generation, and also the number of EVs feeding into/from the grid.

In [6] a mixed-integer linear programming problem is formulated as a double-level stochastic model. At the first level, optimal decisions are made for the day-ahead market at hourly resolution. At the second level, the decision is taken regarding the optimal operation of conventional power plants and batteries, as soon as the output from renewable energy sources and day-ahead energy market prices are known. The objective is to maximize a generation company's expected profit. CVaR-based management is used to control risks related to scenarios with low profit. A VPP is assumed to participate in both day-ahead and balancing energy markets to mitigate the risks using the CVaR algorithm. In the balancing market, the VPP corrects its deviations from the day-ahead schedule by buying energy in case of shortage and selling in case of excess.

The authors of [7] use an MILP-based robust optimization technique for handling uncertainties to allow VPP to self-schedule, i.e., permit VPP to decide on the unit commitment of its DERs, as well as sale and purchase offers to the market. They have also presented a real trading strategy of a VPP through arbitrage modeling. The overall objective of the approach is for a VPP to obtain maximum profit in the market.

A novel strategy to use wind power as an alternative way to decrease the market cost for VPP's when flexible resources are not available is investigated in [8]. The VPP's scheduling and bidding problem is formulated as a two-level stochastic optimization program to reduce VPP's expected market cost during its participation in said markets. The problem is simplified to a certain level such that local optimization algorithms like linear search could be used.

The model discussed in [9] evaluates the physical characteristics of VPP like maximum capacity and ramping constraints with regards to the uncertainty associated with wind power output and demand. The authors proposed a novel distributionally robust optimization model. Its objective is to determine VPP bidding offers for the unit commitment by optimally selecting the necessary parameters. This model uses moment information, i.e., mean and covariance of the unknown parameter. A two-stage model is formulated: the first stage is aimed at minimizing the worst-case expected total net cost (including ramping costs, no-load costs, and fuel costs), while the second stage is formulated to minimize the total operating cost while adhering to dispatch and storage constraints. This model takes into account the real-world dispatching scenario, where the operator takes real-time decisions after knowing the output of uncertain parameters like load and renewable output. This model is solved using linear-decision rules.

The authors of [10] propose a bi-level stochastic programming approach for evaluating the optimal bidding strategy of a VPP that consists of a CHP providing district heating (CHP-DH). The first stage of the method is used to bid VPP energy in the day-ahead market, while the second stage is used for settling the deviation from the day-ahead commitment in the imbalance market. Three different strategies presented are explored: (1) CHP's scheduled output is not adjusted at all, (2) CHP is re-dispatched for the whole day based on the assumption that renewable energy output is known, but considers uncertainty in imbalance prices, and (3) in the second stage, the VPP is re-dispatched at each time step, depending on the renewable output as well as the imbalance price.

The VPP model proposed in [11] participates in two markets, an internal (VPP's own generation and demand) and external (by the ISO). A fuzzy optimization technique is used to model uncertainties associated with renewable sources in the internal day-ahead market before the VPP bids into the external market. The objective of VPP's internal day ahead market optimization takes into consideration uncertainties related to generation, ISO day-ahead prices, and the load curtailment/shift offers. The VPP then accordingly submits bids to the external day-ahead market and implements the dispatch. Mixed-integer nonlinear programming is used. In real-time operation, based on the deviations from the commitment schedule, customers submit bids in the internal market. The deterministic formulation is used to address the real-time nature of VPP's internal balance market optimization, the output of which tells the amount of energy needed to be purchased from or sold to the ISO market. The output is then used to decide on and submit bids to the external balance market by the VPP.

The authors of [12] propose a stochastic adaptive robust model for the dispatch of VPP in different markets, such as day-ahead, real-time, and carbon trading markets. VPP in this paper is specifically assumed to be consisting of PV and central air-conditioning systems. Since the market price predictions are highly accurate, the stochastic programming approach is used to tackle the uncertainties associated with any deviations that happen in the actual market price from the forecasted values. The adaptive robust portion of the model is aimed at addressing the uncertainty associated with PV generation. A binding scenario identification approach to solve the above-mentioned dispatch problem is used.

The overall aim of [13] is to investigate the impacts of the presence of energy storage devices (ESD) in the VPP's mix and to study their contribution towards improving generation system adequacy. A bi-level network flow-based linearized optimization formulation is proposed, to develop a probabilistic model of VPP that consists of distributed generation, loads, and ESD, in addition to considering the topology of the network and associated constraints. The first stage of the algorithm decides the operation strategy of the VPP including the charging and discharging of ESD. The second stage is used to obtain optimal dispatch schedules for DER and ESD using a fair energy allocation strategy.

In [14] a bi-level optimization is formulated that seeks to maximize the virtual power plant profits, while the system operator optimizes the day-ahead scheduling. The VPP aims to maximize its profits by avoiding resorting to balancing energy services; to do that it has to properly account for the load and distributed generation stochasticity, as also the behavior of competing energy actors bidding in the day-ahead market. In this sense, the VPP is acting as a price-maker and not as per the typical paradigm as a price-taker, thus, overcoming successfully the effects of the uncertainty of system and assets behavior. Energy-wise, incorporating demand response programs improves the profitability of the VPP.

The authors of [15] attempt to assess the adjustment in unit dispatching within a virtual power plant that loses unexpectedly a significant amount of power. The said VPP has been committed to a specific output from the day-ahead market and must deliver as promised. A decision tree methodology outputs re-dispatching plans that ensure a fast response to the power loss in a first stage and an optimal (in terms of cost) re-dispatch in the second stage. Multiple levels of optimality allow for the VPP to decide whether to re-dispatch its own resources or resort to balancing markets.

In [16] a scheduling algorithm leverages forecasting with neural networks to handle the uncertainty of wind and solar resources of a VPP in a week-ahead planning horizon. The VPP also comprises hydrogen and thermal generation and is expected to serve a specific demand profile. The scheduling can be adjusted as the meteorological data inputted to the forecasting tools is updated in real time.

The day-ahead scheduling of a VPP in a market with price uncertainty is discussed in [17]. Although scenario reduction is used to handle the volatility of electricity and fuel prices, the researchers propose extension of the methodology to the account for the behavior of solar and wind resources. The work relies on the flexibility offered by heating and electricity storage that comprise the VPP beyond the renewable resources.

As demand response may be sourced from multiple different electricity customers within an area, a community or even a building, the VPP paradigm may ideally serve their aggregation. A VPP of resistance heaters is considered in [18]. In this case a forecast at the aggregated level of the heaters and based on meteorological data is used to anticipate the behavior of the overall VPP heating load. The authors also refine the expected behavior with a model of heat exchanges and losses. The results concern the use of weather forecasts as they are more accurate the shorter the term ahead.

In [19] energy and reserves of a VPP are scheduled in a day ahead electrical and thermal markets. The scheduling accounts for the uncertainty of wind and photovoltaics generation with a point estimate method. The method enhances the role and profitability of VPPs in reserves markets, even if the volatile resources are underperforming.

Integrating an intermittent source, a conventional generator, and a storage unit within a VPP is proposed in [20] in order to describe a more generalized version of the paradigm. The wind power uncertainty is modeled via scenarios and the overall scheduling as a 2-stage optimization. The VPP first dispatches the units within it and then, according to actual wind generation and market prices, joins the balancing market to adequately match its day ahead schedule.

The researchers focus on the uncertainty introduced by multiple wind generators of a VPP in [21] and propose a non-linear probabilistic model of the correlation among them. They then approximate this model using machine learning tools and implement a distributed optimal VPP scheduling that requires minimal information exchange among the control points. The aim is to maximize VPP profit by selling the available power at the given market price, while avoiding penalties for any shortfalls.

10.3 Taxonomy

It is valuable to classify the works that describe how VPPs handle uncertainty, understand dominant trends, the methods which are considered more valuable, and assess any gaps in the research.

10.3.1 Sources of uncertainty in VPP scheduling

VPP control (and dispatching thereof) is designed to handle a high degree of uncertainty related to generation, demand, and pricing while aiming to maximize energy profits. Generation uncertainties involve the stochastic nature of renewable generation attributed to varying levels of solar irradiation, wind flow, and others. Multiple demand-related uncertainties play an important role in VPPs planning and operation stages like uncertainties associated with customer's energy demand and load profiles [22], electric vehicle driving schedules [23], the effect of seasonality on residential electricity and heating demands [24]. Market-pricing related uncertainties are, usually, deviations from forecasted day-ahead market prices [24], uncertainties in reserve capacity and reserve energy prices [25], and imbalance prices [26]. Other uncertainties can be attributed to technical characteristics of components like degradation of batteries [22], degradation of PV system performance due to factors like snow, dirt, soil deposition, shading [16], and also emissions. Another aspect of uncertainty, although not inherent to any VPP, is the health of grid assets. Forced and planned outages are a parameter that may well affect VPP operation even though it is exogenous to it. The work in [27] offers an example on how forecasting uncertainty of wind power is a parameter of a stochastic unit commitment to improve the procurement of spinning reserves in light of uncertainties which also include failures of grid assets.

With the above points in mind, we categorize the works presented in the previous section based on the focus to properly model the uncertainty of each asset considered in the context of VPP control and dispatching. (See Table 10.1.)

Table 10.1 Asset of VPP or electrical grid modeled for its uncertainty.

DER	[1–5,9–16,19,20]
Grid	-
Load	[1,4,14,18,19]
Policy, Market Framework & Pricing	[1,2,4,5,8,10,12,17,19]

10.3.2 Modeling the uncertainty affecting VPP control

Given the described range of assets that may comprise a VPP, their uncertainty needs to be modeled accordingly. We may identify three broad categories of uncertainty modeling approaches. One is focused around risk management. Risk concerns are added and quantified as a component of a VPP's optimization objective. VaR and CVaR are the two most popular risk assessment measures used in electricity markets [28,29]. The second category is a diverse set of methods that model uncertainty through scenarios. The scenarios are account for in the VPP dispatching according to the expectation of their occurrence. The third category includes in general all methods inspired by heuristics and artificial intelligence methods. The angle of the methods in this category is to model the uncertainty in a more abstract manner, identify common characteristics across assets and program the VPP to respond in an agnostic manner.

For the reader's reference we here list the most typical methods from the aforementioned categories: point estimate method [30], sequential Monte Carlo with chance-constrained programming [21], dimension-adaptive sparse grid interpolation and copula theory [31], decision trees [15], tabu-search/PSO (TS-PSO) [14], and biogeography-based optimization [19].

We group the studies presented earlier in Table 10.2 as per the uncertainty modeling preferred in each study.

Table 10.2 Uncertainty modeling for optimal VPP control/dispatching.

VaR, CVaR	[4,6,14]
Scenario-based	[2,3,5,7,9,10,12,13,17,19,20]
Heuristics	[11,15,18]
Other	[8,21]

10.3.3 Control objectives of optimal VPP dispatching

A VPP's optimal modeling governs its planning, unit commitment, and bidding strategies in the energy market. Optimization objectives can be formulated to achieve multiple different purposes. Indicatively, some of them are maximizing VPP's expected profit or income [32] or to minimize its operational costs [24], maximizing retail or customer income [33], minimizing energy losses [34], minimizing voltage deviations, overload failures and maintaining grid balance [23], optimally sizing and placing DERs including batteries [14], penalizing carbon emissions to minimize CO_2 emissions [25], as well as optimizing batteries' charging and discharging schedules [35].

An aspect that has not been given proper attention in optimal VPP control is the grid topology in terms of how the VPP's dispatching might threaten grid stability and cause security concerns. Indicative, the authors of [36] discuss the effect of wind generation uncertainty on cascading failures. Stochastic optimal power flow is used to assess these effects. It is shown that higher wind power penetrations imply a higher risk of cascading failures, due to the uncertainty affecting especially weaker lines in power systems.

A practical aspect of VPP operation that has also been researched very little is demand response. Demand response allows the VPP and its customers to profit by various mechanisms like shifting peak loads to times when market prices are low or energy arbitrage or shedding flexible customer loads when the market prices are expensive. Although most of the currently operation VPPs around the world comprise mostly such interruptible loads, research has focused very little on how to best control them as VPP assets. A representative work as an example on the matter is [37], where a stochastic energy management model is proposed. Hourly interruptible loads may be shed from a variety of customers, considering interruption costs for different customers to determine optimal stochastic load interruption of said assets.

This leads us to group the studies in VPP optimal control as in Table 10.3, which describes what the implicit aims for the optimal VPP control are, i.e., which asset is the focus of control or improved behavior – not the objective function per se.

Table 10.3 Aims of optimal VPP scheduling.

Generation Operation	[8,10,16,20]
Battery/Storage Operation	[13,20]
Placement of Assets	[15,18]
Demand response	-
Profitability	[2,4,5,9,11,12,14,17,19,21]
Grid Security	-
Serving Demand	[18]

10.3.4 Timeframes and market participation of optimal VPP scheduling

The most typical market structures are largely split in two time horizons of planning and procurement, the forward or look-ahead markets and the spot or imbalance markets. The forward markets seek to procure the necessary resources that will serve the forecasted load demand in years and up to minutes ahead of time. The spot markets are the markets that seek to resolve either excess or deficit of power almost in real-time.

From the previous distinction, there also arises the definition of energy and ancillary services as products in the electricity markets. Energy is typically the main product offered and cleared in the forward markets. It involves the commitment of power in specifically defined timeframes (i.e., energy) by each asset. As the forward market relies on forecasts, it is evident that imbalances are expected to occur in real time. To this end, the forward markets also seek to procure balancing reserves that are, essentially, the most typical ancillary service product of the forward markets. Reactive power, black-start capability, and others are also considered as ancillary services, but there is no market maturity or technical framework that would allow VPPs to actively bid for them and offer them.

A separate framework of power actuation is the dynamic scheduling. Dynamic optimization is not driven by a specific market structure, but rather by real-time signals to which the controllable assets respond. An example is the work in [38], which proposes a novel dual horizon rolling scheduling framework for the management of active distribution systems, based on dynamic AC optimal power flow.

As from the preceding analysis, we define the classifications in Tables 10.4 and 10.5.

Table 10.4 VPP scheduling horizon.

Forward market	[2–5,9,13,14,16–19]
Spot market	[21]
Forward & Spot Markets	[6,10–12,20]
Dynamic Optimization	–

Table 10.5 Electricity market products in the VPP scheduling.

Energy	[2–4,8,12–14,16–18]
Ancillary Services	–
Energy & Ancillary Services	[4–6,9–11,19,20]

10.3.5 Forecasting methods to handle VPP uncertainties

It is remarkable to note that with the exception of [16] no other study considers the value or effect of the use of forecasting tools in VPP planning. We will briefly note some crucial works in load forecasting as also forecasting of renewable energy. In general, forecasting can be classified into parametric and non-parametric techniques. Parametric techniques known as statistical techniques could be time-series technique [39], linear regression [29], autoregressive moving average [40], and stochastic time series [34]. Non-parametric techniques include artificial neural networks, support vector machines, adaptive neuro-fuzzy inference system, fuzzy and genetic algorithm kernel-based methods [41], and others.

10.4 Conclusions and path forward

This chapter discusses a wide variety of sources of uncertainty that affect the operation and optimal scheduling of a VPP, and the methods proposed by researchers to address, quantify, and control it. We have explored in detail why uncertainty is a crucial concern in electric power systems and especially for the potential and profitability of VPPs. More than 20 works have been published on how VPPs can operate under uncertainty as actors in modern electricity markets and as parts of large interconnected or smaller grids. After briefly summarizing the most prominent of these works, we have classified them considering their focus, aims, scopes and methodologies. Based on this latter analysis, we can draw the following conclusions on where research should focus in the coming years, as there are clear gaps that need to be addressed.

There has been no research on the matter of *how grid topology might affect the operation or market involvement of a VPP*. It is clear that VPPs do not assume any control or visibility over any part of the electrical grid they are connected to, however, any grid failures or reconfiguration might affect either the commitments or operating integrity of a VPP. One of the most typical operators' plans in day-ahead scheduling is the security constrained dispatch of generating resources [42]. This dispatch assumes that at least one, if not two, transmission grid assets have failed, yet the system must withstand such an event and retain full operational functionality. In this sense, VPP's scheduling should account for either the uncertainty of the grid security or for the uncertainty of those VPP assets affected by grid security concerns. Another aspect in grid topology that touches on VPP scheduling is the optimal placement of assets. Typically, VPPs have been considered to aggregate existing resources with no consideration of which resources would best serve its profitability or other relevant aims within its grid operation. These may further be extended to the optimal placement of new assets in parts of an existing or developing grid.

Management and flexibility of loads, as expressed through the grid service of *demand response, is a crucial contributor to VPP operation and design that has yet to be thoroughly considered and analyzed*. Many practical applications of VPPs are predominantly focused on how to optimize demand response of residential and commercial loads. Most of the demand response strategies fall under two categories, incentive-based and price-based. Some researchers have proposed the concept of active customer participation in the market where a VPP operates an internal market and allows the customers to submit their day-ahead load curtailment and load shift offers. A VPP then uses these offers along with the necessary predictions for generation and market prices to submit its bid to the external day-ahead energy market and announces the winning customers. In real-time, to mitigate the impacts of deviations from committed values, the VPP allows its customers to submit their offers and then accordingly its real-time bid in the balancing market. The concept of active end-user participation seems promising and needs to be researched more by the community. The financial and economic impacts of different demand response strategies on VPP as well as its customers are also worth exploring in more detail. Most of the work on demand response has been focused on maximizing profits for the aggregator realizing it, but hardly any work has been done to study the profits or losses incurred by the customers due to demand response. Researchers can model and experiment with different incentive-based strategies like tax rebates or reduced electricity bill amounts to better understand the benefits and drawbacks of these strategies on customers and the VPP. Real-time load shifting and load curtailment are often desired in a VPP to adjust for the deviations in power generation, load profiles, and market prices. In this case as well, it becomes important to better understand and quantify the benefits and losses incurred by the customers, who are willing to curtail their demand in real-time as per VPP's request. Demand response-based strategies should be designed and their impacts evaluated for each of the three different customer categories – residential, commercial, and industrial. All three types of customers have different demand requirements and schedules. It would be interesting to see if implementing demand response strategies specific to a demand type would increase the VPP and customer profit or not. Further, devising demand response schemes specifically for incentivizing electric vehicles within VPPs for altering their charging schedule is an interesting and relatively less explored research area.

In terms of methodologies and models, *scenario-based formulations seem to have dominated the approaches to account for uncertainty concerns within or affecting a VPP*. Further research should be focused on using chance constrained optimization and methods based on CVaR, VaR, and similar techniques. Recent research in convex relaxation of the AC optimal power flow could also allow for more accurate consideration of losses and voltage profile regulation from a VPP. Lastly, the advances in machine learning, artificial intelligence innovations and, most prominently, deep learning have barely been touched in the field, despite the fact that they promise lots of impact and efficiency in solving complex problems of knowledge-based models as those involving VPPs.

References

[1] S. Yu, F. Fang, Y. Liu, J. Liu, Uncertainties of virtual power plant: problems and countermeasures, Applied Energy 239 (2019) 454–470.
[2] M. Peik-Herfeh, H. Seifi, M. Sheikh-El-Eslami, Decision making of a virtual power plant under uncertainties for bidding in a day-ahead market using point estimate method, International Journal of Electrical Power & Energy Systems 44 (1) (2013) 88–98.
[3] L. Ju, Z. Tan, J. Yuan, Q. Tan, H. Li, F. Dong, A bi-level stochastic scheduling optimization model for a virtual power plant connected to a wind–photovoltaic–energy storage system considering the uncertainty and demand response, Applied Energy 171 (2016) 184–199.

[4] S.R. Dabbagh, M.K. Sheikh-El-Eslami, Risk assessment of virtual power plants offering in energy and reserve markets, IEEE Transactions on Power Systems 31 (5) (2015) 3572–3582.
[5] A. Shayegan-Rad, A. Badri, A. Zangeneh, Day-ahead scheduling of virtual power plant in joint energy and regulation reserve markets under uncertainties, Energy 121 (2017) 114–125.
[6] M.A. Tajeddini, A. Rahimi-Kian, A. Soroudi, Risk averse optimal operation of a virtual power plant using two stage stochastic programming, Energy 73 (2014) 958–967.
[7] M. Shabanzadeh, M.-K. Sheikh-El-Eslami, M.-R. Haghifam, The design of a risk-hedging tool for virtual power plants via robust optimization approach, Applied Energy 155 (2015) 766–777.
[8] Q. Zhao, Y. Shen, M. Li, Control and bidding strategy for virtual power plants with renewable generation and inelastic demand in electricity markets, IEEE Transactions on Sustainable Energy 7 (2) (2015) 562–575.
[9] S. Babaei, C. Zhao, L. Fan, A data-driven model of virtual power plants in day-ahead unit commitment, IEEE Transactions on Power Systems 34 (6) (2019) 5125–5135.
[10] J.Z. Riveros, K. Bruninx, K. Poncelet, W. D'haeseleer, Bidding strategies for virtual power plants considering CHPs and intermittent renewables, Energy Conversion and Management 103 (2015) 408–418.
[11] A.T. Al-Awami, N.A. Amleh, A.M. Muqbel, Optimal demand response bidding and pricing mechanism with fuzzy optimization: application for a virtual power plant, IEEE Transactions on Industry Applications 53 (5) (2017) 5051–5061.
[12] G. Sun, et al., Stochastic adaptive robust dispatch for virtual power plants using the binding scenario identification approach, Energies 12 (10) (2019) 1918.
[13] A. Bagchi, L. Goel, P. Wang, Adequacy assessment of generating systems incorporating storage integrated virtual power plants, IEEE Transactions on Smart Grid 10 (3) (2018) 3440–3451.
[14] E.G. Kardakos, C.K. Simoglou, A.G. Bakirtzis, Optimal offering strategy of a virtual power plant: a stochastic bi-level approach, IEEE Transactions on Smart Grid 7 (2) (2015) 794–806.
[15] P. Moutis, N. Hatziargyriou, Decision trees aided scheduling for firm power capacity provision by virtual power plants, International Journal of Electrical Power & Energy Systems 63 (Dec 2014) 730–739, https://doi.org/10.1016/j.ijepes.2014.06.038 (in English).
[16] A. Tascikaraoglu, O. Erdinc, M. Uzunoglu, A. Karakas, An adaptive load dispatching and forecasting strategy for a virtual power plant including renewable energy conversion units, Applied Energy 119 (2014) 445–453.
[17] H. Taheri, A. Rahimi-Kian, H. Ghasemi, B. Alizadeh, Optimal operation of a virtual power plant with risk management, in: 2012 IEEE PES Innovative Smart Grid Technologies (ISGT), IEEE, 2012, pp. 1–7.
[18] A. Thavlov, H.W. Bindner, Utilization of flexible demand in a virtual power plant set-up, IEEE Transactions on Smart Grid 6 (2) (2014) 640–647.
[19] A.G. Zamani, A. Zakariazadeh, S. Jadid, Day-ahead resource scheduling of a renewable energy based virtual power plant, Applied Energy 169 (2016) 324–340.
[20] H. Pandžić, J.M. Morales, A.J. Conejo, I. Kuzle, Offering model for a virtual power plant based on stochastic programming, Applied Energy 105 (2013) 282–292.
[21] H. Yang, D. Yi, J. Zhao, F. Luo, Z. Dong, Distributed optimal dispatch of virtual power plant based on ELM transformation, Journal of Industrial and Management Optimization 10 (4) (2014) 1297.
[22] G. Cardoso, et al., Optimal investment and scheduling of distributed energy resources with uncertainty in electric vehicle driving schedules, Energy 64 (2014) 17–30.

[23] W.S. Ho, S. Macchietto, J.S. Lim, H. Hashim, Z.A. Muis, W.H. Liu, Optimal scheduling of energy storage for renewable energy distributed energy generation system, Renewable & Sustainable Energy Reviews 58 (2016) 1100–1107.
[24] G. Gutiérrez-Alcaraz, E. Galván, N. González-Cabrera, M.S. Javadi, Renewable energy resources short-term scheduling and dynamic network reconfiguration for sustainable energy consumption, Renewable & Sustainable Energy Reviews 52 (2015) 256–264.
[25] A. Baziar, A. Kavousi-Fard, Considering uncertainty in the optimal energy management of renewable micro-grids including storage devices, Renewable Energy 59 (2013) 158–166.
[26] M. Karami, H. Shayanfar, J. Aghaei, A. Ahmadi, Scenario-based security-constrained hydrothermal coordination with volatile wind power generation, Renewable & Sustainable Energy Reviews 28 (2013) 726–737.
[27] J. Wang, et al., Wind power forecasting uncertainty and unit commitment, Applied Energy 88 (11) (2011) 4014–4023.
[28] D.E. Olivares, C.A. Cañizares, M. Kazerani, A centralized optimal energy management system for microgrids, in: 2011 IEEE Power and Energy Society General Meeting, IEEE, 2011, pp. 1–6.
[29] H. Moghimi, A. Ahmadi, J. Aghaei, A. Rabiee, Stochastic techno-economic operation of power systems in the presence of distributed energy resources, International Journal of Electrical Power & Energy Systems 45 (1) (2013) 477–488.
[30] E.R. Sanseverino, M.L. Di Silvestre, M.G. Ippolito, A. De Paola, G.L. Re, An execution, monitoring and replanning approach for optimal energy management in microgrids, Energy 36 (5) (2011) 3429–3436.
[31] M. Zdrilić, H. Pandžić, I. Kuzle, The mixed-integer linear optimization model of virtual power plant operation, in: 2011 8th International Conference on the European Energy Market (EEM), IEEE, 2011, pp. 467–471.
[32] A. Zakariazadeh, S. Jadid, P. Siano, Stochastic multi-objective operational planning of smart distribution systems considering demand response programs, Electric Power Systems Research 111 (2014) 156–168.
[33] J. Aghaei, H. Shayanfar, N. Amjady, Joint market clearing in a stochastic framework considering power system security, Applied Energy 86 (9) (2009) 1675–1682.
[34] R. Mena, M. Hennebel, Y.-F. Li, C. Ruiz, E. Zio, A risk-based simulation and multi-objective optimization framework for the integration of distributed renewable generation and storage, Renewable & Sustainable Energy Reviews 37 (2014) 778–793.
[35] J.J. Ming Kwok, N. Yu, I.A. Karimi, D.-Y. Lee, Microgrid scheduling for reliable, cost-effective, and environmentally friendly energy management, Industrial & Engineering Chemistry Research 52 (1) (2013) 142–151.
[36] M.H. Athari, Z. Wang, Impacts of wind power uncertainty on grid vulnerability to cascading overload failures, IEEE Transactions on Sustainable Energy 9 (1) (2017) 128–137.
[37] W. Alharbi, K. Raahemifar, Probabilistic coordination of microgrid energy resources operation considering uncertainties, Electric Power Systems Research 128 (2015) 1–10.
[38] A. Saint-Pierre, P. Mancarella, Active distribution system management: a dual-horizon scheduling framework for DSO/TSO interface under uncertainty, IEEE Transactions on Smart Grid 8 (5) (2016) 2186–2197.
[39] A. Rabiee, M. Sadeghi, J. Aghaeic, A. Heidari, Optimal operation of microgrids through simultaneous scheduling of electrical vehicles and responsive loads considering wind and PV units uncertainties, Renewable & Sustainable Energy Reviews 57 (2016) 721–739.
[40] L. Shi, Y. Luo, G. Tu, Bidding strategy of microgrid with consideration of uncertainty for participating in power market, International Journal of Electrical Power & Energy Systems 59 (2014) 1–13.

[41] A. Botterud, et al., Wind power trading under uncertainty in LMP markets, IEEE Transactions on Power Systems 27 (2) (2011) 894–903.
[42] A. Monticelli, M. Pereira, S. Granville, Security-constrained optimal power flow with post-contingency corrective rescheduling, IEEE Transactions on Power Systems 2 (1) (1987) 175–180.

Risk-averse scheduling of virtual power plants considering electric vehicles and demand response

Omid Sadeghian, Amin Mohammadpour Shotorbani, and
Behnam Mohammadi-Ivatloo
Faculty of Electrical and Computer Engineering, University of Tabriz, Tabriz, Iran

Nomenclature

Indexes
t Index for time intervals
v Index for PVs
w Index for WTs
s Index for ESSs
e Index for PVs
j Index for DGs
g Index for upstream grids
l Index for loads
a, b Index for buses
ω Index for scenarios

Constants and parameters
h Duration of each time interval
$F_{DRP,max}$ Maximum participation percentage in DRP
F_{Sh} The load-shifting price coefficient for shiftable loads in DRP
α, β CVaR parameters
$F^s_{S,min}, F^s_{S,max}$ The factor of minimum/maximum allowed energy stored in ESSs
$E^s_{S,min}, E^s_{S,max}$ The minimum/maximum allowed energy stored in ESSs
$E^s_{S,Cap}$ The capacity of ESSs
$F^e_{EV,min}, F^e_{EV,max}$ The factor of minimum/maximum allowed energy stored in EVs
$E^e_{EV,min}, E^e_{EV,max}$ The minimum/maximum allowed energy stored in the EVs
$E^e_{EV,Cap}$ The capacity of EV's battery
$\lambda^{\omega,t}_P$ Electricity price
F^g_{LMP} Factor of the locational marginal price for main grids
ζ^j, μ^j, τ^j Cost coefficient of DGs
C^j_{SU} Start-up cost of DGs
C^j_{SD} Shut-down cost of DGs
$P^j_{J,min}$ Minimum possible production of DGs
$P^j_{J,max}$ Capacity of DGs
R^j_{UP} Ramp-up of DGs
R^j_{DW} Ramp-down of DGs

$P^s_{S,C,max}$ Maximum charging rate of ESSs
$P^s_{S,D.max}$ Maximum discharging rate of ESSs
$P^e_{EV,C.max}$ Maximum charging rate of EVs
$P^e_{EV,D,max}$ Maximum discharging rate of EVs
ECF Energy–distance conversion factor of EVs (km/kWh)
$D^{e,t}_{EV}$ Distance driven by EVs
$L^{e,t}_{EV}$ Electricity load of EVs from driving in daily trips
$C^{\bar{w}}_{WT}$ Total cost of WTs
$C^{\bar{v}}_{PV}$ Total cost of PVs
C_{Sh} Cost of a consumer to shift its load ($/MWh)
$C_{S,deg}$ Cost of ESS degradation ($/MWh)
$C^{\omega,t}_{EV,dis}$ Cost of discharging EV's battery ($/MWh)
$F^{\omega,t}_{EV,dis}$ Cost coefficient of discharging EV's battery ($/MWh)
M_W Location parameter of WTs
M_V Location parameter of PVs
M_L Location parameter of Loads
M_{ESS} Location parameter of ESSs
M_{EV} Location parameter of EVs
M_J Location parameter of DGs
M_B Connection parameter of the system's buses
$B^{a,b}$ The imaginary part of the admittance of line a–b
$X^{a,b}_{Series}$ Series admittance of line a–b
$Y^{a,b}_{Shunt}$ Shunt admittance of line a–b
$F^{a,b}_{L,max}$ Flow limit of line a–b
$L^{l,t}_P$ The value of loads
π^ω Probability of scenarios
$\eta^s_{S,C}$ Charging efficiency of ESSs
$\eta^s_{S,D}$ Discharging efficiency of ESSs
$\eta^e_{EV,C}$ Charging efficiency of EVs
$\eta^e_{EV,D}$ Discharging efficiency of EVs
$P^{g,Max}_G$ The capacity of the lines between VPP and main grids

Variables
TC The total operational cost of VPP
C_{WT} The total cost of WTs ($/MWh)
C_{PV} The total cost of PV systems ($/MWh)
C_{DR} The total cost of shifted loads ($/MWh)
C_{ESS} The total operation cost of BESS ($/MWh)
C_{EV} The total operation cost of BESS ($/MWh)
C_{DG} The total cost of DGs ($/MWh)
C_{Grid} The total cost of trading with main grids ($/MWh)
$P^{s,t}_{S,C}$ Charging of ESSs
$P^{s,t}_{S,D}$ Discharging of ESSs
$E^{s,t}_S$ Stored energy in ESSs
$P^{s,t}_{EV,C}$ Charging of EVs
$P^{s,t}_{EV,D}$ Discharging of EVs
$E^{s,t}_{EV}$ Stored energy in EV's battery
$P^{j,t}_J$ Generation of DGs

$P_S^{s,t}$ Maximum power of ESSs
$P_W^{w,t}$ Generation of WTs
$P_V^{v,t}$ Generation of PVs
$P_G^{g,t}$ Exchanged power with main grids
$L_{P,DRP}^{l,t}$ The load demand in the presence of DRP
$F_{DRP}^{l,t}$ Participation percentage in DRP
$L_{Sh}^{l,t}$ Shifted value of power demand in DRP
$u^{j,t}$ Binary variable of on-off state of DGs
$\psi_U^{j,t}$ Binary variable of DGs startup
$\psi_D^{j,t}$ Binary variable of DGs shutdown
$T_{on}^{j,t}$ Integer variable of on-duration of DGs
$T_{off}^{j,t}$ Integer variable of off-duration of DGs
T_{MU}^{j} Minimum up-time of DGs
T_{MD}^{j} Minimum downtime of DGs
$\theta^{a,t}$ The power angle of the ath bus at the tth interval
$F_L^{a,b,t}$ The flow of line a–b in the tth interval
δ, η CVaR variables

11.1 Introduction

Developments of renewable energy sources (RESs) such as photovoltaic (PV) systems and wind turbines (WTs) in the electricity system have considerably impacted the system and have brought many challenges since these units are non-dispatchable. The virtual power plant (VPP) concept has been introduced by researchers [1] for the aggregation of power generated by RESs to participate in the electricity market [2]. However, the microgrid concept had been presented before VPP, and there are specific differences between VPP and microgrids. In references [3] and [4], these differences have been discussed in detail.

Compared to microgrids, VPPs integrate a wide variety of energy resources in large geographic areas and participate in the wholesale market, while microgrids exchange power only in the form of retail distribution [5]. Additionally, the presence of one or more main grids is essential for a VPP, while microgrids are mainly optimized to operate in the stand-alone mode and, when operating in the grid-connected mode, they only have a single common point with the main grid. Therefore, the locational marginal price (LMP) of the main grid is critical for VPPs connected to multiple main grids [6]. Moreover, VPPs can be implemented under current regulatory structures and tariffs, whereas microgrids have some limitations such as political and regulatory barriers [7]. Microgrids are mainly limited to interconnections between distributed generation units and load demands for load balance, while VPPs mainly depend on the power trade with main grids for improving the economic operation. Another difference is related to the remote control of elements, while microgrids mainly deal with load supply and continuity of power supply with existence power generating elements

[8]. As a consequence, VPP is mainly an economic concept, whereas microgrid is a technical concept.

VPPs can have different elements such as RESs, including the PV and WT, energy storage systems (ESSs), electric vehicles (EVs), and diesel generators (DGs) [9]. The ESS and EVs can store the surplus energy of RESs to be used when needed, especially in high-price hours. EVs can be used as mobile storages, which can be charged in low-demand hours (due to lower electricity price) based on the EVs' battery constraints and even can be discharged for system benefit in high-demand hours. In a VPP, the direction of the power trade with the main grids is determined based on the LMPs. When the electricity price is low, the power can be purchased from the main grids, and when the electricity price is high, the power can be sold to them. In a specific interval, the power can be simultaneously purchased from some main grids and sold to some other main grids. Demand response programs (DRPs) can also be implemented in a VPP to improve the economic operation of the system.

Many researches have been presented in the literature in the field of VPPs. In [10], the maintenance scheduling of DGs has been performed in the presence of RESs, ESS, and curtailable loads by considering risk management to manage the uncertainty in electricity price. In [8], the optimal allocation of ESS in VPPs is studied, in which the conditional value-at-risk (CVaR) is adopted to control the risk of the allocation problem. Reliability evaluation is performed in this research before and after ESS allocation, which shows considerable reliability improvement after installing ESS. In another research [11], the point estimate method is adopted to model the uncertainties in bidding for the participation of VPP in an electricity market. Mid-term optimal scheduling of VPPs containing RESs, pumped hydro storage, and DG has been investigated in [12] by considering the uncertainty in RESs generation. Moreover, in [13], a risk measure based on robust optimization has been presented for optimal scheduling of VPPs to manage the risk caused by the uncertainty in the electricity market. A model for optimal scheduling of VPPs by considering the air pollutant emission and battery depreciation cost has been presented in [14]. In [15], a planning scheme for VPPs with high penetration of RESs is presented. The uncertainty of VPP output power due to inaccuracy in the forecasting of load demand and wind generation is taken into consideration in this research. Economic assessment of a VPP with 67 dwellings in Western Australia has been evaluated in [16], in which the studied VPP includes PV system, ESS, and heat pump, and also DRP is taken into account. This research considers the uncertainty in load demand. In [17], the authors investigate the optimal scheduling of a VPP in Spain with large- and small-scale distributed generation units. WTs and hydroelectric power plants as the large-scale distributed generation units are directly connected to the grid, while PV systems are located in the load points for self-consumption. In [18], economic evaluation of a PV-based VPP in India has been investigated with the aim of cost minimization by considering peak load and reliability of the system. This research considers various demand profiles to specify the optimal investment to benefit both the system and customers.

Some other researches have been accomplished by researchers for optimal dispatch of VPPs. A bi-level programming has been presented in [19] for optimal scheduling of a distribution system contains multiple VPPs. The upper level minimizes the overall cost of the distribution system, while the lower level maximizes the profit of VPPs. A two-stage scheme is proposed in [20], in which the day-ahead market is scheduled in the first stage. In the second stage, real-time scheduling is investigated by deciding on the generation of dispatchable units based on the specific generation of non-dispatchable units. The uncertainty in the price and RESs generation has been considered in this research. In [21], a bi-level scheme is presented for interaction between under-competition VPPs and the distribution company by using modified game theory. Furthermore, the optimal scheduling of energy-hub-based VPPs in the presence of thermal energy market is investigated in [22], in which the risk management is accomplished to manage the risk of scheduling caused by uncertainty in wind power generation.

VPP scheduling is subjected to uncertainty in some parameters such as electricity price, load variations, and RESs generations, which may result in undesirable total costs. Risk management is an effective tool to overcome this challenge by controlling the risk of the VPP scheduling [10]. By risk management, in addition to the expected cost of the scheduling, the cost of each scenario can be managed. For scenarios that may happen, different values are obtained for total cost of a VPP. The worst-case scenarios for the total cost can be improved by risk management based on the risk level. For risk-natural (risk-seeking) decisions, risk management is ignored, i.e., the risk level is the highest, whereas for risk-averse decisions, the risk level is the lowest. Totally, risk management methods improve the objective value of the worst-case scenarios as far as possible based on the operator decision. Different measures have been presented to manage the risk of the scheduling such as CVaR measure, robust optimization, chance constraint method, information gap decision theory, etc. Among them, CVaR is a well-known reliable and applicable measure to appropriately manage the resulting risk caused by uncertain parameters of the VPP. By using this risk measure, in addition to the cost of the worst-case scenario, the cost of all scenarios and their improvement amount compared with the risk-natural decision can be observed and managed as well.

Risk management has been performed in some researches to manage the risk of uncertainties. In [23], optimal dispatch of a VPP, including WT and DG with uncertainties and responsive loads, is investigated. In this research, risk management is accomplished by adopting the information gap decision theory (IGDT) to control the risk due to uncertainties. In [24], risk-averse optimal scheduling of VPPs has been accomplished considering DRP and uncertainty in RESs. This research considers PV, WT, DG, and ESS, and a main grid in the studied VPP. In [25], the authors have proposed a stochastic framework based on the medium-term trading strategy for the integration of a VPP with its neighboring VPPs. An efficient risk management approach based on the concept of first-order stochastic dominance constraints is adopted to manage the risk of the uncertainty in market price. A two-stage optimal dispatch model for a VPP with distributed generation units, including WT and ESS, is presented in [26] for the participation of VPP in the day-ahead and the real-time markets. The particle swarm optimization algorithm is adopted to solve the resulted model. Additionally, the CVaR measure is employed to control the risk of the profit uncertainty affected by uncertainty in electricity price, load demand, and wind speed.

In this chapter, optimal scheduling and operation of VPPs are investigated. The VPP under study contains PV, WT, DG, ESS, and parking lots. Additionally, the VPP can exchange power with the main grids for benefit. The scheduling is performed with the aim of cost minimization. The uncertainty in the electricity market is taken into account. To manage the risk caused by scheduling, CVaR is employed. The problem is formulated as a mixed-integer non-linear programming (MINLP) model, which is implemented in the general algebraic model system (GAMS) environment and solved by using a suitable solver. Numerical simulation is adopted to prove the model's effectiveness and the obtained results are discussed.

The remainder of this chapter is organized as follows: Section 11.2 gives the problem formulation, in which the VPP is modeled and the cost function is formulated. Uncertainty and risk-based optimal scheduling are discussed also in this section. Section 11.3 presents the numerical simulations to validate the effectiveness of the proposed model for optimal scheduling of the VPP. Finally, the conclusions are given in Section 11.4.

11.2 Problem formulation

The studied VPP is presented in Fig. 11.1. The system load is supplied by RESs, including PVs and WTs, DGs, ESS, and EVs. The power trade with the main grids is performed based on the electricity price and risk level. DRP is also considered in this research to improve the operational cost of the system. The uncertainty in electricity price is also taken into account, in which a risk management measure, i.e., CVaR, is employed to manage the risk associated with VPP scheduling. ESSs and EVs can be charged during low-price hours via RESs, DGs, and the main grid. The stored energy in ESSs and EVs can be discharged in high-price hours to supply the demand or even to exchange power with the main grid.

11.2.1 Objective function

In this study, cost minimization is targeted for the studied VPP. The total cost includes the cost of PV, WT, ESS, EV, DGs, DRP, and also the power exchanged with the main grid. The cost of power purchased from the main grid is positive, while the cost of power sold to the main grid is negative. The objective function is written as

$$TC = \sum_{\omega} \pi^{\omega} \sum_{t} (C_{PV} + C_{Wind} + C_{DR} + C_{ESS} + C_{EV} + C_{DG} + C_{Grid}).$$

(11.1)

The variables $C_{PV}, C_{Wind}, C_{DR}, C_{ESS}, C_{EV}, C_{DG}, C_{Grid}$ are calculated as follows:

Risk-averse scheduling of virtual power plants considering electric vehicles and demand response 233

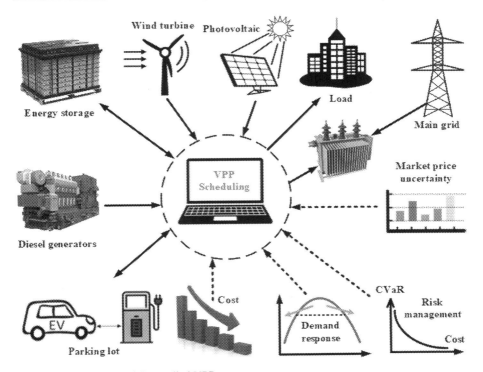

Figure 11.1 Diagram of the studied VPP.

$$C_{PV} = \sum_t \sum_v hC_V^v P_V^{v,t}, \tag{11.2}$$

$$C_{Wind} = \sum_t \sum_w hC_W^w P_W^{w,t}, \tag{11.3}$$

$$C_{DR}^{\omega,t} = \sum_t \sum_l hC_{Sh}^{\omega,t} L_{Sh}^{l,t} \quad \& \quad C_{Sh}^{\omega,t} = F_{Sh}\lambda_P^{\omega,t}, \tag{11.4}$$

$$C_{ESS} = \sum_t \sum_s hC_{S,deg}^s (P_{S,C}^{s,t} + P_{S,D}^{s,t}), \tag{11.5}$$

$$C_{EV}^{\omega} = \sum_t \sum_e h(\lambda_P^{\omega,t} P_{EV,C}^{e,t} + C_{EV,dis}^{\omega,t} P_{EV,D}^{e,t}) \quad \& \quad C_{EV,dis}^{\omega,t} = F_{EV,dis}\lambda_P^{\omega,t}, \tag{11.6}$$

$$C_{DG} = \sum_j h(\zeta^j (P_J^{j,t})^2 + \mu^j P_J^{j,t} + \tau^j)h + \sum_j C_{SU}^j \psi_U^{j,t} + \sum_j C_{SD}^j \psi_D^{j,t}, \tag{11.7}$$

$$C_{Grid} = \sum_t \sum_g hF_{LMP}^g \lambda_P^{\omega,t} P_G^{g,t}. \tag{11.8}$$

11.2.2 Scheduling constraints

- Energy storage system

ESSs are used to meet the intermittent and uncertain generation of RESs (such as PVs and WTs) by storing the surplus power generated by RESs to be used when power is needed.

The state of charge (SoC) of ESSs is obtained as follows [8]:

$$E_S^{s,t} = E_S^{s,t-1} + [(\eta_{S,C} P_{S,C}^{s,t}) - (\frac{1}{\eta_{S,D}} P_{S,D}^{s,t})]. \quad (11.9)$$

The following constraints are considered due to the depreciation of ESSs:

$$E_{S,\min}^s \leq E_S^{s,t} \leq E_{S,\max}^s, \quad (11.10)$$
$$E_{S,\min}^s = F_{S,\min}^s E_{S,Cap}^s, \quad (11.11)$$
$$E_{S,\max}^s = F_{S,\max}^s E_{S,Cap}^s. \quad (11.12)$$

The charging/discharging rates of ESS are limited as:

$$0 \leq (\eta_{S,C}^s P_{S,C}^{s,t}) \leq u_{S,C}^{s,t} P_{S,C,\max}^s, \quad (11.13)$$
$$0 \leq (\frac{1}{\eta_{S,D}} P_{S,D}^{s,t}) \leq u_{S,D}^{s,t} P_{S,D,\max}^s. \quad (11.14)$$

In each interval, the ESS can only have one state of charging or discharging:

$$u_{S,C}^{s,t} + u_{S,D}^{s,t} \leq 1. \quad (11.15)$$

- Electric vehicles

Although EVs impact the load demand and security of electricity systems, they can be used as mobile storages to improve the reliability, cost, and emission of the electricity system [27]. The EVs are charged when they are parked and can even be discharged based on the constraint of the required SoC for daily travel to benefit the electricity system by vehicle-to-grid technology. The stored energy in EVs is expressed as [28]

$$E_{EV}^{e,t} = E_{EV}^{e,t-1} + [(\eta_{EV,C}^e P_{EV,C}^{e,t}) - (\frac{1}{\eta_{EV,D}^e} P_{EV,D}^{e,t})] - L_{EV}^{e,t}. \quad (11.16)$$

The following limitations are considered to prevent the premature degradation of EVs' batteries:

$$E_{EV,\max}^e \leq E_{EV}^{e,t} \leq E_{EV,\max}^e, \quad (11.17)$$
$$E_{EV,\min}^e = F_{EV,\min}^e E_{EV,Cap}^e, \quad (11.18)$$
$$E_{EV,\max}^e = F_{EV,\max}^e E_{EV,Cap}^e. \quad (11.19)$$

The following constraints are considered for limitation of the charging/discharging rate of EVs:

$$0 \leq (\eta_{EV,C}^e P_{EV,C}^{e,t}) \leq u_{EV,C}^{e,t} P_{EV,C,\max}^e, \quad (11.20)$$

$$0 \leq (\frac{1}{\eta_{EV,D}^e} P_{EV,D}^e) \leq u_{EV,D}^{e,t} P_{EV,D}^{e,\max}. \tag{11.21}$$

Based on the following limitations, in each interval, only one state of charging, discharging, or travel is possible for EVs:

$$u_{EV,C}^{e,t} + u_{EV,D}^{e,t} + u_{EV,T}^{e,t} \leq 1, \tag{11.22}$$

where binary $u_{V,T}^{v,t}$ shows the travel state, in which EVs cannot be charged or discharged in this state. The binary variable of travel state ($u_{EV,T}^{s,t}$) is 1 when the EVs consume the energy stored in the battery and are not in the parking lot. This parameter is obtained as

$$u_{EV,T}^{e,t} = \begin{cases} 0, & D_{EV}^{e,t} = 0, \\ 1, & D_{EV}^{e,t} > 0, \end{cases} \xrightarrow{D_{EV}^{e,t} \geq 0} u_{EV,T}^{e,t} = sign(D_{EV}^{e,t}). \tag{11.23}$$

The distance that can be driven by using specific energy is

$$D_{EV}^{e,t} = ECF(L_{EV}^{e,t}). \tag{11.24}$$

Therefore, the load of EVs (required energy for daily travel) is

$$L_{EV}^{e,t}[\text{kWh}] = \frac{D_{EV}^{e,t}[\text{km}]}{ECF[\text{km/kWh}]}, \tag{11.25}$$

where ECF represents the energy–distance conversion factor of EVs.

- **Wind turbine and photovoltaic system**

The power generated by RESs (WTs and PVs) is directly affected by weather conditions, i.e., wind speed and solar radiation, respectively. In this research, the output power is limited to the nominal capacity of RES [29] since the power scenarios for RESs have been employed rather than the scenarios of wind speed and solar radiation, and hence, the scheduling in this chapter is based on the power generation.

- **Demand response program**

In this research, a DRP is implemented to evaluate its impact on the overall cost of the system. DRPs are implemented in electricity systems due to their advantages, such as cost minimization, reliability improvement, and emission reduction [27]. The time-of-use (ToU) DRP is considered in this research for shiftable loads [30]. The shiftable loads are referring to the loads that can be decreased (curtailed) from peak-load hours by an independent system operator and moved to light-load hours. In this study, it is considered that just a fraction of loads can be shifted by payment to the customers based on the optimal scheduling of VPP. This concept is a subset of price-based DRPs. The system load in the presence of DRP is

$$L_{P,DRP}^{l,t} = L_P^{l,t} + L_{Sh}^{l,t}. \tag{11.26}$$

The shifted load in each period is expressed as

$$L_{Sh}^{l,t} = \frac{F_{DRP}^{l,t}}{100} L_P^{l,t}. \qquad (11.27)$$

The shifted loads in each time interval are limited as

$$\left| F_{DRP}^{l,t} \right| \leq F_{DRP,\max}. \qquad (11.28)$$

The following constraint is considered to limit the total amount of the shifted load over a daily period to zero:

$$\sum_t L_{Sh}^{l,t} = 0. \qquad (11.29)$$

- DGs

The DG units have different constraints. Maximum and minimum generation limits are expressed as

$$u^{j,t} P_{J,\min}^j \leq P_J^{j,t} \leq u^{j,t} P_{J,\max}^j. \qquad (11.30)$$

The states related to the startup and shutdown of DGs are determined by (11.31) and (11.32). Eq. (11.31) is used to determine the change of state (startup or shutdown), while Eq. (11.32) means that the two mentioned states cannot happen simultaneously:

$$\psi_U^{j,t} - \psi_D^{j,t} = u^{j,t} - u^{j,t-1}, \qquad (11.31)$$

$$\psi_U^{j,t} + \psi_D^{j,t} \leq 1. \qquad (11.32)$$

Ramp-up and ramp-down constraints of DGs are considered as:

$$R_{UP}^j - (P_J^{j,t} - P_J^{j,t-1}) \geq 0, \qquad (11.33)$$

$$R_{DW}^j - (P_J^{j,t-1} - P_J^{j,t}) \geq 0. \qquad (11.34)$$

The following limitations are considered for minimum uptime and minimum downtime of DGs:

$$T_{on}^{j,t} = T_{on}^{j,t-1} u^{j,t} + u^{j,t}, \qquad (11.35)$$

$$(T_{on}^{j,t-1} + 1 - T_{MU}^j)(u^{j,t-1} - u^{j,t}) \geq 0, \qquad (11.36)$$

$$T_{off}^{j,t} = T_{off}^{j,t-1}(1 - u^{j,t}) + (1 - u^{j,t}), \qquad (11.37)$$

$$(T_{off}^{j,t-1} + 1 - T_{MD}^j)(u^{j,t} - u^{j,t-1}) \geq 0. \qquad (11.38)$$

- System

The power exchanged with the upstream grids is limited as

$$-P_{G,\max}^g \leq P_G^{g,t} \leq P_{G,\max}^g. \qquad (11.39)$$

The DC power flow equation can be expressed as follows:

$$\sum_j M_J^{a,j} P_J^{j,t} + \sum_s M_S^a (P_{S,D}^{s,t} - P_{S,C}^{s,t}) + \sum_e M_{EV}^e (P_{EV,D}^{e,t} - P_{EV,C}^{e,t})$$
$$+ \sum_v M_V^{a,v} P_V^{v,t} + \sum_w M_W^{a,w} P_W^{w,t} + \sum_l M_L^{a,l} P_{CL}^{l,t}$$
$$+ \sum_g M_G^{a,g} P_G^{g,t} - \sum_l M_L^{a,l} P_L^{l,t} = \sum_b B^{a,b} (\theta^{a,t} - \theta^{b,t}) \quad \forall a, t, \quad (11.40)$$

where $B_{a,b}$ is the admittance parameters, which is formulated based on the concept of admittance matrix [8] and can be calculated as follows:

$$jB^{a,b} = \begin{cases} \sum_{k=1}^{A} M_B^{a,k} \left(\dfrac{1}{jX_{Series}^{a,k}} + \dfrac{1}{2} Y_{Shunt}^{a,k} \right), & a = b, \\ -\dfrac{1}{jX_{Series}^{a,b}}, & a \neq b. \end{cases} \quad (11.41)$$

The power lines flows are obtained as

$$F_L^{a,b,t} = B^{a,b} (\theta^{a,t} - \theta^{b,t}) \quad \forall a, b, t, \quad (11.42)$$

and are limited as

$$-F_{L,max}^{a,b} \leq F_L^{a,b,t} \leq F_{L,max}^{a,b} \quad \forall a, b, t. \quad (11.43)$$

11.2.3 Uncertainty

The operation of a VPP is risky due to the uncertain power generation of VPP affected by the uncertainty in distributed generation units and market price. In this research, the uncertainty in market price in VPPs is considered [31,32].

11.2.3.1 Scenario generation

For optimizing the VPP scheduling, proper scenarios of market uncertainty are generated in the following.

The generated scenarios for the electricity price are listed in Table 11.1. This table presents the price coefficients and related probabilities. The scenarios include a wide range of prices to effectively evaluate the uncertainty in the electricity price.

11.2.3.2 Risk management

The optimal scheduling of VPPs and power trade with the main grids is affected by market price, which is usually uncertain in markets. The uncertainty in the market price impacts the operation of units and, consequently, in the scheduling of VPPs. Risk management is an efficient tool to decrease risks due to scheduling. Various risk

Table 11.1 Price coefficient and probabilities of the scenarios.

Scenario	1	2	3	4	5	6	7	8
Price coefficient	0.5902	0.7490	0.9077	1.0665	1.2252	1.3840	1.5428	1.7015
Probability	0.0224	0.1444	0.3017	0.2772	0.1599	0.0637	0.0222	0.0085

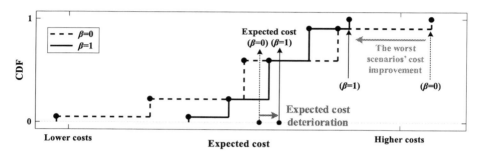

Figure 11.2 Concept of risk management by CVaR.

measures have been proposed for risk assessment in an electricity market [33]. CVaR is the most well-known measure which is widely adopted by researchers to improve the objective of a specific percentage of the worst-case scenarios according to the probability of scenarios (α is a parameter, $\alpha \in (0, 1)$) without adding discrete variables to the problem. The β parameter ($\beta \in (0, 1)$) is the risk level such that $\beta = 0$ and $\beta = 1$ indicate the risk-neutral and risk-averse decisions, respectively. A conceptual diagram of risk management based on the cumulative distribution function (CDF) is depicted in Fig. 11.2. As the figure shows, by increasing the β value, the objective values of the worst scenarios are improved (marked by the green arrow; light gray in print version). Although this achievement increases the VPP expected cost (marked by the red arrow; dark gray in print version), it is beneficial since the worst-case scenarios are improved. Generally, risk management makes the cost of the scenarios closer to each other in order to remove the low expected costs. The new objective function is defined as (11.44) [34]. For $\beta = 0$, the objective (11.44) is identical to the objective (11.1), which is the risk-seeking decision, while for $\beta = 1$ the objective (11.44) is equal to CVaR, which is the risk-averse decision. Selecting the risk level depends on the operator's decisions. However, a compromise between CVaR and the expected cost of the VPP can be performed by the decision-maker:

$$Min.(1-\beta)Cost + (\beta)CVaR. \qquad (11.44)$$

The CVaR is obtained as

$$CVaR = \eta + \frac{1}{1-\alpha}\sum_{\omega}\pi^{\omega}\delta^{\omega}. \qquad (11.45)$$

The variables η and δ depend on each other as

$$\left\{ \sum_{t=1}^{T} \left[\left\{ \begin{array}{l} \sum_w C_W^w P_W^{w,t} + \sum_v C_V^v P_V^{v,t} + \sum_l C_{CL} P_{CL}^{l,t} \\ + \sum_g F_{LMP}^g \lambda_P^{\omega,t} P_G^t + \sum_j \left(\zeta^j \left(P_J^{j,t} \right)^2 + \mu^j P_J^{j,t} + \tau^j \right) \\ \sum_s (P_{S,D}^{s,t} - P_{S,C}^{s,t}) + \sum_e (P_{EV,D}^{e,t} - P_{EV,C}^{e,t}) \\ + \sum_j C_{SU}^j \cdot \psi_U^{j,t} + \sum_j C_{SD}^j \cdot \psi_D^{j,t} \end{array} \right\} h \right] \right\}$$

$$-\eta \leq \delta^\omega, \qquad (11.46)$$

where the real positive variable δ represents the distance of the undesirable objective values of scenarios from η. Additionally, η is a CVaR variable and takes value among scenarios' costs.

11.3 Case study

To demonstrate the performance of the proposed model, both a 6-bus and an 18-bus system are considered. The problem was formulated as the MINLP model [35]. The numerical simulation is carried out in GAMS software [36] (version 27.3) on a system with an Intel Core i5 (Quad-Core 5th generation) processor with 2.50 GHz and 6 GB of RAM. The SCIP solver [37] is employed in GAMS to solve the optimization problem.

11.3.1 Case I: 6-bus system

Firstly, the 6-bus system is adopted. The configuration of this system is illustrated in Fig. 11.3. This system includes 4 loads. The share of the total load at each hour is presented in Fig. 11.4. The hourly profile of the load demand and the electricity price are illustrated in Figs. 11.5(a) and 11.5(b), respectively.

The data of DG, ESS, and EVs are presented in Table 11.2. In the studied VPP, 300 EVs are considered to be managed in the parking lot for optimal scheduling of charging/discharging.

Fig. 11.6 depicts the per-unit generation of PV [38] and WT [39]. Their capacity is 5 MW.

The distance driven by EVs is depicted in Fig. 11.7. The energy–distance conversion factor of EVs (*ECF*) is considered as 6 km/kWh. As mentioned in the previous section, the EVs cannot charge or discharge in the intervals that the EVs consume electricity because they are out of parking for daily travels.

Table 11.3 lists the obtained results for the optimal scheduling of the VPP. As shown in this table, the DG, RESs including PV and WT, static and mobile energy

Table 11.2 Characteristics of the studied VPP's elements.

DG	ζ	μ	τ	$P^j_{J,\min}$	$P^j_{J,\max}$	R^j_{UP}	R^j_{DW}	MUT^j	MDT^j	C^j_{SU}	C^j_{SD}
	1.561	37.963	118.909	2	7	2.5	2.5	2	2	67	55
ESS	$E^s_{S,Cap}$	$F^s_{S,\min}$	$F^s_{S,\max}$		$P^s_{S,C,\max}$	$P^s_{S,D,\max}$	$\eta^s_{S,C}$	$\eta^s_{S,D}$		$E^{s,0}_S$
	20	0.05	0.095		5	5	0.95	0.95		2
EV	$E^e_{EV,Cap}$	$F^e_{EV,\min}$	$F^e_{EV,\max}$		$P^e_{EV,C,\max}$	$P^e_{EV,D,\max}$	$\eta^e_{EV,C}$	$\eta^e_{EV,D}$		$E^{e,0}_{EV}$
	25	0.1	0.9		10	10	0.95	0.95		5

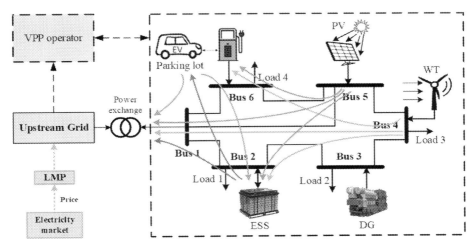

Figure 11.3 Configuration of the 6-bus system (the studied VPP for Case I).

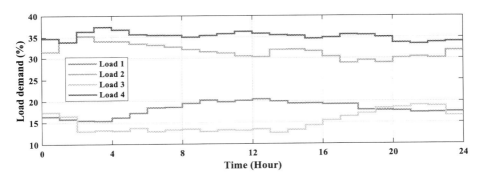

Figure 11.4 The share of the total load of the system.

storage including ESS and EVs, and the main grid are taken into consideration for economic load supply. Most of the time, the power is purchased from the main grid, while in certain hours (14, 15, and 16), the power is sold to the main grid (due to its higher price) for cost improvement. This table also shows that the ESS and EVs are charged in low-price hours and discharged in high-price hours.

Fig. 11.8 illustrates the SoC of ESS and the charging/discharging power of ESS. As expected, the ESS is charged during the low-demand hours (due to the lower price of electricity) and discharged during the high-demand hours (due to the higher price of electricity). It can also be observed that the SoC of ESS has remained in the predetermined range of SoC (5% and 95%, i.e., 1 MWh and 19 MWh).

The SoC of EVs and the charging/discharging power of the parking lot is illustrated in Fig. 11.9. The EVs are charged during the hours with a low price of electricity and discharged during the hours with a high price of electricity. This figure also shows that the SoC of EVs has remained in the predetermined range of SoC (10% and 90%, i.e.,

Table 11.3 Optimal generation scheduling for the considered period for risk-neutral decision (i.e., $\beta = 0$).

Time (Week)	Power-only unit (MW)	PV (MW)	WT (MW)	ESS Charge (MW)	ESS Discharge (MW)	Parking lot Charge (MW)	Parking lot Discharge (MW)	Exchanged power (MW)
1	2.500	–	4.017	–	–	–	–	2.285
2	2.471	–	3.850	2.895	–	–	–	4.655
3	2.000	–	3.959	5.000	–	2.526	–	8.600
4	2.000	–	4.545	5.000	–	3.000	–	8.237
5	2.109	–	4.582	5.000	–	–	–	6.015
6	2.861	–	4.448	–	–	–	–	1.663
7	4.761	–	3.278	–	–	–	–	2.542
8	4.964	–	2.927	–	–	–	–	3.968
9	5.779	0.763	2.913	–	–	–	–	3.685
10	6.845	1.514	2.542	–	–	–	–	2.941
11	5.971	2.099	2.371	–	–	–	–	3.685
12	6.617	2.405	1.777	–	–	–	–	3.544
13	7.000	4.796	2.640	–	5.000	–	0.658	-5.824
14	7.000	4.224	2.755	–	2.100	–	–	-1.079
15	7.000	1.578	2.808	–	5.000	–	3.000	-4.680
16	7.000	0.865	3.079	–	5.000	–	–	-1.311
17	6.686	0.115	2.806	–	–	–	–	4.592
18	5.591	–	2.378	–	–	–	–	5.519
19	6.115	–	3.112	–	–	–	–	3.914
20	4.161	–	3.575	–	–	–	–	4.058
21	4.893	–	4.172	–	–	–	–	2.468
22	4.808	–	4.247	–	–	–	–	2.285
23	4.583	–	4.462	–	–	–	–	1.849
24	3.864	–	4.404	–	–	–	–	1.403

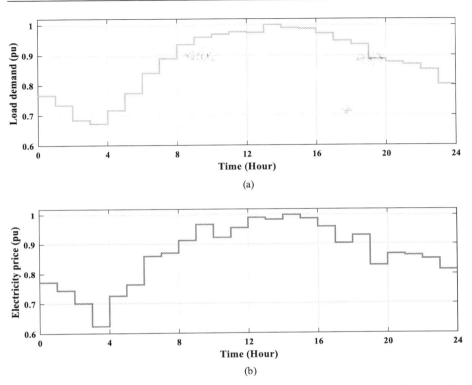

Figure 11.5 Hourly demand and electricity price in Case I: (a) system demand, (b) electricity price.

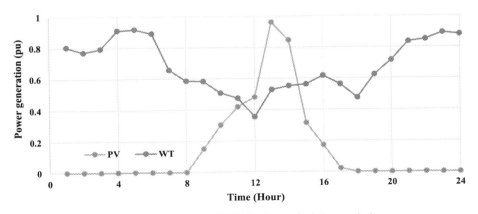

Figure 11.6 Per-unit generation of PV and WT for the studied time period.

0.75 MWh and 6.75 MWh). For each EV, the SoC range is 2.5 kWh and 22.5 kWh, respectively.

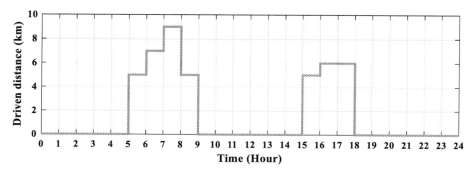

Figure 11.7 Driving pattern of EVs.

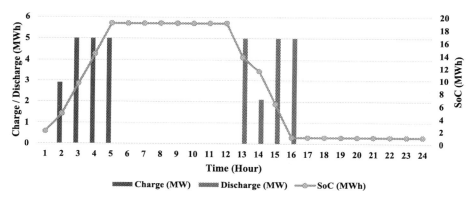

Figure 11.8 The SoC and charging/discharging values of ESS.

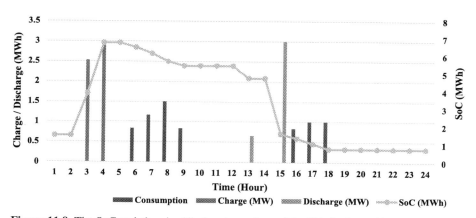

Figure 11.9 The SoC and charging/discharging values of the EVs in the parking lot.

The expected cost for the case of ignoring risk management is $11533.936. Additionally, the cost of the considered scenarios is listed in Table 11.4.

Table 11.4 The scenarios' cost for risk-neutral decision (i.e., $\beta = 0$).

Scenario	1	2	3	4	5	6	7	8
Cost	10206.321	10692.714	11178.801	11665.194	12151.281	12637.675	13124.068	13610.155

Table 11.5 Expected cost and cost of the worst-case scenario respect to risk level β.

Risk management level	Expected cost ($)	Cost of the worst-case scenario (Scenario 8) ($)	Improvement ($)
$\beta = 0$	11533.936	13610.155	–
$\beta = 0.2$	11545.827	13237.406	372,749
$\beta = 0.4$	11574.334	12964.789	645,366
$\beta = 0.6$	11612.609	12764.493	845,662
$\beta = 0.8$	11643.482	12657.944	952,211
$\beta = 1$	11667.531	12598.326	1,011,829

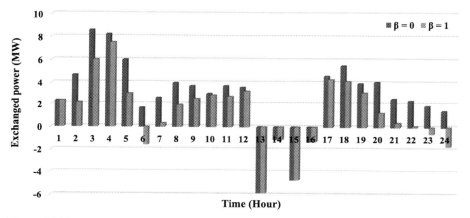

Figure 11.10 The power exchanged with the main grid for both risk-seeking and risk-averse scheduling.

In the following, the CVaR risk measure is employed to control the risks due to scheduling. Table 11.5 lists the obtained results for different risk levels. The expected cost, cost of the worst-case scenario (for this problem, scenario 8), and the improvement value in the cost of the worst-case scenario are presented in Table 11.5. As shown in this table, although the expected cost is increased due to risk management (increasing the β level from 0 to 1), the cost of the worst-case scenario is improved from $13610.155 to $12598.326, which shows an improvement of $1011.829.

The power exchanged with the main grid for both risk-seeking ($\beta = 0$) and risk-averse ($\beta = 1$) decisions is depicted in Fig. 11.10. As seen, the optimal power exchanged with the main grid is influenced by risk management. For the risk-averse decision ($\beta = 1$), to avoid the high prices of electricity and the resulting high costs, the power purchased from the main grid is decreased.

In the following, ToU DRP is implemented for the considered time period in the studied VPP. Table 11.6 shows the improvement trend of the expected cost and cost of the worst-case scenario with respect to the participation percentage in DRP. As

Table 11.6 Expected cost and cost of the worst-case scenario with respect to participation percentage in DRP for $\beta = 1$.

Participation in DRP	Expected cost ($)	Cost of the worst-case scenario ($)	Improvement of the worst-case scenario ($)
0%	11667.531	12598.326	–
5%	11604.368	12493.341	104.985
10%	11541.205	12388.353	209.973
15%	11478.043	12283.364	314.962
20%	11414.880	12178.375	419.951
25%	11351.717	12073.387	524.939
30%	11288.555	11968.398	629.928

this table shows, both mentioned costs are decreased by increasing the participation percentage.

11.3.2 Case II: 18-bus system

For further evaluation of the performance of the proposed scheme for optimal scheduling of VPPs, an 18-bus system is adopted, which is the 33 kV section of the well-known 30-bus system [11]. This system is modified by adding 2 PV systems, 2 WTs, 2 ESSs, 2 parking lots, and 5 DGs. This system is connected to 3 upstream grids with different LMPs. The LMPs are 0.95, 1, and 1.05, respectively, for the upstream grids 1, 2, and 3 located in buses 1, 11, and 16, respectively. (See Fig. 11.11.)

Table 11.7 presents loads of different buses for peak hour. The buses 11, 14, and 16 have no load demand and are connected to the upstream grids.

The hourly demand and electricity price are presented in Figs. 11.12(a) and 11.12(b), respectively.

This system has two parking lots. The driven patterns of the EVs in parking lots 1 and 2 are illustrated in Fig. 11.13.

The optimal charging/discharging values and SoC of ESS 1 and ESS 2 are illustrated in Figs. 11.14(a) and 11.14(b), respectively. As these figures show, the ESSs are charged in low-price hours and discharged in high-price hours. Moreover, the SoC of ESSs is retained in the allowed range to prevent the degradation of ESSs.

Fig. 11.15 presents the optimal SoC of EVs in parking lot 1 (with 300 EVs) and parking lot 2 (with 400 EVs) with related charging/discharging values and energy consumed by EVs. Figs. 11.15(a) and 11.15(b) show the obtained results for EVs in parking lot 1 and parking lot 2, respectively.

The expected cost for the case of ignoring risk management is $32737.332. Additionally, the costs of the considered scenarios are listed in Table 11.8.

In the following, the CVaR risk measure is adopted to control the risk due to scheduling. Table 11.9 lists the obtained results for risk management. The expected cost, cost of the worst-case scenario (for this problem, scenario 8), and the improvement values in the cost of the worst-case scenario are listed in Table 11.9 for different

Table 11.7 The load of the buses for the peak hour (MW).

Bus	1	2	3	4	5	6	7	8	9	10	11	12	13	14	15	16	17	18
Peak load	1.385	2.674	1.480	1.958	0.836	2.149	0.764	2.268	0.525	4.179	0.0	0.764	2.077	0.0	0.836	0.0	0.573	2.531

Table 11.8 The scenarios' cost for risk-neutral decision (i.e., $\beta = 0$).

Scenario	1	2	3	4	5	6	7	8
Cost ($)	22874.244	25183.404	27494.016	29804.628	32113.788	34424.406	36733.56	39044.178

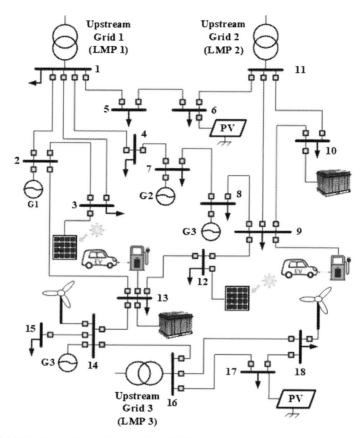

Figure 11.11 Configuration of the modified 18-bus system (the studied VPP for Case II).

risk levels. As can be seen in this table, although the expected cost is increased by risk management (increasing the β level from 0 to 1), the cost of the worst-case scenario is improved from $39044.178 to $33138.258, which shows an improvement of $5905.920.

The ToU DRP is also implemented to evaluate its impact on the cost of VPP. Table 11.10 shows the improvement trend of the expected cost and the worst-case scenario with respect to the participation percentage in DRP. As this table shows, both mentioned costs are decreased by increasing the participation percentage.

11.4 Conclusions

In this chapter, optimal scheduling and operation of the VPP were studied. The studied VPP contains WT, PV, DG, ESS, and EVs. Additionally, a DRP was considered to

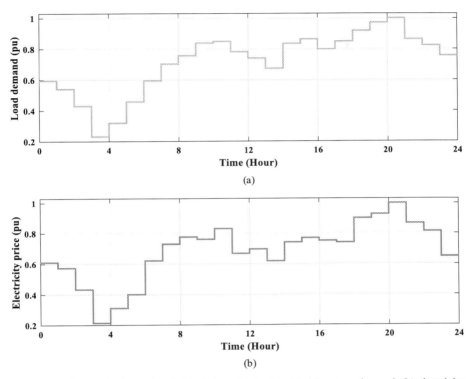

Figure 11.12 Hourly demand and electricity price in Case II: (a) system demand, (b) electricity price.

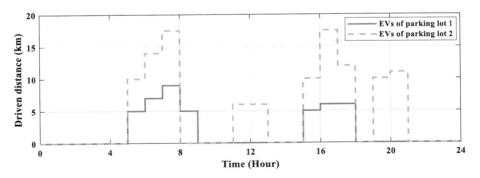

Figure 11.13 Driving patterns of EVs in parking lots 1 and 2.

evaluate its impact on the overall cost. The scheduling was performed to minimize the total cost. Appropriate scenarios were considered to model the uncertainty in the electricity price. The CVaR measure was employed to manage the risk of the high operational costs caused by the uncertainty in the market price. The scheduling problem was

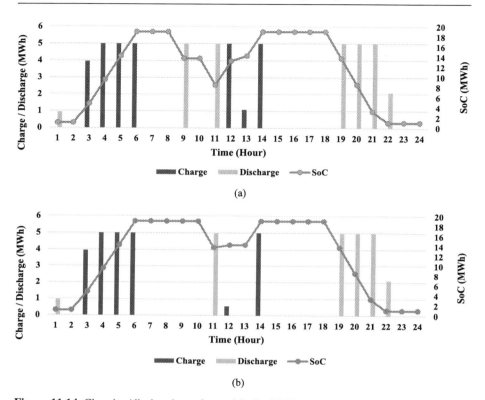

Figure 11.14 Charging/discharging value and SoC of ESSs: (a) ESS 1, (b) ESS 2.

Table 11.9 Short-term generation scheduling for the considered day for different risk levels.

Risk management level	Expected cost ($)	Cost of the worst-case scenario (Scenario 8) ($)	Improvement ($)
$\beta = 0$	32737.332	39044.178	–
$\beta = 0.2$	32763.228	37877.730	1166.448
$\beta = 0.4$	32814.816	36548.250	2495.928
$\beta = 0.6$	32931.942	34837.356	4206.822
$\beta = 0.8$	33132.390	34000.764	5043.414
$\beta = 1$	33138.258	33138.258	5905.920

formulated as an MINLP. The optimization problem was modeled in GAMS software and solved using a suitable solver. Numerical simulation was accomplished to confirm the effectiveness of the proposed scheme and the obtained results were discussed. For evaluating the model, two case studies, including a 6-bus system and the IEEE 18-bus system, were tested. The effectiveness of risk management for reducing the cost of the

Risk-averse scheduling of virtual power plants considering electric vehicles and demand response   253

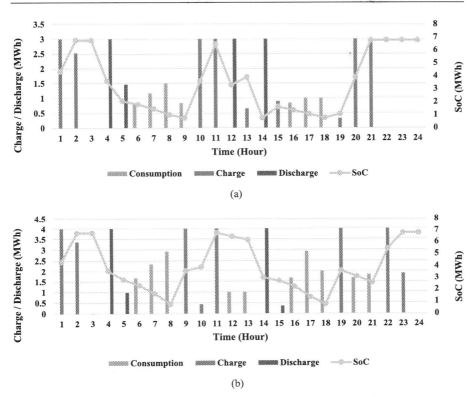

Figure 11.15 Charging/discharging value and SoC of all EVs: (a) parking lot 1 with 300 EVs, (b) parking lot 2 with 400 EVs.

Table 11.10 Improvement trend of the expected cost and cost of the worst-case scenario respect to participation percent in DRP.

Participation in DRP	Expected cost ($)	Improvement of the worst-case scenario ($)
0%	33138.258	–
5%	32474.736	663.522
10%	31678.512	1459.746
15%	31113.684	1997.190
20%	30451.680	2431.602
25%	29725.152	2903.208
30%	29125.080	3301.908

worst-case scenarios was proved for the systems under study. Furthermore, DRP was considered in this research, in which reduced the expected operational cost.

References

[1] K. Dielmann, A. Van Der Velden, Virtual power plants (VPP) – a new perspective for energy generation?, in: Proceedings of the 9th International Scientific and Practical Conference of Students, Post-graduates and Young Scientists – Modern Techniques and Technologies, MTT' 2003, 2003, pp. 18–20.
[2] D. Pudjianto, D. Pudjianto, C. Ramsay, C. Ramsay, G. Strbac, G. Strbac, Virtual power plant and system integration of distributed energy resources, IET Renew. Power Gener. 1 (1) (2007) 10–16.
[3] S.M. Nosratabadi, R.A. Hooshmand, E. Gholipour, A comprehensive review on microgrid and virtual power plant concepts employed for distributed energy resources scheduling in power systems, Renew. Sustain. Energy Rev. 67 (2017) 341–363.
[4] E. Mashhour, S.M. Moghaddas-Tafreshi, A review on operation of micro grids and virtual power plants in the power markets, in: ICAST 2009 – 2nd International Conference on Adaptive Science and Technology, 2009, pp. 273–277.
[5] L. Yavuz, A. Önen, S.M. Muyeen, I. Kamwa, Transformation of microgrid to virtual power plant – a comprehensive review, IET Gener. Transm. Distrib. 13 (11) (2019) 2077–2087.
[6] S. Ghavidel, L. Li, J. Aghaei, T. Yu, J. Zhu, A review on the virtual power plant: components and operation systems, in: 2016 IEEE International Conference on Power System Technology, POWERCON 2016, 2016, pp. 1–6.
[7] S.M. Muyeen, M. Benbouzid, H.M. Hasanien, T. Fernando, Q. Guo, E. Muljadi, Guest editorial: emerging technologies for virtual power plant and microgrid, IET Gener. Transm. Distrib. 13 (11) (2019) 1989–1993.
[8] O. Sadeghian, A. Oshnoei, R. Khezri, S.M. Muyeen, Risk-constrained stochastic optimal allocation of energy storage system in virtual power plants, J. Energy Storage 31 (January 2020) 101732.
[9] S. Seven, G. Yao, A. Soran, A. Onen, S.M. Muyeen, Peer-to-peer energy trading in virtual power plant based on blockchain smart contracts, IEEE Access 8 (2020) 175713–175726.
[10] O. Sadeghian, A.M. Shotorbani, B. Mohammadi-Ivatloo, Generation maintenance scheduling in virtual power plants, IET Gener. Transm. Distrib. 13 (12) (2019) 2584–2596.
[11] M. Peik-Herfeh, H. Seifi, M.K. Sheikh-El-Eslami, Decision making of a virtual power plant under uncertainties for bidding in a day-ahead market using point estimate method, Int. J. Electr. Power Energy Syst. 44 (1) (2013) 88–98.
[12] H. Pandžić, I. Kuzle, T. Capuder, Virtual power plant mid-term dispatch optimization, Appl. Energy 101 (2013) 134–141.
[13] M. Shabanzadeh, M.K. Sheikh-El-Eslami, M.R. Haghifam, The design of a risk-hedging tool for virtual power plants via robust optimization approach, Appl. Energy 155 (2015) 766–777.
[14] Ö.P. Akkaş, E. Çam, Optimal operational scheduling of a virtual power plant participating in day-ahead market with consideration of emission and battery degradation cost, Int. Trans. Electr. Energy Syst. 30 (7) (2020) 1–20.
[15] F. Luo, Z.Y. Dong, K. Meng, J. Qiu, J. Yang, K.P. Wong, Short-term operational planning framework for virtual power plants with high renewable penetrations, IET Renew. Power Gener. 10 (5) (2016) 623–633.
[16] B. Behi, A. Baniasadi, A. Arefi, A. Gorjy, P. Jennings, A. Pivrikas, Cost-benefit analysis of a virtual power plant including solar PV, flow battery, heat pump, and demand management: a Western Australian case study, Energies 13 (10) (2020).
[17] N. Naval, R. Sánchez, J.M. Yusta, A virtual power plant optimal dispatch model with large and small-scale distributed renewable generation, Renew. Energy 151 (2020) 57–69.

[18] H. Sharma, S. Mishra, Techno-economic analysis of solar grid-based virtual power plant in Indian power sector: a case study, Int. Trans. Electr. Energy Syst. 30 (1) (2020).
[19] Z. Yi, Y. Xu, J. Zhou, W. Wu, H. Sun, Bi-level programming for optimal operation of an active distribution network with multiple virtual power plants, IEEE Trans. Sustain. Energy 11 (4) (2020) 2855–2869.
[20] M. Emarati, F. Keynia, M. Rashidinejad, A two-stage stochastic programming framework for risk-based day-ahead operation of a virtual power plant, Int. Trans. Electr. Energy Syst. 30 (3) (2020) 1–13.
[21] A. Zangeneh, A. Shayegan-Rad, F. Nazari, Multi-leader–follower game theory for modelling interaction between virtual power plants and distribution company, IET Gener. Transm. Distrib. 12 (21) (2018) 5747–5752.
[22] M. Jadidbonab, B. Mohammadi-Ivatloo, M. Marzband, P. Siano, Short-term self-scheduling of virtual energy hub plant within thermal energy market, IEEE Trans. Ind. Electron. 68 (4) (2020) 3124–3136.
[23] A. Shayegan Rad, A. Badri, A. Zangeneh, M. Kaltschmitt, Risk-based optimal energy management of virtual power plant with uncertainties considering responsive loads, Int. J. Energy Res. 43 (6) (2019) 2135–2150.
[24] M. Vahedipour-Dahraie, H. Rashidizadeh-Kermani, A. Anvari-Moghaddam, P. Siano, Risk-averse probabilistic framework for scheduling of virtual power plants considering demand response and uncertainties, Int. J. Electr. Power Energy Syst. 121 (February 2020) 106126.
[25] M. Shabanzadeh, M.K. Sheikh-El-Eslami, M.R. Haghifam, Risk-based medium-term trading strategy for a virtual power plant with first-order stochastic dominance constraints, IET Gener. Transm. Distrib. 11 (2) (2017) 520–529.
[26] J. Qiu, K. Meng, Y. Zheng, Z.Y. Dong, Optimal scheduling of distributed energy resources as a virtual power plant in a transactive energy framework, IET Gener. Transm. Distrib. 11 (13) (2017) 3417–3427.
[27] O. Sadeghian, M. Nazari-Heris, M. Abapour, S.S. Taheri, K. Zare, Improving reliability of distribution networks using plug-in electric vehicles and demand response, J. Mod. Power Syst. Clean Energy 7 (5) (2019) 1189–1199.
[28] P. Aliasghari, B. Mohammadi-Ivatloo, M. Alipour, M. Abapour, K. Zare, Optimal scheduling of plug-in electric vehicles and renewable micro-grid in energy and reserve markets considering demand response program, J. Clean. Prod. 186 (2018) 293–303.
[29] M.A. Mirzaei, A. Sadeghi Yazdankhah, B. Mohammadi-Ivatloo, Stochastic security-constrained operation of wind and hydrogen energy storage systems integrated with price-based demand response, Int. J. Hydrog. Energy 44 (27) (2019) 14217–14227.
[30] O. Sadeghian, A. Moradzadeh, B. Mohammadi-Ivatloo, M. Abapour, F.P.G. Marquez, Generation units maintenance in combined heat and power integrated systems using the mixed integer quadratic programming approach, Energies 13 (11) (2020) 2840.
[31] E. Mashhour, S.M. Moghaddas-Tafreshi, Bidding strategy of virtual power plant for participating in energy and spinning reserve markets—part II: numerical analysis, IEEE Trans. Power Syst. 26 (2) (2011) 957–964.
[32] M. Shabanzadeh, M.K. Sheikh-El-Eslami, M.R. Haghifam, An interactive cooperation model for neighboring virtual power plants, Appl. Energy 200 (2017) 273–289.
[33] A.J. Conejo, M. Carrión, J.M. Morales, Decision Making Under Uncertainty in Electricity Markets, Springer, New York, NY, USA, 2010.
[34] O. Sadeghian, A. Mohammadpour Shotorbani, B. Mohammadi-Ivatloo, Risk-based stochastic short-term maintenance scheduling of GenCos in an oligopolistic electricity market considering the long-term plan, Electr. Power Syst. Res. 175 (2019) 105908.

[35] J.J. Cochran, et al., MINLP solver software, in: Wiley Encyclopedia of Operations Research and Management Science, 2011.
[36] R.E. Rosenthal, A GAMS tutorial, in: GAMS – A User's Guide, 2007, pp. 5–26.
[37] T. Achterberg, SCIP: solving constraint integer programs, Math. Program. Comput. 1 (1) (2009) 1–41.
[38] T. Niknam, F. Golestaneh, A. Malekpour, Probabilistic energy and operation management of a microgrid containing wind/photovoltaic/fuel cell generation and energy storage devices based on point estimate method and self-adaptive gravitational search algorithm, Energy 43 (1) (2012) 427–437.
[39] A. Alahyari, M. Ehsan, M. Moghimi, Managing distributed energy resources (DERs) through virtual power plant technology (VPP): a stochastic information-gap decision theory (IGDT) approach, Iran. J. Sci. Technol. - Trans. Electr. Eng. 44 (1) (2020) 279–291.

Optimal operation strategy of virtual power plant considering EVs and ESSs

Milad Kabirifar[a], Niloofar Pourghaderi[a], and Moein Moeini-Aghtaie[b]
[a]Sharif University of Technology, Tehran, Iran, [b]Faculty of Energy Engineering Department, Sharif University of Technology, Tehran, Iran

12.1 Introduction

Diverse distributed energy resources (DERs), including conventional distributed generators (DGs), renewable energy sources (RESs), energy storage systems (ESSs), and demand response (DR) resources, have increasing penetration in networks because of their advantages [1]. In addition, due to environmental reasons, the penetration level of electric vehicles (EVs) in distribution systems has increased significantly in recent years [2]. EVs (especially EVs with vehicle-to-grid (V2G) capability [3,4]) are similar to ESSs because of the possibility of scheduling their charging/discharging cycles. The large penetration of EVs may bring challenges to networks because of increasing charging loads and random behavior of their owners [5–7]. Hence, coordination is required for integrating large scale of EVs in the network [2,8–10]. Moreover, diverse DERs independent operation may have a negative influence on the network operation. On the other hand, individual participation of DERs in electricity markets is not practical because of the large number of these resources and their small capacities [11].

Thus, it is required to establish a coordination framework in order to effectively operate the diverse DERs and exploit their advantages through aggregation. The concept of a virtual power plant (VPP) can address the need for the effective integration of DERs in the electricity grid regarding both technical and economic aspects [12]. Based on the European project FENIX [12], "VPP aggregates the capacity of many diverse DERs; it creates a single operating profile from a composite of the parameters characterizing each DER and can incorporate the impact of the network on aggregated DER output." There are two types of VPP, namely commercial virtual power plant (CVPP) and technical virtual power plant (TVPP) [12]. The CVPP considers diverse DERs around the network as some central resources and the actual location of DERs is not taken into account [13–15]. On the other hand, TVPP considers actual location of DERs in the network, as well as network operational constraints [8,16,17].

It is worth noting that renewable energy sources are becoming a crucial part of power systems. These resources have a growing trend in power systems and are counted as one of the main DERs in electricity distribution networks [18]. It is because of the investment policies of countries on green and sustainable energies for the reason of pollutant reduction policies, etc. [19]. The intermittency, variability, and uncertainty of these resources bring new challenges to power system operators [1]. VPP can efficiently improve the role of RESs in power systems by aggregating them, as well as other DERs, including ESSs and EVs [2,8,20–22]. In recent years, ESSs have been increasingly utilized in different sectors of power systems and they are experiencing decreasing investment costs over the last years [23–25]. ESSs can provide numerous advantages in different sectors and they can significantly accommodate the uncertain RESs into the network [20–22,26,27].

Based on the above explanations, VPP can alleviate the uncertain and intermittent nature of RESs and improve their operation by aggregating and coordinating diverse DERs including ESSs and EVs [2,8,17,20]. Furthermore, VPP is able to schedule its resources in order to make its optimal bidding strategy for participating in electricity markets, making profit and improving its operation [2,8,13,20]. In addition, VPP can offer energy blocks and provide ancillary services, and it may have bilateral contracts with industrial, commercial, and residential customers [28,29]. The general structure of VPP is depicted in Fig. 12.1. As it is shown, VPP aggregates diverse DERs including EVs and ESSs to fulfill its objectives through improving its operation and participating in electricity markets, different aspects of which are addressed in the literature [2,4,8,10,20–22,30–32].

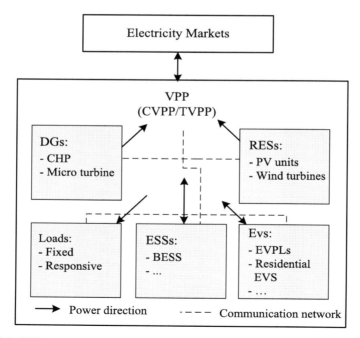

Figure 12.1 VPP structure.

In this chapter, at first an EVs' model, including EVs' operation model, V2G operation, EV parking lots' (EVPLs) model, and associated uncertainties of EVs, is discussed. Then, the ESSs types, associated operation model, and their utilization in the presence of other DERs are considered. After that, the VPP model for integrating EVs and ESSs is proposed based on two centralized and distributed schemes, and finally two examples are evaluated.

12.2 Modeling of EVs

Environmental issues bring incentive for increasing penetration of EVs [2]. The increasing penetration of EVs may bring new challenges; in order to effectively operate the grid in the presence of EVs, the coordination of EVs is required [5–7]. In this regard, in this section the EVs' operation model, V2G operation, parking lots, and associated uncertainties with EVs' operations are discussed.

12.2.1 EVs' operation model

The increasing penetration of EVs may cause reliability issues, congestion, and loss increase. Hence, due to increasing use of EVs in electricity distribution networks, a coordination framework is required. VPP can aggregate EVs, as well as other DERs, in order to take advantage of them and improve its portfolio. EVs can be regarded as mobile ESSs in case of their charge/discharge control. Like ESSs, EVs have potential to improve the operation of VPP. In addition, the aggregation of EVs by VPP can improve the integration ability of renewable energy and alleviate their associated uncertainties [26]. In order to coordinate EVs' operation, VPP should consider the operation model of EVs [17,32]. In this section the charging model of EVs is presented. It should be noted that EVs are not only utilized as the traditional loads and consume the specific amount of energy but also can be exploited as the controllable loads to effectively improve the operation of network and to fulfill VPP's objectives. In addition, exploiting discharging power of EVs can help increase VPP's profit and improve its operation. Utilizing discharge power needs V2G capability which is discussed in Section 12.2.2.

For modeling EVs, some constrains should be satisfied. Some technical limits on the number of charging and discharging cycles of batteries are dictated to customers from battery producers. In this regard the battery's state of health (SOH) should be regarded (see Section 12.3.2). In addition, the EVs owners' welfare constraints and their behavior in using the EVs should be modeled. In this regard the arrival/departure times of EVs to/from charging stations are important factors. It is also known as EVs' access. It should be noted that charging stations can be plugs at homes, commercial buildings, parking lots, and so on.

During the trip interval, when EV is not in the charging mode, the battery cannot be charged. On the other hand, EVs can be charged between arrival and departure times. Hence the set of conditions (12.1) and (12.2) should be met. In addition, the charging limit of EV in the charging mode is adhered by (12.3):

$$E_{e,t}^{Ch\_EV} = 0, \quad \forall t \in T - \left[a^{EV}, d^{EV}\right], \quad (12.1)$$

$$x_{e,t}^{Ch\_EV} \leq 1, \quad \forall t \in \left[a^{EV}, d^{EV}\right], \quad (12.2)$$

$$0 \leq E_{e,t}^{Ch\_EV} \leq \overline{E}_e^{Ch\_EV}, \quad \forall t \in \left[a^{EV}, d^{EV}\right]. \quad (12.3)$$

The EV state of charge (SOC) before departure and after arrival is obtained through (12.4) and (12.5), respectively [33,34]; $E_e^{trip\_EV}$ is the consumed energy during a trip and depends on the driven distance of an EV during its trip. The probabilistic representation of EV's driven distance is presented in Section 12.2.4. We have

$$SOC_{e,t}^{EV} = E_e^{0\_EV} + \sum_{\tau=1}^{t-1} \eta_e^{Ch\_EV} E_{e,\tau}^{Ch\_EV} x_{e,\tau}^{Ch\_EV}, \quad \forall t \leq d^{EV}, \quad (12.4)$$

$$SOC_{e,t}^{EV} = E_e^{0\_EV} + \sum_{\tau=1}^{t-1} \eta_e^{Ch\_EV} E_{e,\tau}^{Ch\_EV} x_{e,\tau}^{Ch\_EV} - E_e^{trip\_EV}, \quad \forall t \geq a^{EV}. \quad (12.5)$$

The SOC should be positive at each time step. In addition, the SOC should be lower than EV's battery capacity. These conditions are respectively addressed in (12.6) and (12.7):

$$SOC_{e,t}^{EV} \geq 0, \quad \forall t \in T, \quad (12.6)$$

$$SOC_{e,t}^{EV} \leq C_e^{EV}, \quad \forall t \in T. \quad (12.7)$$

Before leaving charging station, the charge level of EV should be sufficient for the next trip as follows:

$$E_e^{0\_EV} + \sum_{\tau=1}^{d^{EV}-1} \eta_e^{Ch\_EV} E_{e,\tau}^{Ch\_EV} x_{e,\tau}^{Ch\_EV} \geq E_e^{trip\_EV}. \quad (12.8)$$

In some cases, customers are willing to charge their EVs up to the maximum level. In this condition, the following constraint should be regarded:

$$E_e^{0\_EV} + \sum_{\tau=1}^{d^{EV}-1} \eta_e^{Ch\_EV} E_{e,\tau}^{Ch\_EV} x_{e,\tau}^{Ch\_EV} = C_e^{EV}. \quad (12.9)$$

As noted before, the access to EVs is of importance. Several studies use traffic surveys like national household travel survey (NHTS) to build access patterns of EVs [8]. The behavior of EVs owners has probabilistic nature and it is required to analyze uncertainties arising from EVs owners' behavior (see Section 12.2.4).

12.2.2 EVs' vehicle-to-grid operation

The EVs have the ability to inject their discharge power into the network like energy storage systems. The EVs connection to the grid and ability to inject their discharging power to the grid is called V2G. This ability offers an opportunity to the grid to utilize the EVs like a mobile ESS and use the discharge power when it is required. The utilization of V2G ability of EVs requires necessary infrastructure. The required features include:

- Bidirectional power interface and regulations for buying discharge energy;
- EMS for controlling and measuring;
- Communication protocol in order to transact control signals [35].

Based on [3], three concepts associated with the EVs which are connected to the grid are vehicle-to-home (V2H), vehicle-to-vehicle (V2V), and V2G which is shown in Fig. 12.2. V2H expresses the power exchange between EVs and the home network, while V2V is a local EV community such that EVs can charge/discharge their batteries among themselves. In V2G, the discharging energy of EVs is injected to the power grid through the management and control of VPP [4]. The detailed architecture of V2G and technology of V2G equipment, including unidirectional and bidirectional, are presented in [36].

Based on the above explanations, the vehicles can be discharged when they are parked, before their next use. In addition, parking lots have a great potential to offer V2G. The EVs with V2G capabilities may have positive impact on grid operation (e.g., congestion management, load shifting, power quality improvement, etc.) and also on VPP's performance [31]. In other words, VPP can increase the penetration of RESs in its portfolio and improve its performance in obtaining optimal bidding strategy for participating in electricity markets.

In order to model V2G capability, the discharge of EVs should be modeled. In addition, the EVs owners' welfare constraints and their behavior in using the EVs should be modeled like presented in Section 12.2.4.

Simultaneous charging/discharging of EV when it is connected to grid is prevented in

$$x_{e,t}^{Ch\_EV} + x_{e,t}^{dCh\_EV} \leq 1, \qquad \forall t \in \left[a^{EV}, d^{EV} \right]. \tag{12.10}$$

During the trip time interval, the battery can neither be charged nor discharged, which is adhered through (12.11). Furthermore, charging/discharging limits of EV should be considered by (12.12) and (12.13):

$$E_{e,t}^{Ch\_EV} = 0, E_{e,t}^{dCh\_EV} = 0, \qquad \forall t \in T - \left[a^{EV}, d^{EV} \right], \tag{12.11}$$

$$0 \leq E_{e,t}^{Ch\_EV} \leq \overline{E}_e^{Ch\_EV}, \qquad \forall t \in \left[a^{EV}, d^{EV} \right], \tag{12.12}$$

$$0 \leq E_{e,t}^{dCh\_EV} \leq \overline{E}_e^{dCh\_EV}, \qquad \forall t \in \left[a^{EV}, d^{EV} \right]. \tag{12.13}$$

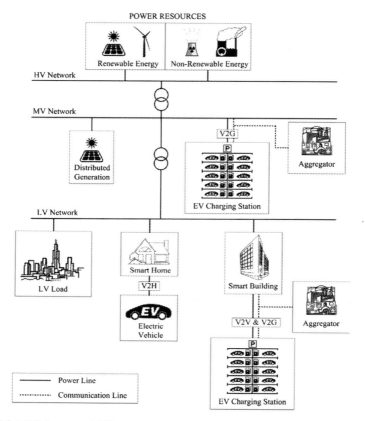

Figure 12.2 V2G framework [3].

The EV's SOC by considering V2G capability is obtained by (12.14) and (12.15) as

$$SOC_{e,t}^{EV} = E_e^{0\_EV} + \sum_{\tau=1}^{t-1} \left(\begin{array}{c} \eta_e^{Ch\_EV} E_{e,\tau}^{Ch\_EV} x_{e,\tau}^{Ch\_EV} \\ -\dfrac{1}{\eta_e^{dCh\_EV}} E_{e,\tau}^{dCh\_EV} x_{e,\tau}^{dCh\_EV} \end{array} \right), \quad \forall t \leq d^{EV},$$

(12.14)

$$SOC_{e,t}^{EV} = E_e^{0\_EV} + \sum_{\tau=1}^{t-1} \left(\begin{array}{c} \eta_e^{Ch\_EV} E_{e,\tau}^{Ch\_EV} x_{e,\tau}^{Ch\_EV} \\ -\dfrac{1}{\eta_e^{dCh\_EV}} E_{e,\tau}^{dCh\_EV} x_{e,\tau}^{dCh\_EV} \end{array} \right) - E_e^{trip\_EV},$$

$$\forall t \geq a^{EV}.$$

(12.15)

As mentioned earlier, the SOC should be positive at each time step. In case of V2G capability, the SOC in Eqs. (12.14) and (12.15) should be greater than the discharge energy (i.e., set of constraints (12.16) should be considered). Furthermore, the SOC should be lower than EV's battery capacity like in Eq. (12.7):

$$SOC_{e,t}^{EV} \geq \frac{1}{\eta_e^{dCh\_EV}} E_{e,t}^{dCh\_EV} x_{e,t}^{dCh\_EV}. \tag{12.16}$$

The charging level of EV should be sufficient for the next trip as follows:

$$E_e^{0\_EV} + \sum_{\tau=1}^{d^{EV}-1} \left(\begin{array}{c} \eta_e^{Ch\_EV} E_{e,\tau}^{Ch\_EV} x_{e,\tau}^{Ch\_EV} \\ -\frac{1}{\eta_e^{dCh\_EV}} E_{e,\tau}^{dCh\_EV} x_{e,\tau}^{dCh\_EV} \end{array} \right) \geq E_e^{trip\_EV}. \tag{12.17}$$

In case EV battery is required to be charged up to the maximum level, the following constraint should be regarded:

$$E_e^{0\_EV} + \sum_{\tau=1}^{d^{EV}-1} \left(\begin{array}{c} \eta_e^{Ch\_EV} E_{e,\tau}^{Ch\_EV} x_{e,\tau}^{Ch\_EV} \\ -\frac{1}{\eta_e^{dCh\_EV}} E_{e,\tau}^{dCh\_EV} x_{e,\tau}^{dCh\_EV} \end{array} \right) = C_e^{EV}. \tag{12.18}$$

12.2.3 EVs' parking lots

The increasing growth rate of EVs necessitates the policy makers to invest in charging stations in different places of cities. The existing parking spaces can be converted to EVPLs by equipping them with charging and V2G infrastructure without requiring new land. The parking lots may be installed in hotels, shopping malls, hospitals, office buildings, universities, airports, and other similar places where EVs are usually parked for some hours.

The EVPLs can be supplied through the grid in order to provide the required charging energy for parked EVs. In addition, they can be equipped with DGs, like photovoltaic (PV) units (namely EV solar parking lots, EVSPLs), conventional microturbines, and ESSs. It is worth noting that parking lots may have the V2G capability. In this case the VPP can exploit the discharging power of numerous EVs which are parked and make profit, as well as improve its operation.

In order to exploit the EVPLs, EVs are required to be parked for large enough periods during a day. It is worth noting that EVPLs can be counted as the aggregators of EVs and also as a bulk ESS. The EVPLs can be equipped with an energy management system (EMS) in order to coordinate the parked EVs based on received signals from VPP or VPP can directly control the EVPLs through centralized control frameworks. In this regard the VPP can control the charging/discharging cycle of EVPLs in order to make profit. If the EVPLs directly control the EVs in a parking lot, the objective of the EVPL owner should be considered by VPP in its coordination model.

The EMS model of EVPLs is proposed in [9,10]. The input required data include:

- Arrival/departure time of EVs,
- SOC of EVs at their arrival time,
- Market data, energy and parking tariffs,
- Data associated with other DERs in EVPLs.

Based on the charging/discharging rate of EVs, the absorbed and injected power from/to grid is limited as follows:

$$P_t^{Ch\_EVPL} + P_t^{g\_EVPL} \leq \gamma^{ch}.n_t^{EV}, \tag{12.19}$$

$$P_t^{dCh\_EVPL} + P_t^{Stor\_EVPL} \leq \gamma^{dCh}.n_t^{EV}, \tag{12.20}$$

where n_t^{EVPL} is the number of parked EVs in EVPL at time t and γ is the rate of charge/discharge of EVPL. It is worth noting that in case of DG or ESS existence in EVPL, $P_t^{g\_DER}$ describes the generated power of DGs in EVPL and $P_t^{Stor\_EVPL}$ is the stored power in the ESS of EVPL.

The SOC of EVPL is as follows. It depends on the SOC of arrived/departed EVs, SOC of the last hour, and exchanged power with grid:

$$SOC_t^{EVPL} = SOC_{t-1}^{EVPL} + SOC_t^{EV,arriv} - SOC_t^{EV,depart}$$
$$+ (P_t^{Ch\_EVPL} + P_t^{g\_EVPL})\eta^{Ch\_EVPL}$$
$$- \frac{1}{\eta^{dCh\_EVPL}}(P_t^{dCh\_EVPL} + P_t^{Stor\_EVPL}). \tag{12.21}$$

The EVs that arrive to and depart from EVPL have the following SOCs:

$$\begin{cases} SOC_t^{arriv} = 0, & SOC_{t-1}^{Nom} \leq SOC_t^{Nom}, \\ SOC_t^{arriv} = SOC_t^{Nom} - SOC_{t-1}^{Nom}, & \text{otherwise}, \end{cases} \tag{12.22}$$

$$\begin{cases} SOC_t^{depart} = 0, & SOC_{t-1}^{Nom} \leq SOC_t^{Nom}, \\ SOC_t^{depart} = \frac{(SOC_{t-1}^{Nom} - SOC_t^{Nom})SOC_t^{EVPL}}{SOC_t^{Nom}}, & \text{otherwise}, \end{cases}$$
$$\tag{12.23}$$

where SOC_t^{Nom} is determined based on capacity and SOC of EVs parked in EVPL using

$$SOC_t^{Nom} = \sum_{e \in E} C_e^{EV}.SOC_{e,t}^{EV}. \tag{12.24}$$

The profit of VPP achieved by controlling EVPL can be defined similar to Eq. (12.46) in Section 12.4.1.

In the above mentioned model, the arrival/departure pattern of EVs and other associated data can be considered the same as in the model presented in Sections 12.2.1 and 12.2.4.

Based on [37], the charging method of EVPLs may be different based on Table 12.1.

In addition, different classes of EVs can be considered in EVPL based on their battery capacity and market share in Table 12.2 [37].

Table 12.1 EVPLs different charging modes.

Charging mode	Charging rate
Slow charging	0.1 BCAP/hour
Quick charging	0.3 BCAP/hour
Fast charging	1.0 BCAP/hour

Table 12.2 Characteristics of different EV classes.

EV class	Maximum battery capacity (kWh)	Energy consumption rate (kWh/mile)	Market share (%)
1	10	0.3790	20
2	12	0.4288	30
3	16	0.5740	30
4	21	0.8180	20

12.2.4 Uncertainties arising from EVs

The main difference of EVs and stationary ESSs is the uncertainty related to EVs owners' behavior. It causes difficulties since EVs are not available at a specified time period. The main parameters of EVs that affect the access probability include arrival/departure time of EVs to/from home, office, parking lots, and other charging stations. On the other hand, the SOC of EVs upon arrival to a charging station is effective for scheduling charging/discharging cycles of EVs and EVPLs. All the mentioned parameters have uncertainty because of the random behavior of the owners. Hence, it is required to model the associated uncertainty.

One of the most effective methods to handle the uncertainties is the stochastic approach [7]. On the other hand, in [5] the fuzzy method is used to model the uncertainty of EVs' charging load in the planning problem, and in [6] for the power consumption model of plugged-in hybrid electric vehicles (PHEVs) during trips the information gap decision theory (IGDT) is used. However, in these frameworks there is no control of EVs and the uncertainty in predicted EVs' charging loads is determined. Based on EVs owners' behaviors, the stochastic approaches are more appropriate for modeling EVs' uncertainties in order to incorporate them in the EVs' coordination problem.

The different EV classes should be considered in a stochastic model. The different classes are determined based on their battery type [38]. The battery capacities can be chosen randomly based on a specific probability distribution in order to generate the scenarios. Two typical battery types that are mostly used are presented in Table 12.3 [39]. In addition, based on the survey "Mobility in Germany," the EVs are classified based on Table 12.4 [30].

The SOC of EVs depends on the daily travel distance and their battery capacity. In [37] and [40] the lognormal distribution is considered to model the random travel distance of EVs owners. In this regard, the daily travel distance is formulated as fol-

Table 12.3 Technical characteristics of two EV battery types [39].

Battery characteristics	Tesla Model S	Nissan Leaf
Battery capacity (kWh)	85	24
Energy consumption (kWh/km)	0.233	0.211
Range (km)	340	150
Charging power (kW)	22	6.6

Table 12.4 Class of EVs based on trip purpose [30].

Trip purpose	Percentage (%)
Leisure	21.9
Work	20.7
Education	1.1
Private business	12.2
Shopping	21.6
Accompanying	9.2

lows:

$$D = \exp\left(\ln\left(\mu_d^2/\sqrt{\mu_d^2 + \sigma_d^2}\right) + N.\ln\left(\mu_d^2/\sqrt{\mu_d^2 + \sigma_d^2}\right)\right), \qquad (12.25)$$

where μ_d and σ_d are respectively the mean and standard deviation of travel distance and N is the standard normal variable. It should be noted that the mean and standard deviation are calculated based on historical data [41,42]. After recognizing the expected driven distance, the expected energy demand of the EV is derived from

$$E_e^{trip\_EV} = \begin{cases} C_e^{EV}, & D_e \geq \overline{D}_e, \\ D_e.E_e^{Cons\_EV}, & D_e \leq \overline{D}_e, \end{cases} \qquad (12.26)$$

where $E_e^{Cons\_EV}$ is the consumed energy of EV per unit of distance and $\overline{D}$ is obtained from

$$\overline{D}_e = \frac{C_e^{EV}}{E_e^{Cons\_EV}}. \qquad (12.27)$$

As noted before the arrival/departure times of EVs are also uncertain. The existing databases (e.g., NHTS [43], in which more than 40,000 users all around the USA are considered), can be used in order to generate stochastic scenarios for arrival/departure of EVs [8,44]. In [37] the Gaussian distribution is used to estimate arrival/departure times as follows:

$$a^{EV} = \mu^{arrival} + \sigma^{arrival}.N_1, \qquad (12.28)$$
$$d^{EV} = \mu^{departure} + \sigma^{departure}.N_2. \qquad (12.29)$$

The μ and σ associated with the arrival and departure of EVs can be obtained from historical data like in [30,43]. Moreover, random variables are calculated by (12.30), in which U_1 and U_2 are uniform random variables:

$$N = \sqrt{-2.\ln(U_1)} \times \cos(2\pi U_2). \tag{12.30}$$

As well as mentioned parameters, the EVPLs numbers of EVs and available capacities in SOC of a parking lot have uncertainty. It is worth noting that the capacity of EVPL is dependent on the number and type of parked EVs in EVPL. The associated uncertainties can be modeled using the stochastic approach through generating scenarios. The generated scenarios can be based on the average traffic behavior of EVs owners [45].

12.3 Modeling of ESSs

Electrical energy storage systems become an important part of power networks and offer various advantages. In order to effectively exploit ESSs, in this section at first ESS types are introduced and then an operation model of ESS is presented. Finally, the ESS utilization in the presence of other DERs is discussed.

12.3.1 Introduction to ESS types

Electrical energy storage systems become an important part of power networks and have various advantages in all generation, transmission, and distribution sections, including power quality improvement, reliability improvement, investment deferral, frequency regulation, energy management, congestion management, uncertainty alleviation, etc. From a customers' point of view, ESSs can be exploited for effective participation in DR programs, bill charge management, backup power, and so on. An overview of ESS applications is as shown in Fig. 12.3 [23].

The increasing trend of ESSs usage in different regions is depicted in Fig. 12.4. On the other hand, the investment cost of different battery ESS types is decreasing. Fig. 12.5 depicts the potential of battery ESSs cost decrease. In addition, from 2010 to 2020 the prices of lithium-ion batteries are illustrated in Fig. 12.6 which shows 88% decrease [25].

Different ESS technologies are typically grouped into six categories which can be used in electricity networks. They are consisting electrical, mechanical, electrochemical, thermochemical, chemical, and thermal as depicted in Fig. 12.7 [47]. The technical maturity of ESS types is as shown in Fig. 12.8.

Based on ESSs characteristics, e.g., power rating and discharge time, they are classified for three applications (Fig. 12.9):

- Uninterruptible power supply (power quality),
- Transmission and distribution grid supports (load shifting),
- Bulk power management.

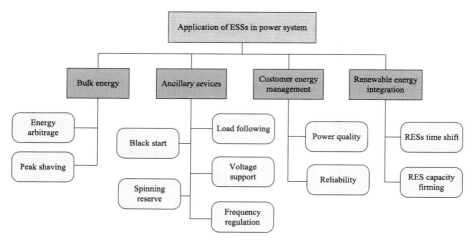

Figure 12.3 The overview of ESS applications.

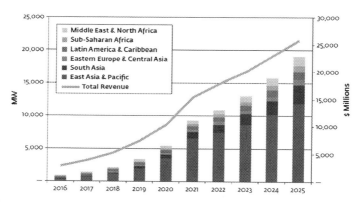

Figure 12.4 Projected annual stationary energy storage deployments, power capacity and revenue by region, emerging markets, 2016–2025 [24].

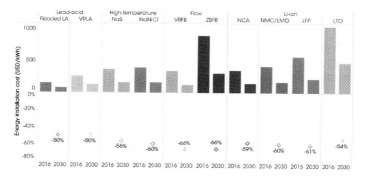

Figure 12.5 Battery ESS installed energy cost reduction potential, 2016–2030 [46].

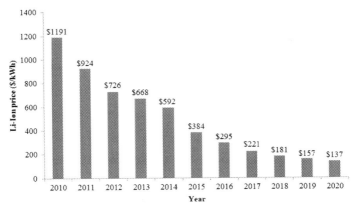

Figure 12.6 Lithium-ion battery prices fell 88% from 2010 to 2020 ($/kWh).

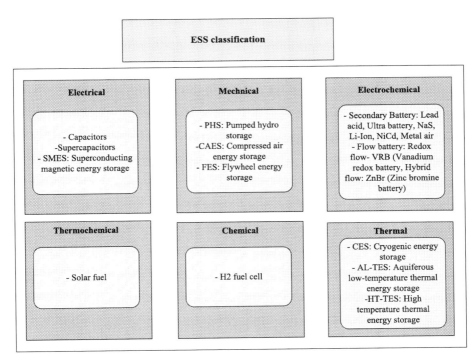

Figure 12.7 ESS types in distribution networks.

The technology mix in installed storage devices between 2011 to 2016 is illustrated in Fig. 12.10.

Based on the above illustrations and reference [48], Li-ion batteries are very capable devices for use in distribution networks for various applications. In addition, the

270　　Scheduling and Operation of Virtual Power Plants

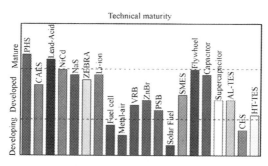

Figure 12.8 Technical maturity of ESS technologies [48].

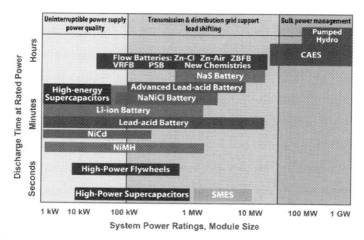

Figure 12.9 Positioning of diverse energy storage technologies per their power rating and discharge times at rated power [46].

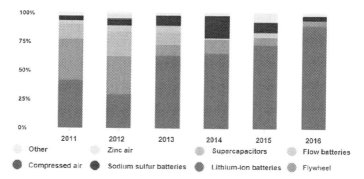

Figure 12.10 Technology mix in storage installations, excluding pumped hydro [49].

lead-acid batteries still are one of the best choices that compromise between performance and cost.

12.3.2 ESSs operation model

In modern power systems, stationary storage systems are able to sell back electricity to the grid. The charging and discharging cycles of ESSs would be determined through VPP or their owner control. The technical constraints associated with the storage systems are presented in the following [50].

Storage systems cannot be charged and discharged simultaneously. Furthermore, the charging and discharging powers have limits. These constraints are addressed in the following equations, respectively:

$$x_{b,t}^{Ch\_ESS} + x_{b,t}^{dCh\_ESS} \leq 1, \qquad \forall b \in B, \ t \in T, \quad (12.31)$$

$$0 \leq E_{b,t}^{Ch\_ESS} \leq \overline{E}_b^{Ch\_ESS}, \qquad \forall b \in B, \ t \in T, \quad (12.32)$$

$$0 \leq E_{b,t}^{dCh\_ESS} \leq \overline{E}_b^{dCh\_ESS}, \qquad \forall b \in B, \ t \in T. \quad (12.33)$$

The stored energy in the ESS should be tracked by considering its SOC. The SOC of ESS based on charging and discharging cycles is presented in Eq. (12.34). It should be noted that the operation efficiency associated with charging and discharging efficiencies (i.e., $\eta_b^{Ch\_ESS}$ and $\eta_b^{dCh\_ESS}$, respectively) should be regarded. This equation can be rewritten as in (12.35).

ESS discharge power at each time step cannot be greater than its SOC at that time step. It is adhered through constraint (12.36). In addition, the SOC should be lower than ESS capacity as given by constraint (12.37):

$$SOC_{b,t}^{ESS} = E_b^{0\_ESS} + \sum_{\tau=1}^{t-1} \left(\eta_b^{Ch\_ESS} E_{b,\tau}^{Ch\_ESS} x_{b,\tau}^{Ch\_ESS} - \frac{1}{\eta_b^{dCh\_ESS}} E_{b,\tau}^{dCh\_ESS} x_{b,\tau}^{dCh\_ESS} \right),$$

$$\forall t \in T, \ b \in B, \quad (12.34)$$

$$SOC_{b,t}^{ESS} = SOC_{b,t-1}^{ESS} + \eta_b^{Ch\_ESS} E_{b,t}^{Ch\_ESS} x_{b,t}^{Ch\_ESS} - \frac{1}{\eta_b^{dCh\_ESS}} E_{b,t}^{dCh\_ESS} x_{b,t}^{dCh\_ESS}, \quad \forall t \in T, \ b \in B, \quad (12.35)$$

$$SOC_b^{ESS} \geq \frac{1}{\eta_b^{dCh\_ESS}} E_{b,\tau}^{dCh\_ESS} x_{b,\tau}^{dCh\_ESS}, \quad (12.36)$$

$$SOC_b^{ESS} \leq C_b^{ESS}. \quad (12.37)$$

In order to avoid unreal solutions, the initial and final SOC during a specified period should be the same. In some applications this equality constraint is considered for a period equal to one day (24 hours) [51]:

$$SOC_{Initial}^{ESS} \left(= E_b^{0\_ESS} \right) = SOC_{Final}^{ESS}. \quad (12.38)$$

The BESS is one of the most popular and applicable storage systems, especially in electricity distribution networks. The daily cycle pattern of BESS is an important factor and directly affects the life cycle of a battery. In this regard the SOH of the battery should be regarded. SOH represents capacity deviation due to battery cycle aging and temperature effects. SOH is the ability of a battery to store energy relative to its nominal capacity. It is represented as the ratio of the real maximum capacity of the battery to the nominal battery capacity,

$$SOH^{ESS} = E_{real}^{ESS}/C^{ESS}. \tag{12.39}$$

The real maximum capacity of a battery includes linear and non-linear terms. The linear terms are associated with daily charging/discharging cycles, while non-linear terms represent the effect of daily cycles along with temperature on battery operation and are estimated using mathematical models [52]. In [27] it is assumed that linear and non-linear terms strongly depend on the daily charging pattern of the battery. Hence, under the condition that the battery operates in the nominal temperature range, the numbers of charging/discharging cycles directly contribute to the degradation of battery capacity [53]. As a result, unnecessary charging/discharging of battery ESSs should be prevented. In other words, the SOH of a battery should be considered in the optimization models.

The degradation model of a Li-ion battery is provided in [54]. It defines the capacity fade ratio of y using (12.40). Then the degradation model is obtained by (12.41):

$$y^{ESS} = 1 - SOH^{ESS} = \frac{C^{ESS} - E_{real}^{ESS}}{C^{ESS}}, \tag{12.40}$$

$$y^{ESS} = A \exp\left(-\frac{E_a}{RT}\right) k^z, \tag{12.41}$$

where k is the number of cycles, T is the temperature in kelvin, R is the constant of gas in J/(mol.K), z is the power law factor, A is constant, and E_a is the activation energy in J/mol.

12.3.3 ESSs utilization in the presence of other DERs

ESS is one of the solutions to accommodate uncertainties and variability caused by high integration of RESs in the grid. By integrating ESS into the grid, intermittent energy produced by renewable energies could be stored during low demand periods, and in the peak periods it can be injected into grid. Moreover, ESS can also provide other benefits such as energy management, frequency regulation, and improve the reliability and power quality of the service [23].

As a result, utilizing the ESSs along with other DERs (especially RESs with high intermittency, uncertainty, and variability) reduces power output fluctuations. In other words, in the case of aggregating DERs by VPP, VPP is able to increase its profit in balance and real time (RT) markets by scheduling charge/discharge cycles of ESSs according to its bidding strategy and the generation of RESs. In this case, VPP exploits the flexibility of ESSs [55]. Furthermore, distributed and mobile ESSs at the power distribution side provide an effective energy management capability for VPP that leads to VPP's operation improvement, network loss reduction, peak shaving, load leveling, etc., and it is worth noting that ESS has a crucial role in supplying power in remote areas especially when renewable based generating units are utilized.

One of the challenges in power systems in case of reliability issues, power fluctuation and occurring events associated with adverse weather conditions is curtailing renewable-based generation and demand. In these conditions ESS can avoid load/generation curtailment and reduce imposed interruption costs to the system. References [27] and [56] have proposed a strategy for optimal integration of BESSs to improve the load and DG hosting ability of the utility grid. In this regard the following strategy is used in case of generation and load curtailment:

$$P_b^{Ch\_ESS} > 0, \quad P_b^{dCh\_ESS} = 0 \quad \text{if generation curtailment} > 0, \quad (12.42)$$

$$P_b^{Ch\_ESS} = 0, \quad P_b^{dCh\_ESS} > 0 \quad \text{if load curtailment} > 0. \quad (12.43)$$

On the other hand, executing DR programs, as an effective DER, brings plenty of preferences. Utilization of ESS by customers not only affects the cooperation and role of customers in DR programs but also offers capability to manage their electricity bills.

In [26] the output generation of PV units and wind turbines is smoothed by exploiting BESSs through the fluctuation rate definition as follows:

$$r^{PV\_W} = f\left(\frac{\overline{P}^{PV} + \overline{P}^W - \underline{P}^{PV} - \underline{P}^W}{P_{rated\_W} + P_{rated\_PV}}\right), \quad (12.44)$$

$$r^{hybrid} = f\left(\frac{\overline{P}^{hybrid} - \underline{P}^{hybrid}}{P_{rated\_hybrid}}\right). \quad (12.45)$$

Eq. (12.44) represent the fluctuation rate of PV and wind units' generation, while in (12.45) fluctuation is obtained by considering the effect of ESSs (hybrid means the combination of PV, wind, and BESS).

The smoothing control of BESS was handled according to the real time feedback control of the BESSs' SOC and power fluctuation rates. In this regard, the rate of power change at each time step and charging/discharging power for each BESS are required to be calculated.

It is worth noting that scheduling ESSs, as well as other DERs, requires EMS. The detail formulation of coordination between DERs through the concept of VPP is presented in Section 12.4.

12.4 VPP operation strategy modeling in the presence of EVs and ESSs

Based on [12], VPP aggregates DERs around the network and makes a single operating profile by utilizing DERs characteristics. VPP optimizes its operation through optimal scheduling of its DERs in order to improve its performance and make profit in the electricity markets. To effectively utilize DERs, VPP should address the challenges of EVs and ESSs growing penetration by aggregating and coordinating their operation. By aggregating EVs and ESSs, VPP is able to effectively manage the charging/discharging cycles of these resources along with optimum management of other DERs and utilize their potentials in order to obtain its optimal operation strategy and fulfill its objectives. The VPP's optimal operation strategy can be obtained by optimal management of aggregated DERs through centralized or distributed schemes which are discussed in the following. It is worth noting that VPP should satisfy DERs technical constraints, as well as EVs owners' preferences. EVs owners can utilize their vehicles as efficient storage devices in the periods that the vehicles are parked in the parking lot (e.g., for billing costs' management). In addition, VPP can utilize the EVs by direct or indirect control in order to fulfill its objectives (e.g., improving its operation and increasing its profit in electricity markets). Thus, VPP should consider the preferences of EVs' owners to obtain its optimal operation strategy. In other words, VPP optimally schedules diverse DERs dispersed in the network by considering their technical characteristics, as well as EVs owners' preferences (and network operational constraints, in case of TVPP), in order to attain its optimal operation strategy and fulfill its objectives.

12.4.1 Centralized scheduling scheme

In a centralized scheme, VPP schedules aggregated DERs by gathering the required data and constraints through an optimization framework [8,10,20–22]. A general centralized scheduling scheme of VPP is depicted in Fig. 12.11. In this framework, VPP gathers the required data associated with aggregated DERs including EVs and ESSs. In case of TVPP, the data of the underlying network should also be taken into account. After that, VPP optimizes its objective by defining an optimization problem. The objective of VPP can be minimizing costs, improving network and DERs operation, and maximizing its profit in electricity markets. Based on [12] and [13], VPP usually optimizes the operation of DERs under its supervision in order to obtain its optimal bidding strategy and maximize its profit in electricity markets (e.g., day-ahead (DA) energy, real time, ancillary services, balancing markets, etc.).

In the optimization framework, VPP considers the operation model of EVs, EV-PLs, ESSs, and other DERs besides modeling the uncertainties of EVs access, load, RESs, and so on. By incorporating an efficient solution approach, the optimal schedule of DERs, including optimal charging/discharging patterns of EVS, EVPLs, and ESSs, is obtained. In addition, the optimum amount of VPP's objective and optimum operational state of the network (in case of TVPP) are determined. It should be noted that

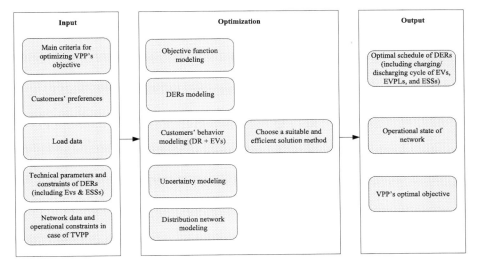

Figure 12.11 VPP's optimal operation strategy framework using centralized scheme.

VPP can control and coordinate DERs in a control center through EMS. In this condition VPP requires remote control equipment and communication system [35,36]. The communication system should be bidirectional in order for VPP to receive and monitor the current status of DERs and also send control signals to DERs.

In the following the general optimization model of VPP is presented.

12.4.1.1 VPP's objective function

As mentioned earlier, VPP usually aims at maximizing its profit in electricity markets, for example, in the DA energy market. The typical objective function of VPP in the DA market is presented as

$$Objective = \sum_{t \in T} \Bigg(\sum_{e \in E} \left(P_{e,t}^{Ch\_EV} \lambda_t^{TOU} - P_{e,t}^{dCh\_EV} \lambda_t^{V2G} \right)$$
$$+ \sum_{pl \in PL} \left(P_{pl,t}^{Ch\_EVPL} \lambda_t^{TOU} - P_{pl,t}^{dCh\_EVPL} \lambda_t^{V2G} \right)$$
$$+ \sum_{b \in B} \left(P_{b,t}^{Ch\_ESS} \lambda_t^{Ch\_ESS} - P_{b,t}^{dCh\_ESS} \lambda_t^{dCh\_ESS} \right)$$
$$- P_t^{Market} \lambda_t^{DA} - \sum_{g \in G} C_g^{MT} - Penalty_t \Bigg). \quad (12.46)$$

The term C_g^{MT} is the generation cost of conventional DGs (e.g., microturbine and combined heat and power (CHP) unit) which is formulated in (12.68); P_t^{Market} is the transacted power between VPP and market, which is obtained based on the amount of VPP's generation and load at each time step. The term penalty is related to agents' optimal costs deviation, like EVs owners whose optimal cost formulation is presented in the following. The model of diverse DERs should also be considered in the optimization problem which is discussed in Section 12.4.1.3.

12.4.1.2 EVs owners' optimal costs

In some cases, VPP is required to consider EVs owners' optimal costs [32]. In other words, VPP calculates the EVs owners' optimal costs and, when it controls EVs to fulfill its objectives, pays a penalty to EVs owners equal to the deviation of their imposed costs from the calculated optimal cost.

EVs' optimal cost ($C_e^{Opt\_EV}$) is calculated through an independent optimization problem which minimizes the following cost function [32]. In this regard, the EV owner minimizes his net costs, including charging costs (it is assumed that the EV owner pays for its charging energy using TOU price, presented in the first part of Eq. (12.47)) minus its revenue obtained from selling its discharging power at V2G price, presented in the second part of

$$C_e^{Opt\_EV} = \sum_{t \in T}(P_{e,t}^{Ch\_EV}\lambda_t^{TOU} - P_{e,t}^{dCh\_EV}\lambda_t^{V2G}), \quad \forall e \in E. \qquad (12.47)$$

12.4.1.3 DERs' modeling

Diverse DERs, including conventional DGs, RESs, responsive loads, EVs, EVPLs, and ESSs, can be aggregated by VPP. The model of all aggregated DERs including technical specifications, operating constraints, and existing uncertainties should be considered in VPP optimal operation strategy problem.

The model of DGs and RESs, as well as stochastic modeling of RESs regarding uncertainties, is presented in [7]. In addition, the responsive loads including customers' appliances and heating, ventilation, and air conditioning (HVAC) systems are addressed in references [8,17,44].

In order to model EVs, they can be modeled as uncontrollable or controllable load. In the first model, the EVs load profile has a fixed pattern and its associated uncertainty can be considered. In this model usually three time periods, including transition, charge, and rest time, are considered [57]. During the transition time, EVs consume energy from their batteries. During charge time, EVs consume power from the grid to charge their batteries, and finally, during the rest time, EVs neither consume power from the batteries nor the grid. When the EVs are controlled directly or indirectly by VPP, the charging and discharging (in case of V2G capability) cycles can be determined based on the specified goals. In the case of EVs control, the constraints (12.1) to (12.9) (if only the charging of EVs can be controlled) or constraints (12.10) to (12.18) (if VPP exploits V2G capability and controls both the charge and discharge cycles) should be considered in VPP's optimization model.

In order to consider the EVPLs in the VPP optimal operation strategy problem, the constraints (12.19) to (12.24) introduced in Section 12.2.3 should be regarded. Furthermore, constraints (12.31) to (12.41) should be added to the framework for incorporating ESSs by VPP.

12.4.1.4 Network modeling

TVPP considers the actual location of DERs in its optimal operation strategy problem. To address this fact, VPP should model network operational constraints in its framework. The main operational constraints of the network are listed below [58]:

$$S_n^G = S_n^L + S_n^T, \qquad \forall n \in N, \quad (12.48)$$

$$S_n^G = \sum_{g \in G} S_g^{MT} + \sum_{\pi \in P} S_\pi^{PV} + \sum_{\omega \in \Omega} S_\omega^W, \qquad \forall n \in N. \quad (12.49)$$

Eq. (12.48) guarantees power balance at each bus of the network. It should be noted that in case of CVPP, the power balance equation is regarded for the entire network, not for each bus. The S_n^T is the power flowing out of bus n. For a bus with k link connections, S_n^T is represented as follows:

$$S_n^T = \sum_{m=1}^{k} S_{nm}, \quad (12.50)$$

$$S_{nm} = V_n[(V_n^* - V_m^*)(g_{nm} - jb_{nm})], \quad (12.51)$$

where g_{ij} and b_{ij} are the real and imaginary parts of admittance of the line between bus n and bus m. The following operating limits should be adhered, which respectively relate to buses' voltages and line flows' limits:

$$\underline{V} \le |V_n| \le \overline{V}, \quad (12.52)$$

$$|S_{nm}| \le \overline{S}_{nm}. \quad (12.53)$$

12.4.2 Distributed scheduling scheme

By increasing deployment of DERs, including battery ESS and EVs, as well as DGs and responsive loads around distribution network, a distributed network of decentralized DERs is formed. The VPP aims at aggregating and coordinating these distributed resources. As explained in the previous section, VPP can centrally coordinate and schedule the operation of its resources by utilizing the communication infrastructure. However, the centralized approach brings some challenges, including jeopardizing customers' privacy, dealing with huge mass of data, challenging computation time, communication network congestion, and so on.

In order to address the aforementioned challenges, distributed frameworks can be exploited by VPP in order to establish coordination and obtain its optimal operation strategy [59]. In addition, the distributed nature of DERs makes the distributed approaches very useful for scheduling the operation of DERs. Based on [60], distributed optimization techniques are based on the augmented Lagrangian decomposition and decentralized solution of the Karush–Kuhn–Tucker (KKT) necessary conditions for local optimality. The first technique includes dual decomposition, alternating direction method of multipliers (ADMM), analytical target cascading, and the auxiliary problem principle. The second technique includes optimally conditions decomposition, consensus, and innovation.

One of the most applicable distributed methods is ADMM [26]. In this method the main problem is divided into some sub-problems, and the solution of the main problem is obtained by solving sub-problems using the distributed facilities. The general structure of the problem is explained in the following.

The objective function of the main problem is decomposed as

$$\min_{x,y} \quad f(x) + g(y) \tag{12.54}$$

$$\text{Subject to} \quad Ax + By = c. \tag{12.55}$$

In (12.54), $f(x)$ and $g(y)$ are local or sub-problems of the main problem, in which x and y are the decision variables of each sub-problems, respectively. In ADMM method, the augmented Lagrangian (i.e., $L_\rho(x, y, \lambda)$) is defined as follows:

$$L_\rho(x, y, \lambda) = f(x) + g(y) + \lambda^T (Ax + By - c) + \frac{\rho}{2} \|Ax + By - c\|_2^2. \tag{12.56}$$

In (12.56), λ is the Lagrangian multiplier and ρ is called the penalty coefficient. The main problem in (12.54) and (12.55) can be solved based on the following procedure:

$$x^{k+1} = \arg\min_{x} L_\rho(x, y^k, \lambda^k), \tag{12.57}$$

$$y^{k+1} = \arg\min_{y} L_\rho(x^{k+1}, y, \lambda^k), \tag{12.58}$$

$$\lambda^{k+1} = \lambda^k + \rho(Ax^{k+1} + By^{k+1} - c). \tag{12.59}$$

The convergence condition of the algorithm is

$$\left\| Ax^{k+1} + By^{k+1} - c \right\| \leq \varepsilon. \tag{12.60}$$

Based on the above explanations, ADMM can be implemented in EMS of VPP. In other words, VPP schedules the DERs under its supervision in a distributed manner in order to obtain its optimal operation strategy.

12.5 Examples for VPP optimal operation strategy considering EVs and ESSs

In this section two examples are covered to investigate the problem of VPP's optimal operation strategy by considering EVs and ESSs. In the first example, VPP aggregates and coordinates EVs along with renewable-based generators and network loads in order to optimize its operation and participate in electricity markets, as well as uses the other advantages of EVs like environmental pollution decrease, etc. [2]. The second example optimizes the operation and bidding strategy of VPP through scheduling ESSs and other DERs [20].

12.5.1 Bidding strategy of VPP by modeling the integration of EVs

As mentioned earlier, the EVs play a crucial rule in the near future distribution networks, and they can be considered as mobile storage systems. Hence, the merits of electric vehicles can be considered in two categories: first, they can be considered as a storage device, and second, VPP can utilize the discharge power using V2G infrastructures. In addition, the charging power of the EVs can be controlled like other responsive loads. In this example the large-scale EVs are considered in the VPP's optimization model [2]. In this regard, the VPP aggregates and coordinates EVs along with renewable-based generators and network loads in order to participate in electricity markets and use the other advantages of EVs like environmental pollution decrease, etc. The detailed charging model of 4 different EVs, including taxi, bus, personal car, and business car, is presented in order to explore the impact of EVs on the operation of VPP. It is worth noting that the charging time, charging power, and access probability of EVs are regarded in the model. Furthermore, the operation of the VPP in three markets is investigated.

a) System model
– General model

In the model, VPP aggregates and coordinates EVs and renewable-based generations to control its resources in order to participate in electricity markets. It is assumed that VPP has access to the status of the resources and also it can send control signals to resources. The VPP framework is as shown in Fig. 12.12.

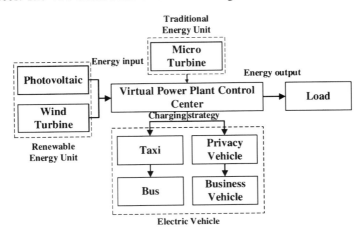

Figure 12.12 The VPP framework [2].

The generating resources include microturbines (MTs), photovoltaic units, and wind turbines. The EMS is the kernel of VPP's control center. It schedules the power flows in the network by controlling the generation of MTs and charging power of EVs. In addition, it determines the transactions with electricity markets based on wind and PV units' generation forecast.

The VPP is a price-taker agent and participates in the DA, RT, and balancing markets. In the DA market, each unit sends its forecasted output to VPP, and VPP determines the bidding strategy in the DA market and also forecasts the biding strategy in the RT market based on the forecasted price of the market. The RT market is cleared every hour. The output of wind and photovoltaic units is more accurate in RT, and VPP determines its bidding based on the latest market price. The balancing market runs after the RT market in order to compensate the deviation in the real and forecasted power of renewable-based generators and loads.

– EVs charging model

EVs are counted as flexible loads which can be utilized in order to alleviate the uncertainty associated with renewable sources. Four EVs' types, including personal cars, business cars, buses, and taxis, are considered. Each EV type has different operation characteristics in comparison with others which are represented by the charging capacities, charging time, charging locations, and the probability of access. The associated parameters of EVs' types are presented in Table 12.5 [2]. The charging power is categorized into slow, normal, and fast charging.

The characteristics of each EV type are summarized in the following:

– Electric bus. The charging power is 0.75 of the capacity between 10:00 and 16:30 and 0.25 of capacity in the period from 22:00 to 5:30. The capacity of buses' batteries is assumed to be 150 kW. The associated probability distribution function of SOC and access of battery is considered to be Normal and Bernoulli, respectively.
– For business cars, only slow charging is used, which is equal to 0.25 of the capacity. The associated capacity is equal to 45 kW.
– The charging time of a taxi is usually on the rest time, and a taxi is charged using the fast method. The probability distribution functions are similar to those of buses.
– Because of various charging patterns of personal cars, two different charging methods are assumed, namely working days and weekends. During working hours, normal charging is utilized. Under the condition of no charging facilities in work places, the owners should charge their vehicles at charging station or at home.

In order to consider the effect of EVs in VPP operation, the number of EVs and their probability of charging are chosen as shown in Table 12.6 [2].

The charging powers of EVs using Monte Carlo simulation are compared in Fig. 12.13 for personal EVs.

The effect of EVs' charging loads on the network load profile is depicted in Fig. 12.14. As it can be seen, the EVs charging profile has a negative impact on system operation without control and coordination on their charging.

b) Mathematical formulation

The mathematical formulation of the VPP problem in order to consider EVs for its optimal operation and participation in the three mentioned market is presented in this part. It is worth noting that the formulation is just presented for VPP participation in the DA market. The detailed formulation of VPP for participation in the RT and balance markets can be found in [2].

Table 12.5 Charging characteristics of different EV types.

EV type	Major charging period	Charging power (kW)	Charging time limit	Access probability	SOC probability	Charging location
Bus	10:00–16:30	112.5	–	Bernoulli(1.1,1.1)	Normal(0.4,0.1)	Charging station
	22:00–5:30	37.5	–	Normal(23,1)	Normal(0.4,0.1)	Charging station
Business car	19:00–6:00	11.25	–	Normal(20,1.5)	Normal(0.5,0.1)	Charging station
Taxi	1:00–5:00	22.5	120 min	Bernoulli(2,4)	Normal(0.3,0.1)	Charging station
	11:00–14:00	56.25	60 min	Bernoulli(2,4)	Normal(0.3,0.1)	Charging station
Personal car (working day)	9:00–17:00	11.25	–	Normal(9,0.5)	Normal(0.5,0.1)	Company
	19:00–7:00	11.25	–	Normal(20,0.5)	Normal(0.5,0.1)	Home
	19:00–21:00	22.5	80 min	Uniform(19,21)	Normal(0.6,0.1)	Shopping mall
	19:00–23:00	56.25	–	Normal(20,0.8)	Normal(0.3,0.1)	Charging station
Personal car (weekend)	10:00–22:00	22.5	80 min	Normal(15,1.5)	Normal(0.5,0.1)	Shopping mall
	17:00–6:00	11.25	–	Normal(20,1.5)	Normal(0.5,0.1)	Home
	16:00–23:00	56.25	–	Normal(17,0.8)	Normal(0.3,0.1)	Charging station

Table 12.6 Charging probabilities and number of different EV types.

EV type	Number	Charging period	Charging probability
Bus	150	10:00–16:30	0.4
		22:00–5:30	0.6
Business car	150	19:00–6:00	1
Taxi	250	1:00–5:00	0.5
		11:00–14:00	0.5
Personal car (working day)	550	9:00–17:00	0.2
		19:00–7:00	0.5
		19:00–21:00	0.1
		19:00–23:00	0.2
Personal car (weekend)		10:00–22:00	0.3
		17:00–6:00	0.4
		16:00–23:00	0.3

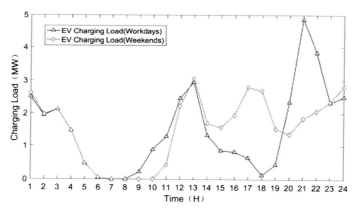

Figure 12.13 Charging pattern of personal EVs [2].

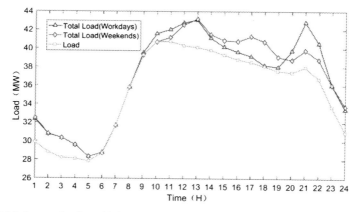

Figure 12.14 System load profile by considering EVs [2].

The objective of the VPP in the DA market is maximizing its profit by aggregating renewable sources and loads and utilizing EVs. The main equations are described below.

The objective is to maximize profit in the DA market as follows:

$$\max \sum_{t \in T} \left(R_t^{W\_DA} + R_t^{PV\_DA} + R_t^{MT\_DA} - (C_t^{DA} + C_t^{MT}) \right), \quad (12.61)$$

$$R_t^{W\_DA} = P_t^{W\_DA} \lambda_t^{DA} + P_t^{W\_RT} \lambda_t^{RT}, \quad (12.62)$$

$$R_t^{PV\_DA} = P_t^{PV\_DA} \lambda_{p,t}^{DA} + P_t^{PV\_RT} \lambda_t^{RT}, \quad (12.63)$$

$$R_t^{MT\_DA} = P_t^{MT\_DA} \lambda_t^{DA}, \quad (12.64)$$

$$P_t^{Market} = (L_t^{DA} + L_t^{EV}) - (P_t^{W\_DA} + P_t^{W\_RT} + P_t^{PV\_DA} + P_t^{PV\_RT} + P_t^{MT\_DA}), \quad (12.65)$$

$$C_t^{DA} = P_t^{DA} \lambda_t^{DA} + P_t^{RT} \lambda_t^{RT}, \quad (12.66)$$

$$P_t^{DA} + P_t^{RT} = P_t^{Market}, \quad (12.67)$$

$$C_t^{MT} = \begin{pmatrix} (y_t^{MT} - y_{t-1}^{MT}) SC^{MT} + k^{MT} \left(P_t^{MY\_DA} - \underline{P}^{MT} \right) \\ + y_t^{MT} FC^{MT} + \sum_i P_t^{MT\_DA} Q_i C_i \end{pmatrix}. \quad (12.68)$$

Eqs. (12.62) and (12.63) respectively describe the forecasting income of wind and photovoltaic units. Eq. (12.65) shows the power balance of VPP in which P_t^{market} is the transaction power between VPP and market. The costs associated with VPP forecasts in the DA market and the operation cost of MTs are obtained using (12.66) and (12.68), respectively. The cost of MT has four terms, including start-up cost, variable cost, fixed cost, and related costs to environmental pollution of MT.

The technical parameters of MTs and coefficients of environmental pollution penalty are depicted in Tables 12.7 and 12.8, respectively. It should be noted that C_i is equal to the sum of environmental value and penalty in Table 12.8.

The constraints are summarized in the following inequalities:

$$P_t^{W\_DA} + P_t^{W\_R} \leq P_t^{WF\_DA}, \quad (12.69)$$

$$P_t^{PV\_DA} + P_t^{PV\_RT} \leq P_{,t}^{PVF\_DA}, \quad (12.70)$$

$$P_t^{MT\_DA} \leq \overline{P}^{MT}, \quad (12.71)$$

$$P_{t+1}^{MT\_DA} - P_t^{MT\_DA} \leq RR^{MT}. \quad (12.72)$$

As mentioned before, it is assumed that VPP is a price-taker agent. In order to express the limits of transaction in a market, the step transaction cost is regarded as follows:

$$C_t^{DA} = P_t^{DA} \delta^{DA} \lambda_t^{DA} + P_t^{RT} \lambda_t^{RT}, \quad (12.73)$$

$$C_t^{RT} = P_t^{DA} \delta^{DA} \lambda_t^{DA} + P_t^{RT} \delta^{RT} \lambda_t^{RT}, \quad (12.74)$$

Table 12.7 MT technical parameters.

Maximum power	Minimum power	K^{MT}	Startup/shut-down costs	Fixed cost	Ramp rate
5.67 MW	2.5 MW	$6.31/MW	$30	$30	3 MW/hr

Table 12.8 Coefficients of environmental pollution penalty.

Parameters	NO_X	CO_2	CO	SO_2
Emission Q_i (kg/MWh)	0.6188	184.0829	0.1702	0.000928
Environmental value E_{id} ($/kg)	1	0.002875	0.125	0.75
Penalty P_i ($/kg)	0.25	0.0125	0.02	0.125

Table 12.9 Coefficient of penalty.

Transaction power (MW)	ζ
$\lvert P_t^{Market} \rvert < 5$	0
$5 < \lvert P_t^{Market} \rvert < 10$	5%
$10 < \lvert P_t^{Market} \rvert < 15$	10%
$15 < \lvert P_t^{Market} \rvert < 20$	15%
$\lvert P_t^{Market} \rvert > 20$	20%

where δ^{DA} and δ^{RT} are coefficients of step transaction costs which are obtained using equations

$$\delta^{DA} = \delta^{RT} = 1 + \zeta, \left(P_t^{Market} > 0 \right), \tag{12.75}$$

$$\delta^{DA} = \delta^{RT} = -\zeta, \left(P_t^{Market} < 0 \right). \tag{12.76}$$

In addition, the penalty coefficient of ζ is obtained based on Table 12.9.

The optimal dispatch of EVs improves VPP operation and also reduces peak–valley difference of the total load profile. The dispatching cost of EVs is obtained using (12.77). In this equation β is the subsidy coefficient which is associated with the charging time before and after control and is assumed to be 0.02/h,

$$C_{dispatch}^{EV} = \sum_{t \in T} \beta P_t^{EV} \lambda_t^{DA}. \tag{12.77}$$

In order to solve the proposed problem, the artificial bee colony algorithm is utilized.

c) Numerical results

The simulation results of evaluated example are presented here. The actual and forecasted DA and RT prices are depicted in Fig. 12.15.

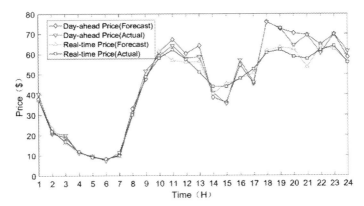

Figure 12.15 Prices in RT and DA markets [2].

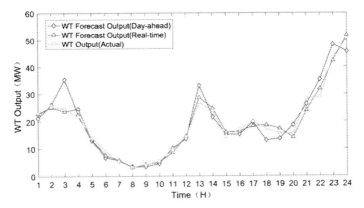

Figure 12.16 Generated power of wind turbines [2].

The outputs of wind and photovoltaic units are depicted in Fig. 12.16 and 12.17, respectively.

The bidding strategy of VPP by utilizing wind, photovoltaic, and microturbine units is depicted in Fig. 12.18. The wind and photovoltaic units' output power is utilized completely because of the low marginal costs. The microturbines compensate the difference of short-term demand and supply of the VPP, and the output depends on the load, renewable generation, and forecasted price.

The exchanged power between VPP and the market and the effect of step transaction cost on it are presented in Fig. 12.19.

The impact of EVs on the VPP's operation is evaluated here. The associated effect of EVs on the exchanged power is illustrated in Fig. 12.20. As it can be seen, the peak demand of VPP increases and VPP should purchase more power from the market. For this reason, the income decreases from $29814 to $29250.

In order to investigate the influence of EVs penetration on VPP operation, the number of EVs increases based on the proportions presented in Table 12.10. As it can be

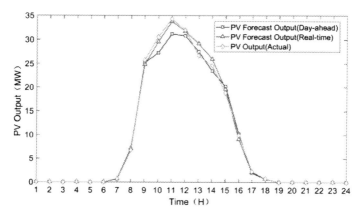

Figure 12.17 Generated power of photovoltaic units [2].

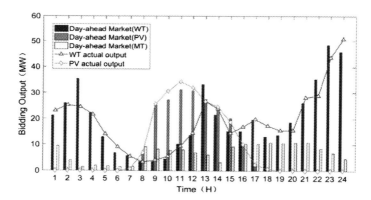

Figure 12.18 VPP bidding strategy in the DA market, neglecting EVs [2].

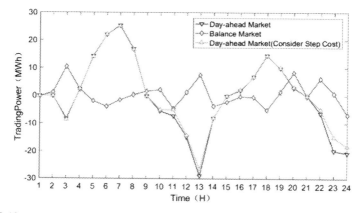

Figure 12.19 Exchanged power between VPP and the market [2].

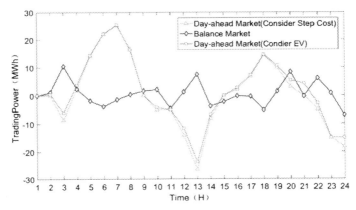

Figure 12.20 Exchanged power between VPP and the market considering EVs integration [2].

Table 12.10 Effect of EVs penetration on VPP revenue.

EVs numbers	0	1000	4000	6000	8000	10000
VPP revenue ($)	29814	29250	26524	24024	21057	17741

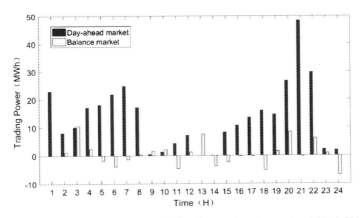

Figure 12.21 Exchanged power in DA and balancing market, integrating 10000 EVs [2].

seen there, the income decreases as the number of EVs increases because of the increasing network load. The exchanged power in the DA and balancing markets by aggregating 10000 EVs is depicted in Fig. 12.21.

The effect of different EVs on the load profile is illustrated in Fig. 12.22. It is assumed that numbers of EVs from each type are the same and equal to 1000.

The effect of different EV types on VPP revenue is shown in Table 12.11.

The effect of EVs on VPP's profit can be sorted as follows: bus, personal car, business car, and taxi. Hence, the optimal dispatch of buses and personal cars is more effective. As it is mentioned before, VPP dispatches EVs in order to maximize profit. The effect of EVs control by VPP is shown in Table 12.12.

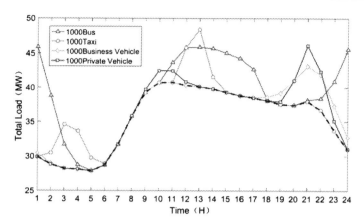

Figure 12.22 Load profile considering different EV types [2].

Table 12.11 Effect of each EV type on VPP revenue.

Types of EVs	1000 buses	1000 taxis	1000 business cars	1000 personal cars
VPP revenue ($)	28709	29908	29117	29015

Table 12.12 Effect of EVs dispatch on VPP operation.

EVs numbers	Parameter	Before control	After control	Comparison
1000 EVs	Peak–valley difference	14.72 MW	14.09 MW	-0.63 MW
	Bidding revenue	$29250	$29532	+$282
	Charging cost	$1774	$1615	-$159
4000 EVs	Peak–valley difference	28.48 MW	20.98 MW	-7.5 MW
	Bidding revenue	$26524	$28081	+$1557
	Charging cost	$7096	$6328	-$768
8000 EVs	Peak–valley difference	47.71 MW	28.94 MW	-18.77 MW
	Bidding revenue	$21057	$25142	+$4085
	Charging cost	$14193	$11394	-$2799

Based on the results, the following points are concluded:

- EVs control can reduce peak–valley difference because of the coordination between EVs and other DERs.
- In addition, the benefit for VPP increases by dispatching.
- The charging cost decreases by coordination of charging.

The equivalent loads profiles before and after EVs control for 4000 EVs are depicted in Fig. 12.23.

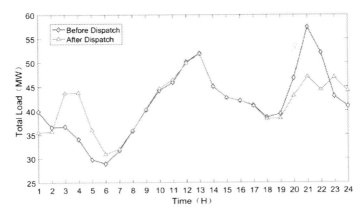

Figure 12.23 Equivalent load profiles before and after EVs control for 4000 EVs [2].

By controlling EVs charge, the load profile will be reshaped in order to improve VPP's operation and bidding strategy. Hence, EVs aggregation and coordination increase VPP benefit and can also increase the penetration level of renewable-based generation in the network.

12.5.2 Optimal VPP operation in the presence of ESSs

In this example the VPP's optimal operation and bidding strategy through the optimal schedule of ESSs in order to achieve maximum benefits in electricity markets is achieved [20]. In this regard, the VPP attains its optimal operation in order to participate in electricity markets. The overall framework is depicted in Fig. 12.24 which shows VPP participation in the electricity market and aggregated DERs including ESSs, PVs, wind turbines, and loads.

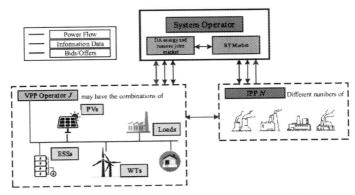

Figure 12.24 VPP framework indicating DERs aggregation to participate in the electricity market [20].

The VPP aims at minimizing total costs by optimal scheduling of its resources including ESSs. The associated costs of VPP in the DA market are described in (12.78), in which C_t^{deg} is batteries' degradation cost:

$$C^{VPP} = \sum_{t \in T} P_t^{DA} \lambda_t^{DA} + C_t^{\text{deg}}. \tag{12.78}$$

The model of ESSs should be considered in VPP's optimization framework. In this regard, the costs associated with batteries' degradation due to numerous charging/discharging cycles are modeled in

$$C_t^{\text{deg}} = \left| \frac{m}{100} \right| \frac{\sum_{t \in T} \max\left[0, SOC_{t-1}^{ESS} - SOC_t^{ESS}\right]}{C^{ESS}} \lambda^{ESS} C^{ESS}, \tag{12.79}$$

where m is the linear approximation factor of battery life which is assumed to be 0.0017 based on current technologies. The batteries' SOC is calculated through (12.80)–(12.82). In addition, the charge/discharge powers are limited in (12.83) and (12.84), respectively:

$$SOC_t^{ESS} = SOC_{t-1}^{ESS} + \eta^{Ch\_ESS} E_t^{Ch\_ESS} - \frac{1}{\eta^{dCh\_ESS}} E_t^{dCh\_ESS}, \tag{12.80}$$

$$SOC_1^{ESS} = SOC_{24}^{ESS} = 0.2 C^{ESS}, \tag{12.81}$$

$$0.2 C^{ESS} \leq SOC_t^{ESS} \leq 0.8 C^{ESS}, \tag{12.82}$$

$$0 \leq E_t^{Ch-ESS} \leq x_t^{Ch\_ESS} \overline{E}^{Ch\_ESS}, \tag{12.83}$$

$$0 \leq E_t^{dCh-ESS} \leq x_t^{dCh\_ESS} \overline{E}^{dCh\_ESS}. \tag{12.84}$$

The corresponding framework is modeled as a mixed-integer linear programming (MILP) problem which is solved using CPLX solver of MATAB. Two case studies are considered to investigate the results; in Case I, VPP ignores uncertainties in its optimization model, while in Case II, the uncertainties in the net load are modeled by a robust approach. The ESS installed capacity is equal to 120 MW and 120 MWh, and its charging/discharging efficiency is 0.95.

The scheduled ESS charge/discharge in comparison with the net load is depicted in Fig. 12.25. In addition, ESS SOC in comparison with the DA energy price is illustrated in Fig. 12.26.

12.6 Conclusion

In this chapter, the model of EVs and ESSs is presented. In this regard, EVs' operation model, V2G operating mode, EVPL considerations, and EVs' associated uncertainties are discussed. Furthermore, ESS types, ESSs operation model, and utilization of ESSs in the presence of other DERs are thoroughly considered. By obtaining EVs and ESSs characteristics and their operation model, the optimal operation model of VPP

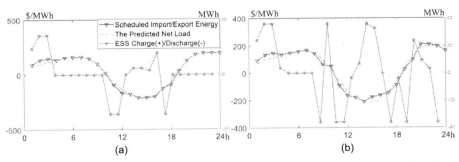

Figure 12.25 The scheduled ESS charge/discharge in comparison with net load: (a) Case I, (b) Case II [20].

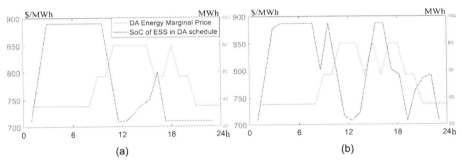

Figure 12.26 ESS SOC in comparison with DA energy price: (a) Case I, (b) Case II [20].

by aggregating and coordinating EVs and ESSs through centralized and distributed schemes is presented. In the model customers' behaviors and uncertainties related to EVs are regarded in order to extract the EVs model. It is worth noting that the utilization of V2G capability of the EVs by VPP can make EVs to be counted as mobile storage devices. As a result, VPP can make profit in different electricity markets by aggregating and coordinating diverse DERs, including EVs and ESSs. Furthermore, the operational aspects of VPP improve, and utilizing EVs and ESSs can increase the penetration level of RESs by alleviating associated uncertainties and variability of these resources. Finally, the role of VPP in aggregating large-scale EVs and ESSs is considered through evaluating two examples.

Nomenclature

Indices (sets)
$t, \tau(T)$ Time steps
$e(E)$ Electric vehicles
$e(E)$ Energy storage systems
$n, m(N)$ Network buses
$\omega(\Omega)$ Wind turbines
$\pi(P)$ Photovoltaic units

$g(G)$ Conventional DGs
$pl(PL)$ EVs parking lots
i Pollutant gases of microturbine

Parameters
$\overline{E}_e^{Ch\_EV}, \overline{E}_e^{dCh\_EV}$ Maximum charging and discharging energy of EV
a_e^{EV}, d_e^{EV} Arrival /departure time of EV to/from charging station
$E_e^{0\_EV}$ Initial charging level of EV
$\eta_e^{Ch\_EV}, \eta_e^{dCh\_EV}$ Charging and discharging efficiency of EV's battery
$E_e^{trip\_EV}$ Required energy of EV for next trip
C_e^{EV} Maximum capacity of EV
$n_{pl,t}^{EV}$ Number of EVs in EVPL
$\gamma_{pl}^{ch}, \gamma_{pl}^{dCh}$ Rate of charge/discharge of EVPL
$SOC_{pl,t}^{EV,arriv}$ SOC of arrived EVs to EVPL
$SOC_{pl,t}^{EV,depart}$ SOC of departed EVs from EVPL
$\eta_{pl}^{Ch\_EVPL}, \eta_{pl}^{dCh\_EVPL}$ Charging and discharging efficiency of EVPL
$SOC_{pl,t}^{Nom}$ Nominal SOC of EVs in EVPL
D_e Travel distance of EV
$E_e^{Cons\_EV}$ Consumed energy of EV per unit of distance
$\overline{D}_e$ Maximum travel distance of EV
$\overline{E}_b^{Ch\_ESS}, \overline{E}_b^{dCh\_ESS}$ Maximum charging/discharging rate of ESS
$\eta_b^{Ch\_ESS}, \eta_b^{dCh\_ESS}$ Charging/discharging efficiency of ESS
$E_b^{0\_ESS}$ Initial charging level of ESS
C_b^{ESS} Maximum capacity of ESS
λ_b^{ESS} Unit price of ESS
$\lambda_t^{TOU}, \lambda_t^{V2G}$ Time of use and V2G price
$\lambda_t^{Ch\_ESS}, \lambda_t^{dCh\_ESS}$ Charging/discharging prices of ESSs
$\underline{V}, \overline{V}$ Minimum/maximum voltage
$\overline{S}_{nm}$ Maximum flow of line from n to m
Q_i Pollutant gas emission of MT
C_i Environmental penalty of MT
L_t^{DA} DA market load level
SC^{MT} Start-up cost of MT
k^{MT} MT variable cost coefficient
FC^{MT} Fixed cost of MT
$P_t^{WF\_DA}$ Forecasted power of wind turbine in DA market
$P_{,t}^{PVF\_DA}$ Forecasted power of PV unit in DA market
RR^{MT} Ramp rate of MT
λ_t^{DA} DA electricity market price
λ_t^{RT} RT electricity market price

Variables
$E_{e,t}^{Ch\_EV}$ Charging energy of EV
$E_{e,t}^{dCh\_EV}$ Discharging energy of EV
$x_{e,t}^{Ch\_EV}$ Binary variable indicating charging status of EV
$x_{e,t}^{dCh\_EV}$ Binary variable indicating discharging status of EV

$C_e^{Opt\_EV}$ Optimum cost of EV
$SOC_{e,t}^{EV}$ EV's state of charge
$P_{pl,t}^{Ch,EVPL}$, $P_{pl,t}^{dCh,EVPL}$ Charging/discharging power of EVPL
$P_{pl,t}^{g,DER}$ Generated power of DGs in EVPL
$P_{pl,t}^{Stor,EVPL}$ Stored power in the ESS of EVPL
$SOC_{pl,t}^{EVPL}$ EVPL's state of charge
$x_{b,t}^{Ch\_ESS}$ Binary variable indicating charging status of ESS
$x_{b,t}^{dCh\_ESS}$ Binary variable indicating discharging status of ESS
$E_{b,t}^{Ch\_ESS}$, $E_{b,t}^{dCh\_ESS}$ Charging/discharging energy of ESS
$SOC_{b,t}^{ESS}$ ESS's state of charge
SOC_{Final}^{ESS} Final SOC of ESS
SOH^{ESS} ESS's state of health
E_{real}^{ESS} Real maximum capacity of ESS
y^{ESS} Capacity fade ratio of ESS
S_n^G, S_n^L Generation and consumption powers
S_{nm} The flow of line from n to m
V_n Voltage of bus n
$R_t^{W\_DA}$ Wind turbines revenue in DA market
$R_t^{P\_DA}$ PV units revenue in DA market
$R_t^{MT\_DA}$ MT revenue in DA market
C_t^{DA} DA trading costs
$C_t^{MT\_DA}$ Microturbine costs
$P_t^{W\_DA}$ Wind turbine output bidding power in the DA market
$P_t^{W\_RT}$ Wind turbine output bidding power in the RT market
$P_t^{PV\_DA}$ PV unit output bidding power in the DA market
$P_t^{PV\_RT}$ PV unit output bidding power in the RT market
P_t^{Market} VPP transacted power with market
L_t^{EV} EV net load
P_t^{DA} VPP transacted power with the DA market
P_t^{RT} VPP transacted power with the RT market
y_t^{MT} Binary variable indicating utilization status of MT

References

[1] Sumeet Kumar Wankhede, Priyanka Paliwal, Mukesh K. Kirar, Increasing penetration of DERs in smart grid framework: a state-of-the-art review on challenges, mitigation techniques and role of smart inverters, Journal of Circuits, Systems, and Computers (2020) 2030014.

[2] Dechang Yang, et al., Bidding strategy for virtual power plant considering the large-scale integrations of electric vehicles, IEEE Transactions on Industry Applications (2020).

[3] Chunhua Liu, et al., Opportunities and challenges of vehicle-to-home, vehicle-to-vehicle, and vehicle-to-grid technologies, Proceedings of the IEEE 101 (11) (2013) 2409–2427.

[4] Diyun Wu, K.T. Chau, Shuang Gao, Multilayer framework for vehicle-to-grid operation, in: 2010 IEEE Vehicle Power and Propulsion Conference, IEEE, 2010.
[5] Ali Ahmadian, Mahdi Sedghi, Masoud Aliakbar-Golkar, Fuzzy load modeling of plug-in electric vehicles for optimal storage and DG planning in active distribution network, IEEE Transactions on Vehicular Technology 66 (5) (2016) 3622–3631.
[6] Seyed Masoud Moghaddas-Tafreshi, et al., Optimal operation of an energy hub considering the uncertainty associated with the power consumption of plug-in hybrid electric vehicles using information gap decision theory, International Journal of Electrical Power & Energy Systems 112 (2019) 92–108.
[7] Milad Kabirifar, et al., Deterministic and probabilistic models for energy management in distribution systems, in: Handbook of Optimization in Electric Power Distribution Systems, Springer, Cham, 2020, pp. 343–382.
[8] Niloofar Pourghaderi, et al., Commercial demand response programs in bidding of a technical virtual power plant, IEEE Transactions on Industrial Informatics (2018).
[9] Helena M.D. Espassandim, et al., Optimal operation of electric vehicle parking lots with rooftop photovoltaics, in: 2019 IEEE International Conference of Vehicular Electronics and Safety (ICVES), IEEE, 2019.
[10] Mohamed Lotfi, et al., Coordinated operation of electric vehicle parking lots and smart homes as a virtual power plant, in: 2020 IEEE International Conference on Environment and Electrical Engineering and 2020 IEEE Industrial and Commercial Power Systems Europe (EEEIC/I&CPS Europe), IEEE, 2020.
[11] Simone Minniti, et al., Local markets for flexibility trading: key stages and enablers, Energies 11 (11) (2018) 3074.
[12] D. Pudjianto, et al., The virtual power plant: enabling integration of distributed generation and demand, Fenix Bulletin 2 (2008) 10–16.
[13] Evaggelos G. Kardakos, Christos K. Simoglou, Anastasios G. Bakirtzis, Optimal offering strategy of a virtual power plant: a stochastic bi-level approach, IEEE Transactions on Smart Grid 7 (2) (2016) 794–806.
[14] Ali Shayegan Rad, et al., Risk-based optimal energy management of virtual power plant with uncertainties considering responsive loads, International Journal of Energy Research 43 (6) (2019) 2135–2150.
[15] Shuangrui Yin, et al., Energy management for aggregate prosumers in a virtual power plant: a robust Stackelberg game approach, International Journal of Electrical Power & Energy Systems 117 (2020) 105605.
[16] Rahmat-Allah Hooshmand, Seyyed Mostafa Nosratabadi, Eskandar Gholipour, Event-based scheduling of industrial technical virtual power plant considering wind and market prices stochastic behaviors—a case study in Iran, Journal of Cleaner Production 172 (2018) 1748–1764.
[17] Niloofar Pourghaderi, et al., Energy management framework for a TVPP in active distribution network with diverse DERs, in: 2019 27th Iranian Conference on Electrical Engineering (ICEE), IEEE, 2019.
[18] Herlambang Setiadi, et al., Modal interaction of power systems with high penetration of renewable energy and BES systems, International Journal of Electrical Power & Energy Systems 97 (2018) 385–395.
[19] Qiang Wang, Shuyu Li, Zhanna Pisarenko, Heterogeneous effects of energy efficiency, oil price, environmental pressure, R&D investment, and policy on renewable energy-evidence from the G20 countries, Energy 209 (2020) 118322.
[20] Wenjun Tang, Hong-Tzer Yang, Optimal operation and bidding strategy of a virtual power plant integrated with energy storage systems and elasticity demand response, IEEE Access 7 (2019) 79798–79809.

[21] Huy Nguyen-Duc, Nhung Nguyen-Hong, A study on the bidding strategy of the Virtual Power Plant in energy and reserve market, Energy Reports 6 (2020) 622–626.
[22] Arian Zahedmanesh, Kashem M. Muttaqi, Danny Sutanto, A consecutive energy management approach for a VPP comprising commercial loads and electric vehicle parking lots integrated with solar PV units and energy storage systems, in: 2019 1st Global Power, Energy and Communication Conference (GPECOM), IEEE, 2019.
[23] Mohammad Faisal, et al., Review of energy storage system technologies in microgrid applications: issues and challenges, IEEE Access 6 (2018) 35143–35164.
[24] Alex Eller, Dexter Gauntlett, Energy Storage Trends and Opportunities in Emerging Markets, Navigant Consulting Inc., Boulder, CO, USA, 2017.
[25] New industry survey, Bloomberg New Energy Finance, London, 2021.
[26] Xiangjun Li, Shangxing Wang, A review on energy management, operation control and application methods for grid battery energy storage systems, CSEE Journal of Power and Energy Systems (2019).
[27] Nadeeshani Jayasekara, Mohammad A.S. Masoum, Peter J. Wolfs, Optimal operation of distributed energy storage systems to improve distribution network load and generation hosting capability, IEEE Transactions on Sustainable Energy 7 (1) (2015) 250–261.
[28] Yizhou Zhou, et al., Four-level robust model for a virtual power plant in energy and reserve markets, IET Generation, Transmission & Distribution 13 (11) (2019) 2036–2043.
[29] Yizhou Zhou, et al., A linear chance constrained model for a virtual power plant in day-ahead, real-time and spinning reserve markets, in: 2019 IEEE Power & Energy Society General Meeting (PESGM), IEEE, 2019.
[30] Maura Musio, Alfonso Damiano, A virtual power plant management model based on electric vehicle charging infrastructure distribution, in: 2012 3rd IEEE PES Innovative Smart Grid Technologies Europe (ISGT Europe), IEEE, 2012.
[31] Maura Musio, P. Lombardi, Alfonso Damiano, Vehicles to grid (V2G) concept applied to a virtual power plant structure, in: The XIX International Conference on Electrical Machines-ICEM 2010, IEEE, 2010.
[32] Niloofar Pourghaderi, et al., Energy scheduling of a technical virtual power plant in presence of electric vehicles, in: Electrical Engineering (ICEE), 2017 Iranian Conference on, IEEE, 2017.
[33] Mohammad Rastegar, Mahmud Fotuhi-Firuzabad, Matti Lehtonen, Home load management in a residential energy hub, Electric Power Systems Research 119 (2015) 322–328.
[34] Mohammad Rastegar, Mahmud Fotuhi-Firuzabad, Farrokh Aminifar, Load commitment in a smart home, Applied Energy 96 (2012) 45–54.
[35] Bernhard Jansen, et al., Architecture and communication of an electric vehicle virtual power plant, in: 2010 First IEEE International Conference on Smart Grid Communications, IEEE, 2010.
[36] Kang Miao Tan, Vigna K. Ramachandaramurthy, Jia Ying Yong, Integration of electric vehicles in smart grid: a review on vehicle to grid technologies and optimization techniques, Renewable & Sustainable Energy Reviews 53 (2016) 720–732.
[37] M. Hadi Amini, Arif Islam, Allocation of electric vehicles' parking lots in distribution network, in: ISGT 2014, IEEE, 2014.
[38] Françoise Nemry, Guillaume Leduc, Almudena Muñoz, Plug-in hybrid and battery electric vehicles: state of the research and development and comparative analysis of energy and cost efficiency, JRC Technical Notes, 2009.
[39] Mart van der Kam, Wilfried van Sark, Smart charging of electric vehicles with photovoltaic power and vehicle-to-grid technology in a microgrid; a case study, Applied Energy 152 (2015) 20–30.

[40] A.D. Domínguez-García, G.T. Heydt, S. Suryanarayanan, Implications of the smart grid initiative on distribution engineering (final project report – part 2), PSERC Document, 2011, 11-05.
[41] Patricia Sandmeier, Simon Felsenstein, Electric vehicle infrastructure ABB Market potentials Electric vehicle infrastructure, using the Swiss example, ABB Sales Switzerland Business Development, 2009.
[42] Rob Van Haaren, Assessment of electric cars' range requirements and usage patterns based on driving behavior recorded in the National Household Travel Survey of 2009, Earth and Environmental Engineering Department, Columbia University, Fu Foundation School of Engineering and Applied Science, New York, 2011.
[43] 2017 National Household Travel Survey, U.S. transportation department. [Online]. Available: https://nhts.ornl.gov/assets/2017_nhts_summary_travel_trends.pdf.
[44] Milad Kabirifar, et al., Centralized framework to coordinate residential demand response potentials, in: Proceeding of International Smart Grid Conference, Gwangjo, Korea, 2015, pp. 243–248.
[45] Maziar Yazdani-Damavandi, et al., Modeling operational behavior of plug-in electric vehicles' parking lot in multienergy systems, IEEE Transactions on Smart Grid 7 (1) (2015) 124–135.
[46] Pablo Ralon, et al., Electricity Storage and Renewables: Costs and Markets to 2030, International Renewable Energy Agency, Abu Dhabi, UAE, 2017.
[47] Choton K. Das, et al., Overview of energy storage systems in distribution networks: placement, sizing, operation, and power quality, Renewable & Sustainable Energy Reviews 91 (2018) 1205–1230.
[48] Abdulwahab Alhamali, et al., Review of energy storage systems in electric grid and their potential in distribution networks, in: 2016 Eighteenth International Middle East Power Systems Conference (MEPCON), IEEE, 2016.
[49] Luis Munuera, Energy Storage: Tracking Clean Energy Progress, International Energy Agency, 2019.
[50] L. Igualada, C. Corchero, M. Cruz-Zambrano, F. Heredia, Optimal energy management for a residential microgrid including a vehicle-to-grid system, IEEE Transactions on Smart Grid 5 (4) (Jul. 2014) 2163–2172.
[51] Mohammad Rasol Jannesar, et al., Optimal placement, sizing, and daily charge/discharge of battery energy storage in low voltage distribution network with high photovoltaic penetration, Applied Energy 226 (2018) 957–966.
[52] Peng Rong, Massoud Pedram, An analytical model for predicting the remaining battery capacity of lithium-ion batteries, IEEE Transactions on Very Large Scale Integration (VLSI) Systems 14 (5) (2006) 441–451.
[53] Habiballah Rahimi-Eichi, et al., Battery management system: an overview of its application in the smart grid and electric vehicles, IEEE Industrial Electronics Magazine 7 (2) (2013) 4–16.
[54] Chang Liu, Yujie Wang, Zonghai Chen, Degradation model and cycle life prediction for lithium-ion battery used in hybrid energy storage system, Energy 166 (2019) 796–806.
[55] Niloofar Pourghaderi, et al., Energy and flexibility scheduling of DERs under TVPP's supervision using market-based framework, in: Proceeding of 2020 IEEE 4th International Conference on Intelligent Energy and Power Systems (IEPS), IEEE, Istanbul, Turkey, 2020.
[56] Stephen Carr, et al., Energy storage for active network management on electricity distribution networks with wind power, IET Renewable Power Generation 8 (3) (2013) 249–259.

[57] Miguel A. López, et al., Demand-side management in smart grid operation considering electric vehicles load shifting and vehicle-to-grid support, International Journal of Electrical Power & Energy Systems 64 (2015) 689–698.
[58] Allen J. Wood, Bruce F. Wollenberg, Gerald B. Sheblé, Power Generation, Operation, and Control, John Wiley & Sons, 2013.
[59] Hongming Yang, et al., Distributed optimal dispatch of virtual power plant via limited communication, IEEE Transactions on Power Systems 28 (3) (2013) 3511–3512.
[60] Daniel K. Molzahn, et al., A survey of distributed optimization and control algorithms for electric power systems, IEEE Transactions on Smart Grid 8 (6) (2017) 2941–2962.

EVs vehicle-to-grid implementation through virtual power plants

Moein Aldin Parazdeh, Navid Zare Kashani, Davood Fateh, Mojtaba Eldoromi, and Ali Akbar Moti Birjandi
Electrical Engineering Department, Shahid Rajaee Teacher Training University, Tehran, Iran

Nomenclature

Variables
$e_{u,t}^{EP}$ Energy supply by EP u during period t (kWh)
$e_{v,t}^{PHEV}$ Battery energy level of EV v at the end of period t (kWh)
$e_{v,t}^{+}, e_{v,t}^{-}$ Energy transferred to/from EV v during period t (kWh)
$x_{v,t}$ 1 if EV v is charged during period t, 0 otherwise
$d_{v,t}^{CD}, d_{v,t}^{CS}$ Travel distance in charge-depleting and charge-sustaining mode during period t by EV v (miles)
$E_{v,t}^{req}$ Required energy for EV v during period t (kWh)
$\delta_{v,t}$ Required energy for EV v during period t (kWh)
$f(\delta)$ Expected battery replacement cost as a function of DoD
$o_{u,t}, s_{u,t}, z_{u,t}$ 1 if EP u is online, started up or shut down respectively in period t, 0 otherwise
$r_{v,t}$ The cost variable of battery deterioration of EV v at period t

Parameters
V, U, T Sets of EVs, EPs, and time periods
$I_{v,t}$ 1 if EV v is available for charging during period t, and 0 otherwise
I_{t}^{TIE} Energy demand to meet the TIE loads during period t (kWh)
$\bar{G}_{u,t}, \underline{G}_{u,t}$ Maximum and minimum energy supply by EP u during period t, respectively (kWh)
$\bar{P}_v, \underline{P}_v$ Battery maximum and minimum energy capacity of EV v, respectively (kWh)
$P0_v$ Initial energy stored in battery of EV v (kWh)
ρ_v^{+}, ρ_v^{-} Total energy transferable to/from EV v during one time period, respectively (kWh)
τ_u^{+}, τ_u^{-} Minimum up and down times of EP u
η_u^{EP} Efficiency of EP u due to plant side and transformer losses
η_v^{+}, η_v^{-} Battery charge and discharge efficiency of EV v, respectively
$C_{u,t}^{EP}$ Price of obtaining energy from EP u during period t (paid by VPP to EPs) (¢/kWh)
C_t^{gas} Price of gasoline during period t (¢/gallon)
cs_v Average gasoline usage of EV v (gallon/mile)
$d_{v,t}^{total}$ Total travel distance during period t by EV v (miles)
E_v Energy required to run EV v on electricity for one mile (kWh)

13.1 Introduction

Electric vehicles (EVs) are distributed energy sources in future smart grids. The main difference between EVs and other DERs is mobility. EVs can connect to different parts of the network and still use the same quality of service. The integration of electric vehicles and other DER units into power systems under the principle of "fit and forget" is not efficient for the safe and secure operation of power grids [1–3]. Several electricity suppliers and DERs can be brought together to meet the demand load. VPP (virtual power plant), a newly introduced aggregation unit, is responsible for load management and resource planning. It obtains energy from DER and contracts with consumers to power EVs and their residential loads. For this purpose, it creates scale savings in a completely new way [4]. VPPs minimize the total cost, while ensuring efficient use of the energy generated by the DER. They have no large-scale infrastructure requirements and can communicate with the smallest DERs with higher efficiency and flexibility [5]. This new technology is commonly referred to as the "energy Internet" [4]. Some real examples of VPP are presented in [6].

The terms "VPP" and "micronetwork" are usually used interchangeably, except that the micronetwork can be "islanded" from the network, while the VPP must be connected to the network [4]. The collector centrally controls the energy distribution for a set of consumers with TE (time-elastic) and TIE (time-inelastic) loads. The latter, such as refrigeration or heating, must be satisfied as soon as requested, while the former, like EVs, can be programmed arbitrarily during the connection period. For example, the EV owner, after returning from work at night, connects the vehicle to the mains and needs a fully charged battery by morning. If needed, the collector can discharge energy from the EV at any time of the night, provided the battery reaches the required level by morning. Fig. 13.1 shows the VPP energy planning process. In the context of this study, electricity suppliers and DERs are referred to as EPs (energy suppliers).

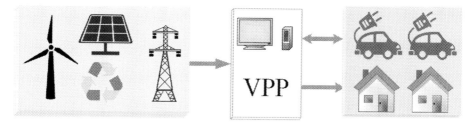

Figure 13.1 VPP energy planning process.

To enable V2G functionality, a series of power electronics converters is needed to connect the battery pack of the EVs to the grid. These converters must have control over both power flow directions to facilitate bidirectional exchange of active power between the AC mains and the DC EV battery. Different topologies are presented in the literature to achieve this feature. In this chapter after a short review on V2G infrastructure, different topologies that can be used in the system are discussed and compared to each other. After that, some common problems in the system such as energy management and battery degradation are modeled. Finally, as a case study, the performance of the system due to some different possibilities is assessed and the results validate the stability and reliability of the system.

13.2 Vehicle-to-grid (V2G)

The evolution of energy related to the development of renewable energy and the rational use of energy is changing the role of the electricity grid, especially the distribution network. The electricity distribution network is currently the main energy interface for renewable sources. This fundamentally changes the task of this system, a task that is increasingly responsible for the operation of interconnection systems between distributed generations and the end user. The rapid growth of distributed generation and the changes in the time and space of production and energy consumption complicate the system too much to be managed with the models used today. This will force the implementation of a new energy management model. One of these networks is the smart grid, which is defined as a modern network that uses advanced communication and control technologies to generate and distribute electricity more efficiently, economically, and safely. These structures represent the backbone of VPPs and require the widespread presence of electrical energy storage systems. EV, and especially its energy storage equipment, is one of the most interesting technical solutions to implement outlet storage and allow the use of these models.

Several models have been developed that describe the future impact of EVs on the grid performance and economics. One of them proposes "vehicle-to-grid" (V2G) technology as a possible solution to affect the stability of the distributed generation development network. In such a model, electric vehicles are considered as both a means of transportation and an energy storage system. EV can be used for multiple network services to stabilize the network and support the exploitation of renewable energy sources. This consideration is based on the analysis of the number of cars per capita, the distribution of space and time of the car, the average usage coefficient, and the correlation of time and place between electrical energy and car use. In fact, vehicles are typically parked 90% of the time and are an idle electrochemical source that can absorb energy to or from the grid when connected. This is the basic concept of the so-called vehicle-to-grid (V2G) technology, in which EV must have three or more elements to provide such services. First of all, it must have a bidirectional power interface to supply or absorb energy. The second essential element is the energy management system (EMS). Such an element must control and measure the energy flows from/to the battery pack. Finally, a communication protocol is needed between the network operator and the EV owner to transfer all useful information to the EMS and evaluate the cost and revenue of the service.

13.3 Bidirectional converters for V2G systems

The first stage of a bidirectional converter circuit, shown in Fig. 13.2, is a bidirectional AC–DC converter (BADC), which is required to modify the power factor correction (PFC), as well as to facilitate bidirectional AC–DC power conversion. The second stage is a bidirectional DC–DC converter (BDC), which provides a voltage match between the EV battery and the DC bus. Although some current research focuses on using an AC–DC power conversion stage to minimize system components, cost, and footprint, much of the literature has reported that two dedicated bidirectional power conversion stages have been used: one AC–DC and one DC–DC for V2G systems. Bidirectional converters usually use two current control methods. The indirect current control method uses PWM control on the AC side to generate sine PWM so that phase and amplitude can be controlled. Its implementation is due to the lack of a simple closed current-control loop, but it has a sluggish transient function and the presence of a DC component on the AC side. These shortcomings are addressed in the direct current control method, which consists of two closed-loop control systems, namely an external voltage loop to stabilize the DC voltage and an internal current loop to control the inductor current. The direct current control method also improves the static and dynamic performance of the system [7–12].

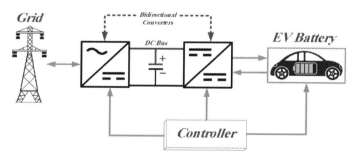

Figure 13.2 Bidirectional AC–DC converter (BADC) connected to the grid.

13.4 Bidirectional AC–DC converter (BADC)

BADC acts as a rectifier that converts AC power to DC power in G2V mode and as an inverter that converts DC power to AC power in V2G mode. As a requirement, the BADC must also comply with power factor correction (PFC) and harmonic injection into and from the network to meet standards such as IEEE Std. 519-2014. An important aspect to watch out for in BADC in G2V mode is the accurate detection of zero-voltage grids, which in the event of failure can lead to a neutral-phase short circuit through the mains converter and distortion of the input voltage and current waveform [13,14]. These distortions occur mainly due to the forward voltage drop across the switches and diodes at zero passes [15]. Ideally, the voltage provided in

V2G mode should be compatible with the mains voltage in terms of voltage level, phase, and frequency. It is desirable that the total harmonic distortion (THD) of the output current in V2G mode be less than 5%.

Since most V2G systems use two dedicated converters, one BADC with BDC, the performance of each converter is clearly influenced by the coupled converter. Thus, a direct comparison of the performance of BADCs without considering the effect of associated BDC is not logical. But in order to evaluate the performance characteristics to accurately compare different BADCs, each of the BADC topologies must be tested with a specific BDC topology, which is not feasible. A comparison of different BADC topologies represented in Table 13.1.

As can be seen from the table, full bridge BADC is the most common topology due to its simple structure and control flexibility. The power factor performance of the unit is easily achievable. However, the harmonic distortion of the current waveform can be a major concern, and alternative control methods must be identified to address this problem. In addition, most full-bridge BADCs in the literature indicate the need for a large electrolytic bus capacitor to function as an intermediate energy storage element, which increases system cost and efficiency, and reduces reliability and power compression. Sine charging of EV batteries, as discussed in the section below, can effectively solve this problem but at the cost of degrading the battery. Single-stage conversion topologies, including matrix converters, eliminate the need for DC-bus electrolytic capacitors as intermediate energy storage elements, but require additional stabilizing circuits to maintain unity power factor operation, while ensuring minimum THD. Although the BADC multilevel topology is scalable and offers less THD without the use of filters, it requires more switching devices compared to the full bridge topology and should therefore only be considered for very high charge/discharge systems. The eight-switch topology in the literature is a good candidate for high-power charger/discharge systems, but requires twice the number of switches compared to a full bridge topology, resulting in lower reliability and higher cost.

13.5 Bidirectional DC–DC converter (BDC)

Bidirectional DC–DC converters (BDCs) are electronic power converters that enable the bidirectional conversion of DC voltage between two levels, with the ability to reverse the direction of current by receiving the appropriate control command [16]. One of the two voltage levels has a higher value compared to the other. The two most important BDC functions in a V2G connection system can be expressed exactly as follows. The first is mode, when the EV battery voltage is converted to DC bus voltage. It is always recommended and expected that the BDC will charge EV batteries after two charging phases – a constant current phase and a constant voltage phase [17,18]. BDCs always operate at high switching frequencies to increase their power density capability. As a result, HF noise is generated due to the rapid switching on and off of devices, which causes EMI in the network and disrupts the performance of other equipment connected to the network. Therefore, effective EMI reduction and control techniques should be considered when designing BDCs.

Table 13.1 Comparison of BADCs developed for V2G applications.

BADC topology	Rated power	DC bus/ Battery voltage range	Number of switching devices	Passive requirements (excluding filter)	Filter requirements	Power factor	THD	Remark
full-bridge	500 W	60–120 V	4 IGBTs	4 diodes	LC on grid-side	1	not known	hard-switched, low efficiency
full-bridge	500 W	60–120 V	4 MOSFETs with body diode	nil	LC on grid-side	0.991	4.3%	no DC-bus capacitor, smaller footprint
full-bridge	3.5 kW	300–340 V	4 IGBTs with freewheeling diode	CBUS	RLC on grid-side	1	not known	no galvanic isolation
full-bridge	3.6 kW	270–360 V	4 IGBTs with freewheeling diode	CBUS	L on grid-side	0.99	< 3%	low THD
full-bridge	3 kW	120 V	4 IGBTs with freewheeling diode	Lg, CBUS	nil	1	4.5%	high THD
full-bridge	3.3 kW	400–450 V	4 MOSFETs with body diode	CBUS, snubber capacitor across each switch	LCL on grid-side	variable	not known	fast response to transient load demands, VAR compensation

continued on next page

Table 13.1 (continued)

BADC topology	Rated power	DC bus/Battery voltage range	Number of switching devices	Passive requirements (excluding filter)	Filter requirements	Power factor	THD	Remark
full-bridge	400 W	120 V	4 IGBTs with freewheeling diode	nil	LC on grid-side	variable	6.15%	VAR compensation, high THD
three-phase full-bridge	20 kW	800 V	6 MOSFETs with body diode	CBUS, grid-side	LCL on grid-side resistances	not known	~3.3%	97–99% efficiency due to SiC devices
eight-switch	30 kW	177–201 V	8 IGBTs with freewheeling diode	Lg, CBUS	LC on grid-side	1	2.72%	can be connected to 3-φ grids, acceptable THD
three-level	18 kW	345.6 V	6 IGBTs with freewheeling diode	Lg, 2 CBUS, 2 diodes	nil	1	2.35%	scalable, small grid-side inductor, more number of switches, high conduction loss
single-stage isolated	3.3 kW	280–430 V	8 MOSFETs with body diode	nil	2 LC, one each on grid and battery-side	0.985 at 1.656 kW	<5% at 1.656 kW	97.8% peak efficiency, complex control

continued on next page

Table 13.1 (continued)

BADC topology	Rated power	DC bus/Battery voltage range	Number of switching devices	Passive requirements (excluding filter)	Filter requirements	Power factor	THD	Remark
single-stage cyclo-converter	33.3 kW	600 V	8 IGBTs with freewheeling diode	nil	L on grid-side	not known	not known	97% efficiency, soft-switched, accurate design of center tapped transformer
non-isolated matrix	10 kW	24–72 V	12 IGBTs with freewheeling diode	nil	3-φ LC on grid-side, LC on battery-side	1	not known	cost-effective, fast response to transient load demands, low utilization of components, complex control, no galvanic isolation

continued on next page

Table 13.1 (continued)

BADC topology	Rated power	DC bus/Battery voltage range	Number of switching devices	Passive requirements (excluding filter)	Filter requirements	Power factor	THD	Remark
isolated matrix	30 kW	500 V	16 IGBTs with free-wheeling diode	Lr, Cr	3-φ LC on grid-side, LC on battery-side	1	< 3% (G2V), < 5% (V2G)	fast response to transient load demands, fast charging, galvanic isolation
isolated matrix	3.6 kW	384 V	12 IGBTs with free-wheeling diode	nil	RLC on grid-side, C on battery-side	1	not known	low switching frequency
isolated matrix	4 kW	200 V	8 switches with free-wheeling diode	Lr, Cr	LC on grid-side	~1	< 4%	cost-effective

Table 13.2 compares the different BDC topologies implemented in V2G systems. Buck increase is the most common non-isolated BDC topology, while full bridge is the most common isolated BDC topology. Non-isolated topologies have smaller footprint, lower weight, lower cost, and easier control due to the lack of isolation transformers. Fewer switching devices increase the use of components in such converters. However, they limit the power density capability due to excessive conduction loss from high current through switching devices and other passive components, resulting in low efficiency of these converters. Therefore, if you want fast charging/discharging capability, cross-connection in unbundled BDCs is preferred. In addition, soft switching techniques should be considered to eliminate switching losses, as control algorithms in non-isolated BDCs do not provide normal ZVS/ZCS. To achieve excellent transient performance, the torture topology can be considered, which also provides scaling flexibility if desired for high power transmission capability.

Isolated BDC topologies offer better safety in the event of internal faults, but at the expense of an HF isolation transformer, which increases footprint, weight, and system cost. Nevertheless, isolation in accordance with international standards is essential, especially when high power transmission occurs. Non-resonant topologies are a desirable option due to the reduction in the number of passive components, which increases the reliability of these converters. In addition, a carefully designed control algorithm can eliminate the need for sabotage or closure circuits to achieve ZVS/ZCS, thus reducing the cost and size of BDC. However, compared to the resonant topology, the flow through switching devices in the non-resonant topology is slightly higher, resulting in higher conductivity loss and lower efficiency. Resonance topologies show very good performance in both currents, but the transient response is often sluggish. In addition, a variable frequency is usually required to ensure ZVS over a wide input voltage range and all load conditions. Although additional inactive elements are required to form the resonant circuit in such transducers, the need for sabotage or closing circuits can be eliminated for soft switching of devices. If there is a demand for further reductions in the size and location of the BDC, sinusoidal charging can be considered, but battery damage is inevitable. Broadband devices can lead to high efficiency, better power density and thermal properties, and can be the best power devices for very high charge/discharge systems. The performance of these devices in non-isolated BDC topologies for V2G applications has not yet been investigated [19].

13.6 Modeling the problem

This section is made up of two subsections: the first introduces the VPP energy management model, and the second deals with the battery degradation cost modeling.

Table 13.2 Comparison of BDCs developed for V2G applications.

BDC topology	Type of feed	Rated power	DC battery voltage range	Number of switching devices	Passive requirements (excluding filter)	Battery-side filter	Efficiency	Remark
buck-boost (with auxiliary switching network)	current-voltage-fed	500 W	60–120 V	2 IGBTs	2 diodes, Lin	C	83%	low efficiency, high charging current ripple
buck-boost (with auxiliary switching network)	dual current fed	500 W	60–120 V	2 IGBTs	2 diodes, Lin	LC	< 85%	low efficiency, charging current ripple
buck-boost	dual voltage-fed	3.5 kW	270–360 V	2 IGBTs with free-wheeling diode	CBUS	LC	> 90%	fewer components
buck-boost	current-voltage-fed	1.2 kW	106–136 V	2 IGBTs with free-wheeling diode	CBUS	LC	not known	fewer components, no experimental verification
interleaved buck-boost	current-voltage-fed	30 kW	177–201 V	4 IGBTs with free-wheeling diode	2 Lin, CBUS	C	not known	high power transfer capability, small inductor sizing and current ripple

continued on next page

Table 13.2 (continued)

BDC topology	Type of feed	Rated power	DC battery voltage range	Number of switching devices	Passive requirements (excluding filter)	Battery-side filter	Effi-ciency	Remark
interleaved buck-boost	current-voltage-fed	400 W	120 V	4 IGBTs with free-wheeling diode	2 Lin	C	> 94%	inverted output, low power output
non-inverted buck-boost	dual voltage-fed	18 kW	345.6 V	5 IGBTs with free-wheeling diode	boost inductor, 5 diodes	C	95.25%	high number of switching devices and diodes
cascaded buck-boost	dual current fed	9 kW	350 V	4 IGBTs	Cin, Lin, CBUS	LC	91.61%	modular structure, excellent transient performance
dual full-bridge	dual voltage-fed	3.3 kW	235–430 V	8 MOSFETs with body diode	CBUS, Llk, snubber capacitor across each switch	CLC	not known	low EMI, low diode reverse recovery problem
dual full-bridge	dual voltage-fed	30 kW	360 V	8 IGBTs with free-wheeling diode	CBUS, Llk	C	not known	reactive power compensation

continued on next page

Table 13.2 (continued)

BDC topology	Type of feed	Rated power	DC battery voltage range	Number of switching devices	Passive requirements (excluding filter)	Battery-side filter	Efficiency	Remark
dual full-bridge	current-voltage-fed	250 W	150–300 V	8 MOSFETs with body diode	Cin, Lin, Llk, snubber capacitor across each switch	C	93%	high number of switching devices, control flexibility, symmetric structure, high conduction losses
half-full-bridge	current-voltage-fed	1 kW	250–450 V	6 MOSFETs with body diode	Cin, 2 Lin, Llk, snubber capacitor across each switch	C	95.8%	control flexibility, symmetric structure, high efficiency
full-bridge resonant	dual voltage-fed	3.5 kW	250–450 V	4 IGBTs with free-wheeling diode for 1° devices, 4 MOSFETs with body diode 2° devices	Cin, 2 Cr, 2 Llk, Lm	C	97.7% (G2V), 98.1% (V2G)	high efficiency, variable frequency operation to ensure ZVS

continued on next page

Table 13.2 (continued)

BDC topology	Type of feed	Rated power	DC battery voltage range	Number of switching devices	Passive requirements (excluding filter)	Battery-side filter	Efficiency	Remark
full-bridge resonant	dual voltage-fed	6.6 kW	250–415 V	8 MOSFETs with body diode	Cr, Llk, Lm, snubber capacitor across each switch, charge-pump capacitor	LC	97.7% (G2V), 97.3% (V2G)	high efficiency, high power density capability, slow transient response
full-bridge resonant	dual voltage-fed	2.5 kW	50 V	4 IGBTs with free-wheeling diode for 1° devices, 4 MOSFETs with body diode 2° devices	Cin, Cr, 2 Llk	C	96%	high efficiency, wide voltage conversion range, low THD
half-bridge resonant	dual voltage-fed	3.3 kW	360 V	4 SiC FETs with body diode	Cin, 4 Cr, 2 Llk, Lm	C	96%	small footprint, high efficiency, high power density capability, complex transformer design

continued on next page

Table 13.2 (continued)

BDC topology	Type of feed	Rated power	DC battery voltage range	Number of switching devices	Passive requirements (excluding filter)	Battery-side filter	Efficiency	Remark
half-bridge resonant	dual voltage-fed	3.3 kW	250–420 V	4 MOSFETs with body diode	4 Cr, 2 Llk	C	97.5% (G2V), 97.3% (V2G)	high efficiency, synchronous rectification
half-bridge resonant	dual voltage-fed	3.3 kW	250–410 V	4 IGBTs with free-wheeling diode	2 Cin, 2 Cr, Llk	LC	95.7% (G2V), 95.4% (V2G)	high efficiency, sinusoidal charging, battery degradation
half-full-bridge resonant	dual voltage-fed	10 kW	400 V	6 SiC FETs with body diode	2 Cr, Llk, Lm	LC	>96% (G2V), >98% (V2G)	ZVS for wide DC-bus voltage range, high efficiency ZVS for wide DC-bus voltage range, high efficiency

13.6.1 VPP energy management model

The objective of the VPP energy management model is to minimize the total cost of energy in order to satisfy the loads and to generate the charging and discharging schedules of the EVs. The parameters and the variables to be used are presented below.

The cost minimization model is as follows:

$$\min \sum_{t \in T} \left[\sum_{u \in U} C_{u,t}^{EP} \times e_{u,t}^{EP} + \sum_{u \in U} C_u^{SU} \times s_{u,t} \right. \\ \left. + \sum_{v \in V} \left(C_t^{gas} \times cs_v \times d_{v,t}^{CS} + \left[f\left(\delta_{v,t}\right) - f\left(\delta_{v,t-1}\right)^+ \right] \right) \right] \quad (13.1)$$

subject to

$$\underline{P}_v \leq e_{v,t}^{PHEV} \leq \bar{P}_v; \quad \forall v \in V, \forall t \in T, \quad (13.2)$$

$$e_{v,t}^{PHEV} = e_{v,t-1}^{PHEV} + \eta_v^+ \times e_{v,t}^+ - \frac{1}{\eta_v^-} \times e_{v,t}^- - E_{v,t}^{req}; \quad \forall v \in V, \forall t \in T, \quad (13.3)$$

$$e_{v,0}^{PHEV} = P0_v; \quad \forall v \in V, \quad (13.4)$$

$$e_{v,t}^+ \leq \rho_v^+ \times I_{v,t} \times x_{v,t}; \quad \forall v \in V, \forall t \in T, \quad (13.5)$$

$$e_{v,t}^- \leq \rho_v^- \times I_{v,t} \times (1 - x_{v,t}); \quad \forall v \in V, \forall t \in T, \quad (13.6)$$

$$d_{v,t}^{CD} \leq d_{v,t}^{total}; \quad \forall v \in V, \forall t \in T, \quad (13.7)$$

$$d_{v,t}^{CD} \leq \frac{\left(e_{v,t-1}^{PHEV} - \underline{P}_v\right)}{E_v}; \quad \forall v \in V, \forall t \in T, \quad (13.8)$$

$$d_{v,t}^{CS} = d_{v,t}^{total} - d_{v,t}^{CD}; \quad \forall v \in V, \forall t \in T, \quad (13.9)$$

$$E_{v,t}^{req} = d_{v,t}^{tCD} \times E_v; \quad \forall v \in V, \forall t \in T, \quad (13.10)$$

$$\delta_{v,t} = 1 - \frac{e_{v,t}^{PHEV}}{\bar{P}_v}; \quad \forall v \in V, \forall t \in T, \quad (13.11)$$

$$\sum_{u \in U} \eta_u^{EP} \times e_{u,t}^{EP} + \sum_{v \in V} e_{v,t}^- = \sum_{v \in V} e_{v,t}^+ + l_t^{TIE}; \quad \forall t \in T, \quad (13.12)$$

$$o_{u,t} \times \underline{G}_{u,t} \leq e_{u,t}^{EP} \leq o_{u,t} \times \bar{G}_{u,t}; \quad \forall u \in U, \forall t \in T, \quad (13.13)$$

$$o_{u,t} - o_{u,t-1} = s_{u,t} + z_{u,t}; \quad \forall u \in U, \forall t \in T, \quad (13.14)$$

$$\sum_{y=t-\tau_u^+} s_{u,y} \leq o_{u,t}; \quad \forall u \in U, \forall t \in T, \quad (13.15)$$

$$\sum_{y=t-\tau_u^-} z_{u,y} \leq 1 - o_{u,t}; \quad \forall u \in U, \forall t \in T, \quad (13.16)$$

$$x_{v,t}, o_{u,t}, s_{u,t}, z_{u,t} \in \{0,1\}; \quad \forall u \in U, \forall v \in V, \forall t \in T, \quad (13.17)$$

$$e_{u,t}^{EP}, e_{v,t}^{PHEV}, e_{v,t}^+, e_{v,t}^-, E_{v,t}^{req}, d_{v,t}^{CD}, d_{v,t}^{CS}, \delta_{v,t} \geq 0; \quad \forall u \in U, \forall v \in V, \forall t \in T. \quad (13.18)$$

The objective function minimizes the cost of satisfying the residential load and EV travel requirements. The cost components are the cost of energy generation, startup costs of EPs, gasoline prices, and battery degradation cost which is a function of depth of discharge (DoD). The details of this function are presented in the following subsection. Constraints (13.2)–(13.11) are related to EVs. Constraint (13.2) enforces the minimum and maximum charge limits for each EV. Constraint (13.3) is the EV battery storage balance equation between periods. If the battery is charged (discharged), the stored energy level in the battery in the following period is increased (decreased) accordingly. Constraint (13.4) sets the initial battery level of each EV. Constraints (13.5) and (13.6) jointly ensure that charging and discharging do not occur simultaneously in each period and that either can only occur when the EV is connected to the grid. Constraints (13.7) and (13.8) together force the distance traveled in CD mode to be the minimum of the "actual trip distance" and "possible travel distance with the available energy left in the EV battery." Constraint (13.9) sets the CS mode travel distance. Constraint (13.10) calculates the energy required to travel the trip distance and (13.11) calculates the DoD for each period. Constraints (13.12)–(13.16) are related to EPs. Constraint (13.12) is the energy balance equation. The sum of the total energy obtained from EPs and the discharged energy from the batteries equals the supplied energy for the EVs and the TIE loads. Constraint (13.13) ensures that the minimum and maximum capacities of each EP are met, and forces the binary variable out to take value 1 if energy is generated by EP u in period t. Constraint (13.14) sets the startup and shut down binary variables to correct values. Constraints (13.15) and (13.16) enforce the minimum and maximum up and down times of EPs. Finally, (13.17) defines the binary variables and (13.18) forces the non-negativity on the variables.

13.6.2 Theorem

In an optimal solution of the VPP energy management model, simultaneous charging and discharging do not occur for any EV in any period when the binary variable $x_{v,t}$ is omitted from the formulation.

Proof. (By contradiction) Assume that charging and discharging simultaneously occur for EV w during period p in an optimal solution of the above formulation when the binary variable $x_{v,t}$ is omitted. Let $e^+_{w,p} = a > 0$ and $e^-_{w,p} = a > 0$. Then the VPP supplies a unit of energy for the EV w and receives b units of energy from the EV w. Therefore, the net energy generation required is $a - b$ units. On the other hand, due to the charging and discharging efficiencies of the EV w, it actually stores $\eta^+_w \times a$ units of energy and loses $1/\eta^-_w \times b$ units of energy back to the VPP. Then, the net energy stored in the EV w battery is $c = \eta^+_w \times a - (1/\eta^-_w) \times b$. But c units of energy could be directly provided by only charging $1/\eta^+_w \times c$ units of energy, which is equal to $a - (1/\eta^+_w \times \eta^-_w) \times b$. Since $\eta^+_w < 1$ and $\eta^-_w < 1$, we have $(1/\eta^+_w) \times c < a - b$. Thus, the same level of storage for the battery of EV w during period p could be achieved by generating less energy, which means less cost. This contradicts the assumption that the solution with simultaneous charging and discharging is the optimal (i.e., the least-cost) solution.

In other words, because the batteries are not 100% efficient when charging or discharging, it is less costly to charge the required energy directly rather than first charging more energy and then discharging the batteries. Note the non-restrictive underlying assumption that the more energy is obtained from EPs, the more cost the VPP incurs. As a result of this theorem, the binary variable $x_{v,t}$ can be omitted from the energy management model.

Solving the above model gives the least cost that can be achieved by the VPP. Observe that if we solve the model for only a given vehicle v by taking into account the grid as the single EP, then we obtain the least cost of this EV for traveling the desired trip mileage if it was in the national grid domain.

13.6.3 Modeling the battery degradation cost

In order to model the battery degradation cost, we follow a similar methodology to that of Sioshansi and Denholm [20]. The battery of the EV has a limited lifespan and it deteriorates through usage. Therefore, discharging or depleting the battery shortens its life; and after sufficiently many times, the battery needs to be replaced. Therefore, the VPP incurs a cost each time a battery is used. The lifetime of a EV battery is inversely proportional with the DoD [21] as shown in Fig. 13.3.

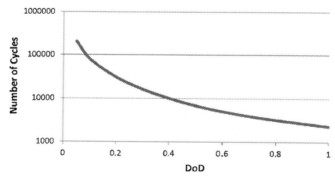

Figure 13.3 Cycle life of EV batteries as a function of DoD [23].

Therefore, the more the energy is discharged, the less the lifespan of the battery is; and this implies more cost for deeper discharging. The nonlinearity in the battery deterioration cost as a function of DoD can be approximated with a piecewise linear function. An example battery degradation function for a battery that costs $4189 is depicted in Fig. 13.4. Approximations by a linear function and a 2-piece piecewise linear function are also depicted in this figure.

To handle the linearization of the battery degradation cost function, we adopted the methodology presented in Nemhauser and Wolsey [22]. Consider a piecewise linear function with K pieces, let a_v^i be the DoD values at the break points, and $f(a_v^i)$ be the function values evaluated at each a_v^i for $i \in \{1, ..., K\}$. Let $r_{v,t} = [f(\delta_{v,t}) - f(\delta_{v,t-1})]^+$ be the cost variable of battery deterioration of EV v at period t; and $\lambda_{v,t}^i$ and $y_{v,t}$ be the auxiliary variables used in linearization. Then, in order

EVs vehicle-to-grid implementation through virtual power plants

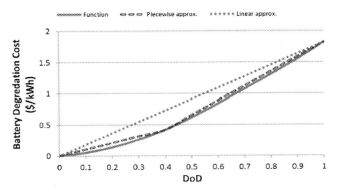

Figure 13.4 Battery degradation cost function and approximating functions [23].

to deal with the $\left[f(\delta_{v,t}) - f(\delta_{v,t-1})\right]^+$ term, we incorporate the following integer linear program into our model:

minimize

$$\min \sum_{t \in T} \sum_{v \in V} r_{v,t} \tag{13.19}$$

subject to

$$r_{v,t} \geq \sum_{i=1}^{K} \lambda_{v,t}^i \times f\left(a_v^i\right) - \sum_{i=1}^{K} \lambda_{v,t-1}^i \times f\left(a_v^i\right); \quad \forall v \in V, \, \forall t \in T, \tag{13.20}$$

$$r_{v,t} \geq 0; \quad \forall v \in V, \, \forall t \in T, \tag{13.21}$$

$$\delta_{v,t} = \sum_{i=1}^{K} \lambda_{v,t}^i \times a_v^i; \quad \forall v \in V, \, \forall t \in T, \tag{13.22}$$

$$\lambda_{v,t}^1 \leq y_{v,t}^1; \quad \forall v \in V, \, \forall t \in T, \tag{13.23}$$

$$\lambda_{v,t}^i \leq y_{v,t}^i + y_{v,t}^{i-1}; \quad \forall v \in V, \, \forall t \in T, \, i = 2, \ldots, K-1, \tag{13.24}$$

$$\lambda_{v,t}^K \leq y_{v,t}^{K-1}; \quad \forall v \in V, \, \forall t \in T, \tag{13.25}$$

$$\sum_{i=1}^{K} \lambda_{v,t}^i = 1; \quad \forall v \in V, \, \forall t \in T, \tag{13.26}$$

$$\sum_{i=1}^{K-1} y_{v,t}^i = 1; \quad \forall v \in V, \, \forall t \in T, \tag{13.27}$$

$$\lambda_{v,t}^i \geq 0; \quad \forall v \in V, \, \forall t \in T, \tag{13.28}$$

$$y_{v,t}^i \in \{0, 1\}; \quad \forall v \in V, \, \forall t \in T, \, i = 1, \ldots, K-1. \tag{13.29}$$

If the $\sum_{t \in T} \sum_{v \in V} \left[f(\delta_{v,t}) - f(\delta_{v,t-1}) \right]^+$ term in the objective function of the VPP energy management model is replaced with $\sum_{t \in T} \sum_{v \in V} r_{v,t}$, and constraints (13.20)–(13.29) are appended to (13.2)–(13.18), we obtain an MILP for the energy management problem. The larger the number of the pieces, the better the approximation is of the nonlinear form.

13.7 Case study

In this study, as mentioned before, due to better performance, structure simplicity, acceptable efficiency and cost, interleaved PFC and dual active bridge are selected as bidirectional AC to DC and bidirectional DC to DC converter, respectively. In Fig. 13.5, the block diagram of the single-phase structure is given [24]. Bidirectional interleaved PFC is used by shaping the current of the grid in sinusoidal shape that is in phase with grid voltage to achieve a unity power factor and reduce the harmonics injected to the power grid.

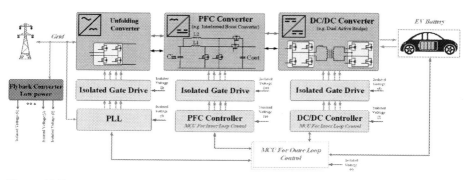

Figure 13.5 Architecture of the single phase battery charger.

Following the PFC stage, there is a DC/DC stage that provides galvanic isolation and generates the output in a way that follows the battery charging pattern. Galvanic isolation plays an important role in order to maintain safety and regulatory standards for the converter. Advantages of this converter are the low number of components used for having two full bridges, soft switching, which is possible without adding auxiliary components, multiple possible modulation techniques, and the bipolar flux swing in the transformer, allowing for a higher power capability compared to the half-bridge topologies. Each voltage source makes a square waveform on each side of the leakage inductor, and the phase shift between these two determines the amount and the direction of the power flow. A disadvantage of this converter is that large input and output capacitors are needed for a low ripple at both sides. Furthermore, the simple phase shift modulation method can lose soft switching for light loads, and with this modulation method the conduction loss is high due to circulating power. There are more sophisticated current modulation techniques available which can improve the soft-switching range and can reduce the circulating reactive power; this will require a more complex controller compared to the other methods. This converter could also work in resonant mode, by placing a resonant tank in-between. Due to the advantages of the operation, the converter can be operated under a higher frequency and efficiency and achieve a wider soft-switching range, but additional resonant components also bring extra size and costs. This converter will have better performance and voltage conversion ratio if a variable switching frequency is permitted, but the variable frequency operation and moving outside a specific frequency range will increase the losses. So, the frequency of this converter should be fixed and the voltage transfer ratio should be set near the turns ratio.

Seamless mode transition between the charging and discharging modes is one of the requirements of the system, so when VPP chooses one charging station to stop charging the battery and instead transfer power to the grid (in case of peak power shaving or frequency regulation, etc.), the converter could follow the rules and reverse its power flow direction without any difficulty. In Fig. 13.6, the simulation result of this mode transition of the abovementioned converter is shown. At first the converter is working in the charging mode, when the current and voltage of the grid are in phase. Suddenly the central control unit of VPP decides that the power must transmit back to the grid due to an emergency. It can be seen that seamless mode transition occurs and there is no extra harmonics injected to the grid. Another requirement for this charger is to handle different current levels due to the variation of the battery voltages in different EVs. In order to maintain maximum available power in every condition, a current loop control, followed by a voltage loop control, is provided to do this task. So, this converter, regardless of the main charging/discharging strategy that is decided by central VPP controlling unit, could interact with different battery voltages in different EVs.

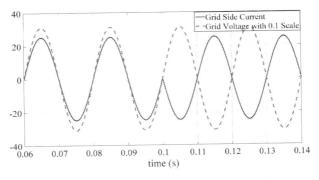

Figure 13.6 Simulated voltage and current of the grid in mode transition.

As an example, in the case which is shown in Fig. 13.7, an EV is parked at the charging station 1 for a long time, so its battery is almost full. Another EV just parked at the station 2 to charge its battery and continue its trip. So, it needs to be charged with maximum power. The VPP central unit decides to deliver some portion of this

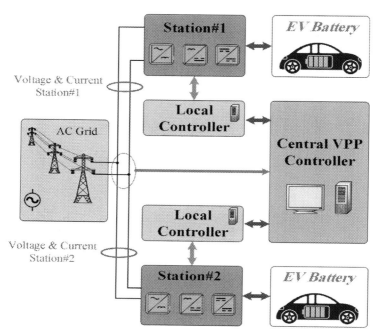

Figure 13.7 Example case, station 2 is being charged by station 1 and the grid.

power from station 1 and the rest of the power from the grid. So immediately station 1 starts discharging and delivering the power to the grid. And the EV at station 2 starts charging with full capacity.

The waveforms of voltage and current of both stations 1 and 2 are shown in Figs. 13.8(a) and 13.8(b), respectively. Such characteristics help the VPP to share the loads with different DERs and improve the stability and reliability of the whole system.

In order to investigate the effect of electric vehicles on power management in a distribution system, the IEEE 69-bus system has been used to investigate the power supply with the help of electric vehicles [25] as shown in Fig. 13.9. In this system, some busbars have several KVA powers that are used by consumers. In Fig. 13.10, gray bar diagrams show the power of crossing lines between busbars and consumers without the use of electric vehicles. These powers are supplied through the power supply.

Assuming that the charged vehicles are connected to high-power busbars, the throughput of the lines between the busbars and the consumers is calculated. This is illustrated with black bar charts. These diagrams show the power absorbed by the same busbars during the connection and discharge of electric vehicles. In this case, the absorbed power from each bus has significantly decreased, so that electric cars in each of these rails have been able to supply several KVA.

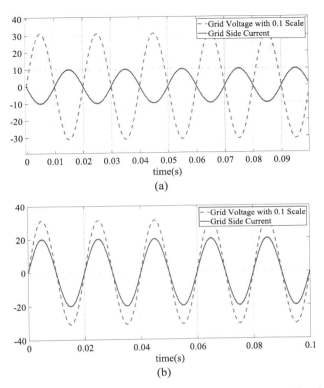

Figure 13.8 (a) Voltage and current of station 1; (b) Voltage and current of station 2.

13.8 Conclusion

Vehicle-to-grid (V2G) is a promising technology that allows idle or parked EVs to act as distributed sources, which can store or release energy at appropriate times, thus allowing the exchange of power between the network and the EV. This increases the total capacity of electricity generation and improves the stability, reliability, and efficiency of the network. Bidirectional electronic converters, namely AC–DC (BADC) and DC–DC (BDC), are commonly used to facilitate G2V and V2G power transmission between the network and the EV battery. A variety of bidirectional converters have been successfully developed and implemented in V2G systems. These converters help achieve high efficiency power conversion, and with the growth of such converters and charging stations will help in the transition from the conventional to electric vehicles, and in the end will lead to a green environment. This chapter reviews the various BADC and BDC topologies implemented in V2G systems. In this study, we consider a group of people who install a VPP into an electricity grid to reduce overall electricity costs through the economy of scale. We assume that the VPP obtains energy from several renewable energy sources.

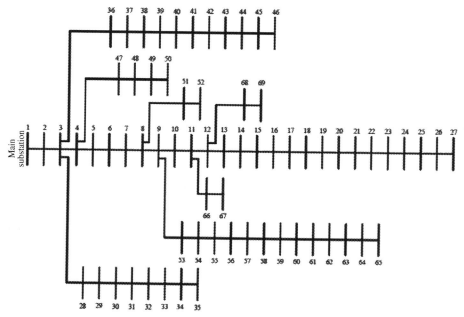

Figure 13.9 IEEE 69-bus system.

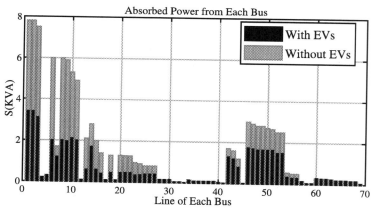

Figure 13.10 Absorbed power from each bus in IEEE 69-bus system.

References

[1] J.A. Peças Lopes, Nikos Hatziargyriou, Joseph Mutale, Predrag Djapic, N. Jenkins, Integrating distributed generation into electric power systems: a review of drivers, challenges and opportunities, Electric Power Systems Research 77 (9) (2007) 1189–1203.

[2] Danny Pudjianto, Charlotte Ramsay, Goran Strbac, Virtual power plant and system integration of distributed energy resources, IET Renewable Power Generation 1 (1) (2007) 10–16.
[3] Kai Strunz, Henry Louie, Cache energy control for storage: power system integration and education based on analogies derived from computer engineering, IEEE Transactions on Power Systems 24 (1) (2008) 12–19.
[4] Peter Asmus, Microgrids, virtual power plants and our distributed energy future, The Electricity Journal 23 (10) (2010) 72–82.
[5] Xi Fang, Satyajayant Misra, Guoliang Xue, Dejun Yang, Smart grid—the new and improved power grid: a survey, IEEE Communications Surveys and Tutorials 14 (4) (2011) 944–980.
[6] Łukasz Bartosz Nikonowicz, Jarosław Milewski, Virtual power plants-general review: structure, application and optimization, Journal of Power Technologies 92 (3) (2012) 135–149.
[7] Fucun Li, Feng Ji, Hongxia Guo, Hao Li, Zhenhua Wang, Research on integrated bidirectional control of EV charging station for V2G, in: 2017 2nd International Conference on Power and Renewable Energy (ICPRE), IEEE, 2017, pp. 833–838.
[8] Z.U. Zahid, Z.M. Dalala, R. Chen, B. Chen, J. Lai, Design of bidirectional DC–DC resonant converter for vehicle-to-grid (V2G) applications, IEEE Transactions on Transportation Electrification 1 (3) (Oct. 2015) 232–244.
[9] Z. Wang, Y. Zhang, S. You, H. Xiao, M. Cheng, An integrated power conversion system for electric traction and V2G operation in electric vehicles with a small film capacitor, IEEE Transactions on Power Electronics 35 (5) (May 2020) 5066–5077.
[10] J. Singh, R. Tiwari, Cost benefit analysis for V2G implementation of electric vehicles in distribution system, IEEE Transactions on Industry Applications 56 (5) (Sept.-Oct. 2020) 5963–5973.
[11] S.H. Hosseini, R. Ghazi, H. Heydari-Doostabad, An extendable quadratic bidirectional DC–DC converter for V2G and G2V applications, IEEE Transactions on Industrial Electronics 68 (6) (June 2021) 4859–4869.
[12] S. Semsar, T. Soong, P.W. Lehn, On-board single-phase integrated electric vehicle charger with V2G functionality, IEEE Transactions on Power Electronics 35 (11) (Nov. 2020) 12072–12084.
[13] Saranga Weearsinghe, Duleepa J. Thrimawithana, Udaya K. Madawala, Modeling bidirectional contactless grid interfaces with a soft DC-link, IEEE Transactions on Power Electronics 30 (7) (2014) 3528–3541.
[14] Hua Han, Yonglu Liu, Yao Sun, Hui Wang, Mei Su, Single-phase current-source bidirectional converter for V2G applications, Journal of Power Electronics 14 (3) (2014) 458–467.
[15] Mei Su, Hua Li, Yao Sun, Wenjing Xiong, A high-efficiency bidirectional AC/DC topology for V2G applications, Journal of Power Electronics 14 (5) (2014) 899–907.
[16] Manu Jain, Bi-directional DC–DC converter for low power applications, PhD Thesis, Concordia University, 1998.
[17] J.G. Pinto, Vítor Monteiro, Henrique Gonçalves, João Luiz Afonso, Onboard reconfigurable battery charger for electric vehicles with traction-to-auxiliary mode, IEEE Transactions on Vehicular Technology 63 (3) (2013) 1104–1116.
[18] Arun Kumar Verma, Bhim Singh, D.T. Shahani, Grid to vehicle and vehicle to grid energy transfer using single-phase bidirectional AC–DC converter and bidirectional DC–DC converter, in: 2011 International Conference on Energy, Automation and Signal, IEEE, 2011, pp. 1–5.

[19] Angshuman Sharma, Santanu Sharma, Review of power electronics in vehicle-to-grid systems, Journal of Energy Storage 21 (2019) 337–361.
[20] Ramteen Sioshansi, Paul Denholm, Emissions impacts and benefits of plug-in hybrid electric vehicles and vehicle-to-grid services, Environmental Science & Technology 43 (4) (2009) 1199–1204.
[21] M. Duvall, M. Alexander, Batteries for Electric Drive Vehicles–Status 2005, Tech. Rep., Electric Power Research Institute, 2005.
[22] Laurence A. Wolsey, George L. Nemhauser, Integer and Combinatorial Optimization, vol. 15, Wiley-Interscience, 1988.
[23] Okan Arslan, Oya Ekin Karasan, Cost and emission impacts of virtual power plant formation in plug-in hybrid electric vehicle penetrated networks, Energy 60 (2013) 116–124.
[24] Xun Gong, Jayanth Rangaraju, Taking charge of electric vehicles-both in the vehicle and on the grid, in: Texas Instruments, Dallas, TX, USA, 2018, pp. 1–13.
[25] Emad Ali Almabsout, Ragab A. El-Sehiemy, Osman Nuri Uç An, Oguz Bayat, A hybrid local search genetic algorithm for simultaneous placement of DG units and shunt capacitors in radial distribution systems, IEEE Access 8 (2020) 54465–54481.

Short- and long-term forecasting
Wind speed forecasting by spectral analysis

14

Hüseyin Akçay and Tansu Filik
Department of Electrical and Electronics Engineering, Eskişehir Technical University, Eskisehir, Turkey

14.1 Introduction

Reducing uncertainty in wind speed has been the focus of research and new developments in the last several decades. Numerous studies on wind speed prediction, which can be vaguely divided into three categories of physical, statistical, and artificial intelligence models, have been reported in the literature. Spatial correlation models lie between statistical and artificial intelligence models. The division between statistical and artificial intelligence models is not sharp as the following brief survey shows. In fact, hybrid wind speed forecasting schemes are becoming an industry norm nowadays.

Physical models require more parameters such as geographic and geomorphic conditions, temperature, and pressure to build forecasting models [36,51]. They produce poor estimation results when used for short-term wind speed forecasting [20], but they are more suitable for long-term predictions, for example, at least for three-hour predictions [26]. A statistical approach usually picks one model structure from linear-in-parameters time-series models [45,47,50,52], nonlinear-in-parameters artificial neural network models [29], wavelet transform models [53,54], and space-time models, i.e., spatio-temporal models [8,55,56]. Kalman filter-based forecasting methods were also proposed [11,35]. Hybrid models unifying autoregressive moving average (ARMA) models or autoregressive integrated moving average (ARIMA) models, Kalman filters, artificial neural networks, wavelet transforms, support vector machines, and artificial bee colony algorithms were proposed recently [12,14,17,30–33]. This area of research is very active and many hybrid estimation algorithms are still being developed. The time resolution of a short-term forecasting model ranges from several minutes to several hours ahead. Fractional ARIMA models were used in [27] to model and forecast wind speeds on one and two days-ahead horizons. To forecast wind speed and direction simultaneously, a vector ARMA model was used in [16].

Statistical approaches are preferred for short-term wind speed forecasting. But they have also found applications in atmospheric weather prediction models. For example, Kalman filtering was applied in [11,35,40] as a posterior data-processing procedure. Linear-in-parameters time series models do not require long data records for model building and analysis [47]. These models are easy to implement and apply to a large class of dynamic systems. A drawback of this model structure is that the performance of a linear time-series model severely deteriorates when some measurements are missing, for example, due to sensor failures and the nature of wind. Degradation in the estimation accuracy when some measurements are missing is usually offset by collecting estimation data over long horizons. Nonlinear model structures, which also need long data records, have been introduced to ameliorate shortcomings of linear model structures. Nonlinear artificial neural network model proposed in [13] uses nine years of wind speed measurements for assessment of probability distributions. Spatio-temporal estimation algorithms proposed in [55,56] utilize longer prediction horizons over multiple sensor locations to build accurate models.

In this chapter, we will use hourly wind speed measurements supplied by the Turkish Meteorological Service to illustrate properties of a short-time wind speed forecasting scheme developed in [5]. Short-term forecasting needs significant penetration levels of the wind energy, which may take a long time to develop [38]. While many forecasting studies have pointed out annual and diurnal trends, only in a few works [22,23,56] diurnal and annual trends were explicitly modeled. Furthermore, any data acquisition procedure is subject to missing values due to sensor failures and still weather. In this chapter, we study the wind speed forecasting framework proposed in [5]. In this framework, first data are de-trended for periodic components. Next, covariance-factorization via a recent subspace method is performed. In the third stage, wind speeds are forecasted by one- and/or multi-step-ahead Kalman filters. In this framework, intermittently or sequentially missed measurements are easily handled, without a significant deterioration in estimation accuracy.

14.2 Wind speed forecasting from long-term observations

In this section, we will review the wind speed forecasting scheme proposed in [5]. This scheme builds input/output models for wind speeds in several stages from wind speed/velocity measurements. The properties of the proposed scheme are illustrated on data sets collected from several stations of the Turkish Meteorological Service in Marmara region. The meteorological stations are shown schematically in Fig. 14.1. This region is known as having one of the highest wind energy potential in Turkey. The considered stations were selected arbitrarily from the available measurement stations. Each station is isolated from the rest. The wind speed and the direction measurements are intermittently and sequentially missing and have been collected over a period of $6\frac{2}{3}$ years between 2008 and 2014. The wind data were acquired at 10 m height. The data sets are available on the web site [19]. In Fig. 14.2, the wind speed measurements over 58,969 hours are plotted for Station BOZ. The data record is not complete. In fact, 1.68% of the wind speed and the direction measurements are missing.

In this section, we will consider the problem of wind speed forecasting from the wind speed measurements with missing values. Wind speed forecasts can be converted into wind power forecasts by using deterministic power curves supplied by wind turbine manufacturers. Wind speed forecasts should also be extrapolated to the turbine operation height by using a known relationship between the height and wind speed.

Short- and long-term forecasting

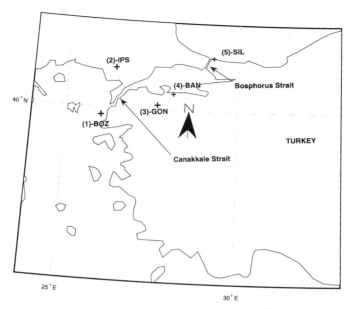

Figure 14.1 The meteorological stations: Bozcaada (BOZ), İpsala (IPS), Gönen (GON), Bandırma (BAN), and Şile (SIL) with N indicating the North.

Standard identification methods cannot be directly applied due to missing data and the enormous size of the data set. An inspection of Fig. 14.2 reveals that the measurements vary wildly, thus making short-term forecasting difficult. By focusing on large data segments, we plan to use stationary random process theory. The wind speed forecasting scheme will be presented in stages in the sequel.

14.2.1 Trend removal

Recall that the wind speed measurements plotted in Fig. 14.2 have been collected over a long period of time. As a result, the measurements comprise a significant amount of deterministic, periodic, and predictable components in the trend. On top of this, some trend components may be time-varying and admit only nonlinear data models. In an effort to reduce the trend-based uncertainty, it has been suggested in several works [56,60] to remove a diurnal pattern

$$y_d[k] = c_0 + \sum_{i=1}^{n_d} c_{1i} \cos(\pi i k/12) + \sum_{i=1}^{n_d} c_{2i} \sin(\pi i k/12) \qquad (14.1)$$

from the wind speed measurements $y[k]$, $k = 1, \ldots, N$. Observe that a diurnal pattern model is a trigonometric function with a period of 24 h and it is useful to model daily trends. A generalization of this idea includes weekly, monthly, and annual patterns

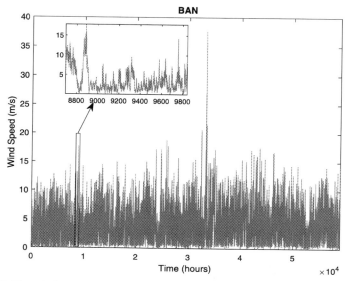

Figure 14.2 The wind speed measurements at Station BAN over 58,969 h. The measurements are intermittently and sequentially missing.

defined respectively by the equations

$$y_w[k] = \sum_{i=1}^{n_w} c_{3i} \cos(\pi i k/84) + \sum_{i=1}^{n_w} c_{4i} \sin(\pi i k/84), \quad (14.2)$$

$$y_m[k] = \sum_{i=1}^{n_m} c_{5i} \cos(\pi i k/365) + \sum_{i=1}^{n_m} c_{6i} \sin(\pi i k/365), \quad (14.3)$$

$$y_a[k] = \sum_{i=1}^{n_a} c_{7i} \cos(\pi i k/4380) + \sum_{i=1}^{n_a} c_{8i} \sin(\pi i k/4380). \quad (14.4)$$

The overall trend and the residuals are then given by the equations

$$y_p[k] = y_d[k] + y_w[k] + y_m[k] + y_a[k], \quad (14.5)$$
$$y_r[k] = y[k] - y_p[k]. \quad (14.6)$$

The trend model (14.5) is linear in the $n_t = 1 + 2n_d + 2n_w + 2n_m + 2n_a$ regression parameters $c_0, c_{11}, \ldots, c_{8n_a}$ which will be estimated by solving an optimization problem.

The above trend model was applied in [5] to the wind speed measurements collected from the five meteorological stations plotted in Fig. 14.1. The regression parameters in the model (14.5) were stacked in a vector $c \in \mathbb{R}^{n_t}$ and estimated by minimizing

Short- and long-term forecasting

the sum of the squares of the residuals

$$J(c) = \sum_{k=1}^{N_e} \chi_y(k) \, y_r^2[k], \tag{14.7}$$

where the indicator function $\chi_y(k)$ is 1 if $y(k)$ is measured and 0 if $y(k)$ is missing and $N_e = 52{,}560$ was chosen as the estimation data length. The data $y[k]$ for $k = 52{,}561, \ldots, 58{,}969$ h, including missing measurements, are reserved to validate the estimated final model. The trend removal process forms the first-stage of the wind speed estimation framework proposed in [5].

In Fig. 14.3, the trend estimate is plotted for Station BAN with the number of harmonics $n_d = 1$, $n_w = 2$, $n_m = 3$, and $n_a = 4$ chosen on a trial-and-error basis. The visible trends in the data plotted in Fig. 14.2 are mostly captured. The periodic components in $y[k]$ manifest themselves as poles on the unit circle in the z-transform of $y[k]$. Since ARIMA models have only one pole at $z = 1$, $y_p[k]$ may be viewed as a generalization of the ARIMA models [31] for the wind speed measurements. The same comment applies to the fractional ARIMA models as well. The numerical experiments with the rest of the stations can be found in [5].

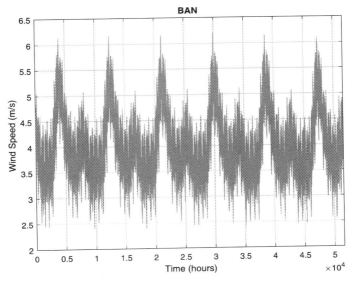

Figure 14.3 The trend estimates for Station BAN.

As an alternative method for the determination of $c \in \mathbb{R}^{n_t}$ by the least-squares formula $c^* = \arg\min J(c)$, consider for a given $\delta > 0$ the following sparsity promoting optimization problem:

$$\hat{c}_0 = \arg\min_{c \in \mathbb{R}^{n_t}} \|c\|_0 \quad \text{subject to} \quad J(c) \leq \delta, \tag{14.8}$$

where $\|c\|_0$ is the ℓ_0 norm of c, that is, the number of the nonzero entries of c. This formulation does not assume n_t known. Nevertheless, it is a nonconvex optimization problem without a guarantee for a global minimum. To overcome numerical issues, we relax ℓ_0 to an ℓ_1 norm and get the following convex variant of (14.8), known in the literature as the basis pursuit denoising:

$$\hat{c}_1 = \arg \min_{c \in \mathbb{R}^{n_t}} \|Wc\|_1 \text{ subject to } J(c) \leq \delta, \tag{14.9}$$

where W is a diagonal positive-definite weight matrix and $\|c\|_1$ is the ℓ_1 norm of c, that is, the sum of the absolute values of the entries of c. The hyperparameter δ shapes the solutions of (14.8) and (14.9). There are many efficient algorithms to solve (14.9). Orthogonal matching pursuit, iteratively reweighted least squares, and iterated shrinkage algorithms are among them. These algorithms are collectively known as compressive sensing identification/approximation algorithms. Exact recovery conditions when $\delta = 0$ and $N_e < n_t$ in (14.8) and (14.9) are interestingly the same and depend on the sparsity level of c denoted by $\|c\|_0$, N_e, n_t, and the regressors chosen [9]. The $n_t < N_e$ case was studied in [10]. The trend filtering problem was studied in [28,44] in compressive sensing frameworks. A compressive sensing algorithm with wavelet transform applied to output measurements was proposed in [54] to estimate wind speeds. The wavelet transform reduces data size by filtering outputs and maintains only stationary artifacts. The transformed data is then modeled by an underdetermined system of linear equations.

Diurnal, weekly, monthly, and annual cycles in the time series data are explicitly modeled by (14.5), whereas in [54], and more generally in the empirical mode decomposition method [24], they are implicitly distributed among intrinsic mode functions. On the other hand, the residuals $y_r[k]$ in (14.6) may be represented by a linear combination of several mode functions with parameters extracted from the solutions of (14.8) and (14.9). The calculation of c by unconstrained minimization of $J(c)$ makes sense if either $y_p[k]$ is orthogonal to $y_r[k]$ or the latter is uncorrelated with the former. Since N_e is very large, it is plausible to assume that $y_p[k]$ is uncorrelated with $y_r[k]$. This assumption is the foundation of spectral analysis elaborated in the following section.

14.2.2 Spectral analysis

Recall that power spectral density function of a zero-mean wide-sense stationary stochastic process $y[k] \in \mathbb{R}$ is defined by

$$S(e^{j\omega}) = \sum_{\tau=-\infty}^{\infty} R_y(\tau) e^{j\omega\tau}, \qquad 0 \leq \omega < 2\pi, \tag{14.10}$$

where $R_y(\tau)$ is the autocorrelation function of $y[k]$,

$$R_y(\tau) = \mathrm{E}\left(y[k+\tau]y[k]\right), \qquad \forall k, \tag{14.11}$$

Short- and long-term forecasting

with $E(x)$ denoting the expected value of a given random variable x. Assuming $y[k]$ is an ergodic process, for each fixed integer τ as $M \to \infty$, with probability one (w.p.1)

$$\frac{1}{M} \sum_{k=1}^{M} y[k+\tau]y[k] \to R_y(\tau). \tag{14.12}$$

Suppose $y[k]$ are specified for $k = 1, \ldots, M$. Then, it is necessary to impose $|\tau| < M$ and replace the above sums with

$$\hat{R}_y^M(\tau) = \frac{1}{M - |\tau|} \sum_{k=1}^{M-|\tau|} y[k+|\tau|]y[k]. \tag{14.13}$$

When τ is restricted to bounded sets, $\hat{R}_y^M(\tau) \to R_y(\tau)$ w.p.1 as $M \to \infty$. An estimator of $S(e^{j\omega})$ based on $\hat{R}_y^M(\tau)$ is proposed as follows:

$$\hat{S}_R^M(e^{j\omega}) = \sum_{\tau=1-M}^{M-1} w^M(\tau)\hat{R}_y^M(\tau)e^{-j\omega\tau}, \quad 0 \leq \omega < 2\pi, \tag{14.14}$$

where $w^M(\tau) \geq 0$ is a symmetric scalar window function satisfying some regularity conditions [43]. Since $\hat{R}_y^M(0)$ is calculated by averaging M terms while $\hat{R}_y^M(M-1)$ merely equals $y[M]y[1]$, $\hat{R}_y^M(\tau)$ is less reliable for large $|\tau|$. The disturbing effect of large lags can be offset by selecting a triangular window function $w^M(\tau) = 1 - M^{-1}|\tau|$, $|\tau| < M$ satisfying the regularity conditions.

When some measurements are missing, it is necessary to modify $\hat{R}_y^M(\tau)$ as described next. Pick a fixed integer $T > 0$ and for each $-T \leq \tau \leq T$ define a set $\mathbb{N}_\tau$ by

$$\mathbb{N}_\tau = \{k : 1 \leq k + \tau \leq M, \ y[k+\tau] \text{ observed}\}. \tag{14.15}$$

Given a finite set X, let $\text{card}(X)$ denote its number of elements. Let $\mathcal{S}_\tau = \mathbb{N}_0 \cap \mathbb{N}_{k+\tau}$ and modify the definition of $\hat{R}_y^M(\tau)$ as follows:

$$\check{R}_y^M(\tau) = \frac{1}{\text{card}(\mathcal{S}_\tau)} \sum_{k \in \mathcal{S}_\tau} y[k+\tau]y[k]. \tag{14.16}$$

Then, from the wide-sense stationarity of $y[k]$, observe that $\check{R}_y^M(\tau) \to R_y(\tau)$ for each fixed τ as $M \to \infty$ provided that $\text{card}(\mathcal{P}_\tau) \to \infty$. See [4] for further details.

We divided the wind speed measurements into L overlapping/non-overlapping blocks and for each τ and averaged $\check{R}_y^M(\tau)$ over the blocks. We estimated the autocorrelation function of Station BAN by plugging $y_r[k]$ in place of $y[k]$ and letting $L = 6$ and $T = 256$ where $N_e = LM$. In Fig. 14.4, the estimation results are plotted.

The results were nearly identical when L was increased from 6 to 48. Likewise, T was varied between 64 and 6000 and the change was not significant. A visual inspection shows that $y_r[k]$ may be modeled by a 1st or 3rd order ARMA model; but the 3rd model is seen to have an oscillatory impulse response. A systematic model order selection procedure, based on the extraction of the so-called observability range space, is presented next.

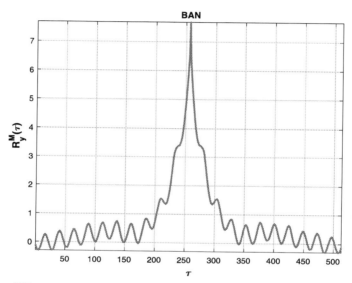

Figure 14.4 $\check{R}_y^M(\tau)$ for Station BAN and $L = 6$.

14.2.3 ARMA modeling of the residuals

For a positive integer T, define a block Hankel matrix $\mathcal{Y}_R$ by

$$\mathcal{Y}_R(i_1, i_2) = \check{R}_y^M(i_1 + i_2 - 1) + \check{R}_y^M(i_1 + i_2 - 1 - T), \quad (14.17)$$

where $1 \leq i_1, i_2 \leq p$ are respectively the block-row and the block-column indices and $T = 2p$. Suppose that $S(e^{j\omega})$, $0 \leq \omega < 2\pi$ has been generated by an nth order linear-shape filter $G(z) = c^T(zI_n - A)^{-1}b + d$, that is, $S(e^{j\omega}) = G(e^{j\omega})G(e^{-j\omega})$ for some nth order stable and minimum-phase transfer function $G(z)$ where I_n denotes the $n \times n$ identity matrix and $p > 2n$. Recall that $S(z)$ can be decomposed uniquely in terms of the spectral summands $H(z)$ and $H(z^{-1})$ as $S(z) = H(z) + H(z^{-1})$ where for some column vector f and a scalar $e > 0$, $H(z) = c^T(zI - A)^{-1}f + e/2$. Let $X^\dagger = (X^T X)^{-1} X^T$ denote the Moore–Penrose pseudo-inverse of a given full-column rank matrix X with X^T denoting the transpose of X. Let $\hat{H}(z)$ denote an estimator of $H(z)$. It is a well-known fact that a spectral factor can be extracted from $\hat{H}(z)$ if and only if $\hat{H}(z)$ is positive-real. If $\hat{H}(z)$ is not positive-real, by modifying the estimates of f and

e denoted by $\hat{f}$ and $\hat{e}$, $\hat{H}(z)$ be deemed positive-real. While spectral factorization is straightforward for single-input/single-output systems and easily carried out by pole-zero factoring, it is a nontrivial task for multi-input/multi-output systems.

In [2], it was shown that Algorithm 1 outlined in the sequel estimates $H(z)$ and a spectral factor up to an orthogonal matrix with probability one from output-only measurements as M grows unboundedly. This algorithm utilizes the regularized and reweighted nuclear norm heuristic developed recently in [3] to ensure positivity of the estimated power spectrum.

Algorithm 1. Subspace algorithm estimating a spectral factor from $\mathcal{Y}_R$.

Input: $\mathcal{Y}_R$
1: Calculate the singular-value decomposition
$$\mathcal{Y}_R = \begin{bmatrix} U^\sharp & U' \end{bmatrix} \begin{bmatrix} \Sigma^\sharp & 0 \\ 0 & \Sigma' \end{bmatrix} \begin{bmatrix} V^\sharp \\ V' \end{bmatrix}$$
where $\Sigma^\sharp$ contains the $2n$ largest singular values.
2: With $U^\sharp$ defined above and J_u and J_d by $J_u = [0 \ I_{p-1}]$ and $J_d = [I_{p-1} \ 0]$, calculate $A^\sharp = (J_d U^\sharp)^\dagger J_u U^\sharp$.
3: Put $A^\sharp$ into the following Jordan canonical form
$$A^\sharp = \begin{bmatrix} \Pi_c & \Pi_{ac} \end{bmatrix} \begin{bmatrix} \Sigma_c & 0 \\ 0 & \Sigma_{ac} \end{bmatrix} \begin{bmatrix} \Pi_c & \Pi_{ac} \end{bmatrix}^{-1},$$
where the eigenvalues of Σ_c and Σ_{ac} are respectively inside and outside the unit circle and let $\hat{A} = \Sigma_c$, $\hat{c}^T = J_f U^\sharp \Sigma_c$ with $J_f = [1 \ 0]$.
4: Set $\hat{e} = \check{R}_y^M(0)/2$ and estimate $\hat{f}$ by solving
$$\begin{bmatrix} \check{R}_y^M(1) \\ \vdots \\ \check{R}_y^M(T-1) \end{bmatrix} = \begin{bmatrix} \hat{c}^T \\ \vdots \\ \hat{c}^T \hat{A}^{T-2} \end{bmatrix} \hat{f}$$
in the least-squares sense.
5: Extract a spectral factor $\{\hat{A}, \hat{b}, \hat{c}, \hat{d}\}$ from $\{\hat{A}, \hat{f}, \hat{c}, \hat{e}/2\}$
Output: $\hat{G}(z) = \hat{c}^T(zI_n - \hat{A})^{-1}\hat{b} + \hat{d}$.

In Fig. 14.5, the power spectrum estimated by Algorithm 1 is plotted for Station BAN. The shape-filter order is one. The jiggles in the middle-frequency band cannot be modeled by a higher-order shape filter since extra poles will cluster around the unit circle and generate filter zeros around the unit circle (approximate pole-zero cancellation). The zeros of a power spectrum are known to be very sensitive to perturbations of the spectrum. ARMA models are parsimonious with respect to AR models. To see this, write $G(z) = n(z)/d(z)$ as an all-pole filter $G(z) = 1/(n^{-1}(z)d(z))$. The truncation of $n^{-1}(z)$ is possible if the zeros of $n(z)$ are not on the unit circle. In [31], AR model orders of 8, 9, and 19 were found in the ARIMA models. AR models are used to train a three-layer ANN. Thus, over-parameterization of spectral factors has a compound effect.

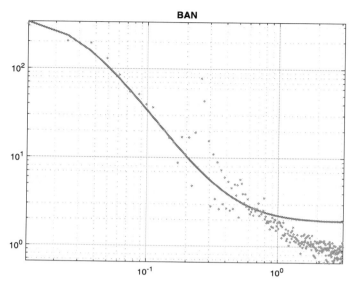

Figure 14.5 The spectral density estimated by Algorithm 1 (solid line) and computed from the autocorrelation function in Fig. 14.4 (dotted line) for Station BAN.

14.2.4 Forecasting by Kalman filters

We collect one- and multi-step ahead predictor formulas in Algorithm 2. For the derivations, see Lemma 8.2 in [57]. In Algorithm 2, $y_p[k]$ refers to the extension of $y_p[k]$ introduced in Section 14.2.1 to the entire time interval $[1, N_e]$. The predictors are stable and $\hat{d}$ is invertible since $\hat{G}(z)$ is minimum-phase. The missing values are recovered from the predictor formulas on setting $\hat{x}[1|0] = 0$ and letting $y[k] = y_p[k]$ until the first measurement is acquired. Thereafter, the prediction filters are run. Due to initialization errors, it may take a while for the predictions to settle. For large N_e, the quadruplet $\{\hat{A}, \hat{b}, \hat{c}^T, \hat{d}\}$ may be considered independent from $\zeta[k]$ on short intervals.

14.2.5 Predictor performance

The predictor performance will be judged by computing the normalized ℓ_q-norms of the i-step-ahead prediction errors defined as

$$\rho_q^K(i) = \left(\frac{\sum_{k=1}^{N_e} |\hat{y}[k+i|k] - y[k+i]|^q}{\sum_{k=1}^{N_e} |y[k]|^q} \right)^{\frac{1}{q}}, \quad i \geq 1, \tag{14.18}$$

where in the sums missing values are omitted. It is known as the normalized i-step-ahead mean-absolute-error (MAE) for $q = 1$ and root-mean-square-error (RMSE) for $q = 2$, respectively. In Fig. 14.6, the RMSE and MAE of the predictors proposed in

Algorithm 2. Prediction by Kalman filters.

Input: Spectral factor realization $\{\hat{A}, \hat{b}, \hat{c}, \hat{d}\}$
1: Write $\hat{G}(z)$ in the innovation form:
$x[k+1] = \hat{A}x[k] + \hat{b}\zeta[k]$,
$y_r[k] = \hat{c}^T x[k] + \hat{d}\zeta[k]$
driven by a zero-mean and unit-intensity white-noise $\zeta[k]$.
2: The one-step-ahead predictor:
$\hat{x}[k+1|k] = (\hat{A} - \hat{b}\hat{d}^{-1}\hat{c}^T)\hat{x}[k|k-1] + \hat{b}\hat{d}^{-1}(y[k] - y_p[k])$,
$\hat{y}[k+1|k] = \hat{c}^T \hat{x}[k+1|k] + y_p[k]$.
3: The multi-step-ahead ($i > 1$) predictors:
$\hat{x}[k+i|k] = \hat{A}^{i-1}\hat{x}[k+1|k]$,
$\hat{y}[k+i|k] = \hat{c}^T \hat{x}[k+i|k] + y_p[k]$.
Outputs: $\hat{y}[k+i|k], i \geq 1$.

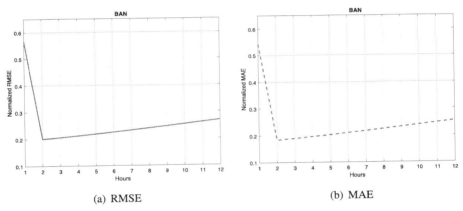

Figure 14.6 The normalized i-step-ahead prediction errors of Algorithm 2 for Station BAN.

Algorithm 2 are plotted versus step size for Station BAN. The errors fall sharply at $i = 2$ and afterward increase slowly as i increases.

14.2.6 Model validation

Recall that we have saved 6409 measurements including missing values for model validation. We calculated the RMSE and the MAE errors for various step sizes with $\hat{G}(z)$ estimated by using the first 52,560 measurements. The results plotted in Fig. 14.7 are very similar to the estimation errors plotted in Fig. 14.6. Thus, $\hat{G}(z)$ is a valid model.

Lastly in this section using the wind speed measurements at Station BAN, we plot the 2, 6, and 12-hour ahead forecasting results in Fig. 14.8 on the first 600 hours of the validation data. In Fig. 14.8(d), the forecasting results are plotted together and zoomed. The proposed scheme provides very good forecasting even for rather

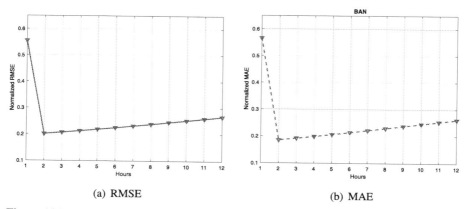

Figure 14.7 The prediction errors of Algorithm 2 calculated on the validation data for Station BAN.

large step sizes. Remarkably, the changes in the wind speed are detected for all step sizes.

14.2.7 Extensions of Algorithms 1–2 to vector-valued measurements

The framework presented above for scalar-valued measurements extends to vector-valued measurements with only a slight change in the notation. More specifically, let $y[k] \in \mathbb{R}^\sigma$ denote the wind speed measurements of σ stations at time k stacked into a vector. For each station, the wind speed and the direction information may also be represented by a velocity vector in $\mathbb{R}^2$ by passing to the Cartesian coordinates. Thus, in the most general case $y[k] \in \mathbb{R}^{2\sigma}$. Let $\|x\|_2$ denote the Euclidean norm of a given vector x. The regression parameters in Section 14.2.1 are calculated by minimizing $\sum_{k=1}^{N_e} \|y_r[k]\|_2^2$. The definition of $\check{R}_y^M(\tau)$ remains unchanged if $\mathbb{N}_\tau$ is redefined as the intersection of the index sets for all stations though this modification increases the number of the missing data. Alternatively, $\check{R}_y^M(\tau)$ can be calculated componentwise to maximally utilize the available data. Algorithm 2 already operates on matrix-valued autocovariance estimates. The predictors do not need any changes. Lastly, it suffices to replace the scalar norms in $\rho_q^K(i)$ with the ℓ_q-norms. This vectorization procedure was applied to planar wind velocity measurements in [6] with $\sigma = 1$.

14.2.8 Persistence predictors

The following family of predictors:

$$\hat{y}[k+i|k] = y[k], \qquad i \geq 1$$

Short- and long-term forecasting

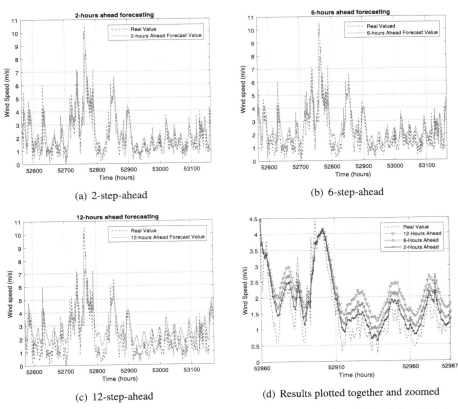

Figure 14.8 The i-step-ahead forecasting results calculated on the first 600 h of the validation data at Station BAN: the measurements (dashed line); the proposed scheme (solid line).

are well-known in the literature on wind energy. They are called persistence predictors and regarded as the industry standards and provide benchmark values to evaluate performance of a candidate predictor. In Fig. 14.9, the prediction errors of the persistence models are plotted versus the step size for Station BAN. They all increase rapidly as the step size increases in contrast to the curves in Fig. 14.7. The numerical results are summarized in Table 14.1 for Algorithm 2 and the persistence predictors with errors calculated on the validation data for i ranging between 1 and 12 for Station BAN. The normalized i-step prediction errors of the persistence predictors are denoted by $\rho_q^{\text{per}}(i)$.

14.2.9 Summary of the numerical results

Based on the numerical results above, we can draw the following conclusions:

1. The proposed scheme has superior performance relative to the persistence predictors for all step sizes, but one.

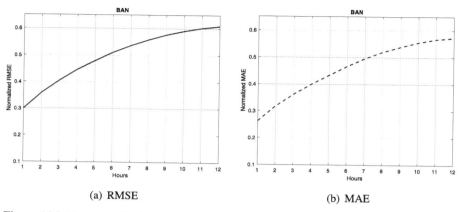

Figure 14.9 The normalized i-step-ahead prediction errors of the persistence predictors.

Table 14.1 The normalized i-step-ahead MAEs and RMSEs of the proposed predictors and the persistence predictors calculated on the validation data set for Station BAN.

i	$\rho_2^{per}(i)$	$\rho_2^{K}(i)$	$\rho_1^{per}(i)$	$\rho_1^{K}(i)$
1	0.299	**0.554**	0.263	**0.565**
2	0.358	0.201	0.318	0.185
3	0.405	0.206	0.362	0.191
4	0.445	0.212	0.399	0.198
5	0.478	0.218	0.433	0.205
6	0.510	0.225	0.466	0.213
7	0.536	0.231	0.496	0.221
8	0.558	0.238	0.520	0.229
9	0.577	0.245	0.539	0.237
10	0.591	0.252	0.555	0.245
11	0.602	0.259	0.567	0.253
12	0.610	0.267	0.573	0.261

2. The proposed one-step ahead predictor is the only predictor outperformed by the persistence predictor in both the RMSE and MAE criteria.
3. The performance improvements achieved by the proposed scheme in either the RMSE or the MAE criterion surpass the performance improvements recorded by a great number of competitive algorithms, for example, [15,22,23,48,60].

14.3 Conclusion

The forecasting scheme studied in this chapter is based on a high-resolution spectral estimator, that is, a parametric spectrum estimator is constructed from ouput-only measurements. In this process, not only the filter poles but also the filter zeros are shaped. The resolution of the spectrum estimator is determined jointly by its pole and zero sets. A numerical study showed that the predictors derived from the spectrum estimator track changes in the wind speed even for very large step sizes. The framework studied is statistics-free, that is, there is no need for prior probability distributions; all one needs to implement this framework are the tools of numerical linear algebra.

The foremost advantage of the studied scheme over most hybrid techniques is the absence of iterative search procedures in the parameter space which are not guaranteed to converge to the optimal values unless suitably started. For example, parameters of the artificial neural networks or the support vector machines are optimized using a genetic or more generally a nonlinear-type optimization algorithm. Most algorithms in the literature are not capable of handling measurements with missing values whereas in the studied scheme this situation is easily tackled since the measurements are assumed to be stationary once the periodic component is removed. Merging subspace and recursive identification methods with compressive sensing techniques needs to be explored. Some preliminary results are available in this direction [6].

In [6], application domain of the forecasting scheme studied in this chapter was enlarged from multi-step ahead only to one- and multi-step-ahead and from long-term observations to long- and mid-term observations. It was observed that the proposed scheme outperforms the persistence and artificial neural network (ANN) predictors by a large margin for all step sizes considered. A second scheme presented in this work was demonstrated to outperform the multi-step-ahead persistence and the ANN predictors. This scheme is a compressive subspace algorithm developed in [7] for models in the innovation form and uses short-term observations. The wind speed estimation algorithm presented in this chapter has a simple structure. Recently, a state-of-the-art hybrid algorithm encompassing empirical mode decomposition, a bagging artificial neural network, k-means clustering, and a new intelligent optimization method was proposed to reduce uncertainty in the wind speeds [1]. In a case study, the proposed method was reported superior to several other methods. Many decomposition techniques are available to extract features [21,25,34,39,41,42,46,53,54,58,61]. Deep learning strategies are being used increasingly in time-series analysis [49] and wind speed forecasting [18,37,59].

References

[1] O. Abedinia, M. Lotfi, M. Bagheri, B. Sobhani, M. Shafie-khah, J.P.S. Catalão, Improved EMD-based complex prediction model for wind power forecasting, IEEE Transactions on Sustainable Energy 11 (4) (2020) 2790–2802.

[2] H. Akçay, Spectral estimation in frequency-domain by subspace techniques, Signal Processing 101 (2014) 204–217.

[3] H. Akçay, S. Türkay, Positive realness in stochastic subspace identification: a regularized and reweighted nuclear norm minimization approach, in: Proceedings of the 14th European Control Conference, Linz, Austria, July 2015, pp. 1754–1759.

[4] H. Akçay, Time-domain identification of rational spectra with missing data, in: Proceedings of the 5th International Conference on Systems and Control, Marrakech, Morocco, May 2016, pp. 445–450.
[5] H. Akçay, T. Filik, Short-term wind speed forecasting by spectral analysis from long-term observations with missing values, Applied Energy 191 (2017) 653–662.
[6] H. Akçay, T. Filik, Wind speed forecasting by subspace and nuclear norm optimization based algorithms, Sustainable Energy Technologies and Assessments 35 (2019) 139–147.
[7] H. Akçay, S. Türkay, Power spectrum estimation in innovation models, Mechanical Systems and Signal Processing 121 (2019) 227–245.
[8] M. Alexiadis, P. Dokopoulos, H. Sahsamanoglou, Wind speed and power forecasting based on spatial correlation models, IEEE Transactions on Energy Conversion 14 (1999) 836–842.
[9] A.M. Bruckstein, D.L. Donoho, M. Elad, From sparse solutions of systems of equations to sparse modelling of signals and images, SIAM Review 51 (2009) 34–81.
[10] E.J. Candés, T. Tao, Decoding by linear programming, IEEE Transactions on Information Theory 51 (2005) 4203–4215.
[11] F. Cassola, M. Burlando, Wind speed and wind energy forecast through Kalman filtering of numerical weather prediction model output, Applied Energy 99 (2012) 154–166.
[12] J.P.S. Catalão, H.M.I. Pousinho, V.M.F. Mendes, Short-term wind power forecasting in Portugal by neural networks and wavelet transform, Renewable Energy 36 (2011) 1245–1251.
[13] A.N. Çelik, M. Kolhe, Generalized feed-forward based method for wind energy prediction, Applied Energy 101 (2013) 582–588.
[14] K. Chen, J. Yu, Short-term wind speed prediction using an unscented Kalman filter based state-space support vector regression approach, Applied Energy 113 (2014) 690–705.
[15] J. Dowell, S. Weiss, D. Hill, D. Infield, Short-term spatio-temporal prediction of wind speed and direction, Wind Energy 17 (2013) 1945–1955.
[16] E. Erdem, J. Shi, ARMA based approaches for forecasting the tuple of wind speed and direction, Applied Energy 88 (2011) 1405–1414.
[17] S.W. Fei, Y. He, Wind speed prediction using the hybrid model of wavelet decomposition and artificial bee colony algorithm-based relevance vector machine, International Journal of Electrical Power & Energy Systems 73 (2015) 625–631.
[18] C. Feng, M. Cui, B.M. Hodge, J. Zhang, A data driven multi-model methodology with deep feature selection for short-term wind forecasting, Applied Energy 190 (2017) 1245–1257.
[19] T. Filik, Wind forecasting, multichannel wind speed and direction data, http://www.eem.anadolu.edu.tr/tansufilik/EEM%20547/icerik/windforecasting.htm.
[20] G. Giebel, R. Brownsword, G. Kariniotakis, M. Denhard, C. Draxl, The state-of-the-art in short term prediction of wind power: a literature overview, second edition, Report ANEMOS.plus Second Edition, Technical University of Denmark, 2011.
[21] Q. Han, H. Wu, T. Hu, F. Chu, Short-term wind speed forecasting based on signal decomposing algorithm and hybrid linear/nonlinear models, Energies 11 (2018) 2976.
[22] A.S. Hering, M.G. Genton, Powering up with space-time wind forecasting, Journal of the American Statistical Association 105 (2010) 92–104.
[23] D.C. Hill, D. McMillan, K.R.W. Bell, D. Infield, Application of auto-regressive models to U.K. wind speed data for power system impact studies, IEEE Transactions on Sustainable Energy 3 (2012) 134–141.
[24] N.E. Huang, Z. Shen, S.R. Long, M.C. Wu, H.H. Shih, Q. Zheng, N.-C. Yen, C.C. Tung, H.H. Liu, The empirical mode decomposition and the Hilbert spectrum for nonlinear

and non-stationary time series analysis, Proceedings of the Royal Society A 454 (1998) 903–995.

[25] Y. Jiang, S. Liu, N. Zhao, J. Xin, B. Wu, Short-term wind speed prediction using time varying filter-based empirical mode decomposition and group method of data handling-based hybrid model, Energy Conversion and Management 220 (2020) 113076.

[26] G. Kariniotakis, I.H-P. Waldl, I. Marti, G. Giebel, T.S. Nielsen, J. Tambke, J. Usaola, F. Dierich, A. Bocquet, S. Virlot, Next generation forecasting tools for the optimal management of wind generation, in: Proceedings of the 9th International Conference on Probabilistic Methods Applied to Power Systems, Stockholm, Sweden, 2006, pp. 1–6.

[27] R.G. Kavasseri, K. Seetharaman, Day-ahead wind speed forecasting using f-ARIMA models, Renewable Energy 34 (2009) 1388–1393.

[28] S.J. Kim, K. Koh, S. Boyd, D. Gorinevsky, ℓ_1 trend filtering, SIAM Review 51 (2009) 339–360.

[29] G. Li, J. Shi, On comparing three artificial neural networks for wind speed forecasting, Applied Energy 87 (2010) 2313–2320.

[30] Y. Liu, J. Shi, Y. Yang, W.J. Lee, Short-term wind-power prediction based on wavelet transform-support vector machine and statistic-characteristics analysis, IEEE Transactions on Industry Applications 48 (2012) 1136–1141.

[31] H. Liu, H.Q. Tian, Y.F. Li, Comparison of two new ARIMA-ANN and ARIMA-Kalman hybrid methods for wind speed prediction, Applied Energy 98 (2012) 415–424.

[32] H. Liu, H.Q. Tian, D.F. Pan, D.F. Li, Forecasting models for wind speed using wavelet, wavelet packet, time series, and artificial neural networks, Applied Energy 107 (2013) 191–208.

[33] D. Liu, D. Niu, H. Wang, L. Fan, Short-term wind speed forecasting using wavelet transform and support vector machines optimized by genetic algorithm, Renewable Energy 62 (2014) 592–597.

[34] L. Liu, T. Ji, M. Li, Z. Chen, Q. Wu, Short-term local prediction of wind speed and wind power based on singular spectrum analysis and locality-sensitive hashing, Journal of Modern Power Systems and Clean Energy 6 (2018) 317–329.

[35] P. Louka, G. Galanis, N. Siebert, G. Kariniotakis, P. Katsafados, I. Pytharoulis, G. Kallos, Improvements in wind speed forecasts for wind power prediction purposes using Kalman filtering, Journal of Wind Engineering and Industrial Aerodynamics 96 (2008) 2348–2362.

[36] P. Lynch, The origins of computer weather prediction and climate modeling, Journal of Computational Physics 227 (2008) 3431–3444.

[37] Z. Ma, H. Chen, J. Wang, X. Yang, R. Yan, J. Jia, W. Xu, Application of hybrid model based on double decomposition, error correction and deep learning in short-term wind speed prediction, Energy Conversion and Management 205 (2020) 112345.

[38] M. Marquis, J. Wilczak, M. Ahlstrom, J. Sharp, A. Stern, J.C. Smith, S. Calvert, Forecasting the wind to reach significant penetration levels of wind energy, Bulletin of the American Meteorological Society 92 (2011) 1159–1171.

[39] X. Mi, S. Zhao, Wind speed prediction based on singular spectrum analysis and neural network structural learning, Energy Conversion and Management 216 (2020) 112956.

[40] L.D. Monache, T. Nipen, Y. Liu, G. Roux, R. Stull, Kalman filter and analog schemes to postprocess numerical weather predictions, Monthly Weather Review 139 (2011) 3554–3570.

[41] T. Niu, J. Wang, K. Zhang, P. Du, Multi-step ahead wind speed forecasting based on optimal feature selection and a modified bat algorithm with the cognition strategy, Renewable Energy 118 (2018) 213–229.

[42] T. Peng, J. Zhou, C. Zhang, Y. Zheng, Multi-step ahead wind speed forecasting using a hybrid model based on two-stage decomposition technique and AdaBoost-extreme learning machine, Energy Conversion and Management 153 (2017) 589–602.
[43] M.B. Priestley, Spectral Analysis and Time Series, Elsevier, London, 2004.
[44] A. Ramdas, R.J. Tibshirani, Fast and flexible ADMM algorithms for trend filtering, Journal of Computational and Graphical Statistics 25 (2016) 839–858.
[45] G.H. Riahy, M. Abedi, Short term wind speed forecasting for wind turbine applications using linear prediction method, Renewable Energy 33 (2008) 35–41.
[46] S.M. Moreno, R.D.S. Gomes, V.M. Cocco, L.C.D. Santos, Multi-step wind speed forecasting based on hybrid multi-stage decomposition model and long short-term memory neural network, Energy Conversion and Management 213 (2020) 112869.
[47] A. Sfetsos, A comparison of various forecasting techniques applied to mean hourly wind speed time series, Renewable Energy 21 (2000) 23–35.
[48] A. Sfetsos, A novel approach for the forecasting of mean hourly wind speed time series, Renewable Energy 27 (2002) 163–174.
[49] Z. Shen, Y. Zhang, J. Lu, J. Xu, G. Xiao, A novel time series forecasting model with deep learning, Neurocomputing 396 (2020) 302–313.
[50] O.B. Shukur, M.H. Lee, Daily wind speed forecasting through hybrid KF-ANN model based on ARIMA, Renewable Energy 76 (2015) 637–647.
[51] G. Sideratos, N. Hatziargyriou, An advanced statistical method for wind power forecasting, IEEE Transactions on Power Systems 22 (2007) 258–265.
[52] D.A. Smith, K.C. Mehta, Investigation of stationary and nonstationary wind data using classical Box–Jenkins models, Journal of Wind Engineering and Industrial Aerodynamics 49 (1993) 319–328.
[53] A. Taşçıkaraoğlu, M. Uzunoğlu, A review of combined approaches for prediction of short-term wind speed and power, Renewable & Sustainable Energy Reviews 34 (2014) 243–254.
[54] A. Taşçıkaraoğlu, B.M. Sanandaji, K. Poolla, P. Varaiya, Exploiting sparsity of interconnections in spatio-temporal wind speed forecasting using wavelet transform, Applied Energy 165 (2016) 735–747.
[55] J. Tastu, P. Pinson, P.J. Trombe, H. Madsen, Probabilistic forecasts of wind power generation accounting for geographically dispersed information, IEEE Transactions on Smart Grid 5 (2014) 480–489.
[56] L. Xie, Y. Gu, X. Zhu, M.G. Genton, Short-term spatio-temporal wind power forecast in robust look-ahead power system dispatch, IEEE Transactions on Smart Grid 5 (2014) 511–520.
[57] M. Verhaegen, V. Verdult, Filtering and System Identification: A Least Squares Approach, Cambridge University Press, NY, 2007.
[58] C. Wang, H. Zhang, P. Ma, Wind power forecasting based on singular spectrum analysis and a new hybrid Laguerre neural network, Applied Energy 259 (2020) 114139.
[59] Y. Zhang, L. Le, X. Liao, F. Zheng, Y. Li, A novel combination of forecasting model for wind power integrating least square support vector machine, deep belief network, singular spectrum analysis and locality-sensitive hashing, Energy 168 (2019) 558–572.
[60] X. Zhu, M.G. Genton, Short-term wind speed forecasting for power system operations, International Statistical Review 80 (2012) 2–23.
[61] Q. Zhu, J. Chen, D. Shi, L. Zhu, X. Bai, X. Duan, Y. Liu, Learning temporal and spatial correlations jointly, IEEE Transactions on Sustainable Energy 11 (2020) 509–523.

Forecasting of energy demand in virtual power plants

Farshad Khavari[a], Jamal Esmaily[a], and Morteza Shafiekhani[b]
[a]Electrical Engineering Department, Shahid Rajaee Teacher Training University, Tehran, Iran, [b]Department of Electrical Engineering, Faculty of Engineering, Pardis Branch, Islamic Azad University, Pardis, Tehran, Iran

Introduction

Inside a smart grid, a new energy production model called a virtual power plant (VPP) emerges [1]. VPPs comprise a plethora of different elements designed to solve local problems, but they need to interact together to behave as a unit [2]. One of the most critical issues for the flexible operation of the VPP is demand forecasting, which allows the VPP to know in advance the amount of power that will be required, so that generation planning is feasible. When we talk about VPP load forecasting, the goal is energy load forecasting on a small (microgrid) scale [3–6]. The objective of the multi-agent system (MAS) demand forecasting is to forecast VPP users' energy demand by disaggregated sectors. To perform this task, the VPP needs to communicate with other agents and collaborate with the other processes to ensure VPP stability [2]. Typically, short-term load forecasting (STLF) consists of hourly prediction of the load for a lead time ranging from one hour to a week. Most studies in electric load forecasting in the past decades have focused on point load forecasting, meaning that at each time point, one value is provided, usually an average. Recently, market competition and requirements to integrate renewable technology have made researchers and operators more interested in load forecasts. Therefore, many solutions have been proposed [7].

A multi-agent system creates a demand forecasting system (DeF) for a VPP whose objective is to predict VPP users' energy demand by disaggregated sectors. To ensure stability and reliability, VPPs need to communicate with agents. These agents include historical control agent (DeF-HcA), smart meter control agent (DeFMcA), smart home/house data control agent (DeF-DcA), data preprocessing agent (DeFDpA), forecasting agent (DeF-FoA), retraining control agent (DeFReA), new smart home/house control agent (DeF-NcA), and external control agent (DeF-EcA). Their task is described below [2]:

    DeF-HcA provides historical data for prediction purposes;
    DeFMcA accurately transmits consumption data;
    DeF-DcA coordinates with DeF-McA and reports to DeFDpA;
    DeFDpA is responsible for data storage and standardization;
    DeF-FoA triggers demand forecasting;

DeFReA controls and retains the forecasting model;
DeF-NcA creates or removes other agents;
DeF-EcA communicates with DeF system and monitors customer interaction between different VPPs.

The principal objective of STLF is to provide load prediction for DERs scheduling and assess the security of system operation. The interactions between VPP's aggregator and short-term load forecasting box are shown in Fig. 15.1. The load forecasting box, based on historical load and weather data, predicts the load profile in a period, then sends it to the VPP's aggregator. VPP's aggregator optimizes its goal function based on forecasted loads.

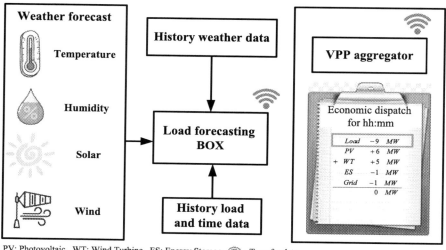

PV: Photovoltaic, WT: Wind Turbine, ES: Energy Storage, 📶 : Transfer data

Figure 15.1 Architecture and functional diagram of virtual power plant and load forecasting box.

Load behavior

Unlike other traded commodities, electricity power cannot be stored. Therefore, unutilized demand could be easily lost, which is the unique feature of the power market and it makes the load prediction task inevitable [8]. Yet, extracting patterns on a given load dataset would be critical and cumbersome. To elaborate on this point, the profile of load in a district in Tehran (Shahe-e-Ray) in 2010 is shown in Fig. 15.2. In this figure, there is a vague behavior, and the behavior of the load has no definite order.

Looking more closely at the load profile, nonetheless, shows a certain order. If we focus on 300 hours (equal to 12 days) of the load profile, it is clear that a rather meaningful connection exists among load profiles on different days (Fig. 15.3). Therefore, if this semantic connection is pre-determined, the load profile can be predicted.

Forecasting of energy demand in virtual power plants

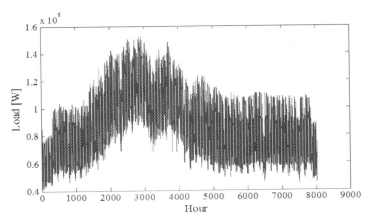

Figure 15.2 A real hourly load profile for a year.

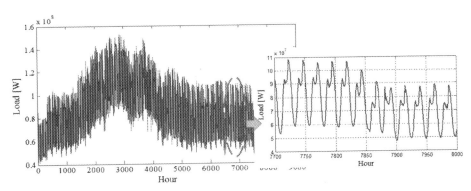

Figure 15.3 Focused load behavior on a weekly basis.

The load profile depends on time series parameters such as type of load, economic conditions, time factor, etc. Weather data could also meaningfully assist the prediction of load data since load consumption and weather statistics are usually correlated. Weather conditions consist of variables such as temperature, wind speed, daylight illumination, humidity, dew point, etc. [9].

- **Type of load**
  On a specific day, the load profiles of industrial, residential, and commercial customers are different. Industrial district's load profile is a rather flat curve. It also shows different temporal patterns since industrial customers consume more energy than residential districts in working shifts. In Fig. 15.4, the load profiles of two districts, named Chahar-Dange and Shahe-e-Ray, are presented for a typical day in 2010. Chahar-Dange is an industrial district and Shahe-e-Ray is a residential district. Chahar-Dange would consume electrical energy from 8 a.m. through 6 p.m. during its working shift.

Therefore, the amount of load that is used in Chahar-Dange is higher than in Shahe-e-Ray. The opposite pattern is evident in Shahe-e-Ray where residents would use more energy later that night.

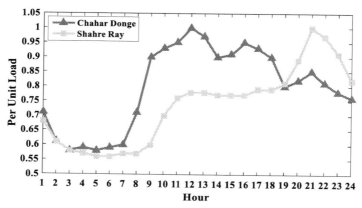

Figure 15.4 Different temporal patterns of load in industrial and residential districts on a daily basis.

- **Economic conditions**

Consumer economic condition is another parameter that can influence the load profile. In Fig. 15.6, the load profiles of two districts, named Gholhak and Shahe-e-Ray, are presented for a typical day in 2010. The majority of people that live in Gholhak are financially stronger than in Shahe-e-Ray. In houses that are located in Gholhak, there are usually an air-conditioning device, washing machine, dishwasher, electrical oven, and different electrical facilities which are more expensive and more likely to be affordable by residents of Gholhak. Therefore, the economic shift could alter the pattern of load usage as Fig. 15.5 shows.

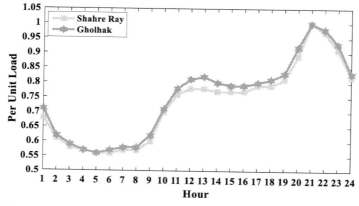

Figure 15.5 The impact of financial power on load consumption on a daily basis.

- **Time factor**

Another important factor that influences the load profile is the time step. Unlike Fig. 15.2, we could evaluate the load profile based on a 24 hour (a day) time step. In this way we could form distinct periods corresponding to seasons. The average of daily load profile in a district in Tehran (Shahe-e-Ray) during 2010 is shown in Fig. 15.6.

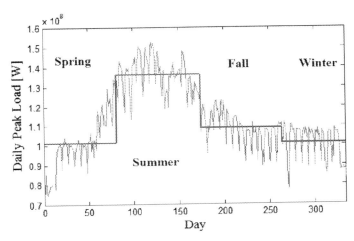

Figure 15.6 The average of daily and seasonal load profile.

In Fig. 15.7, four panels are represented which indicate different weekdays. Load profiles in five weeks are considered in each graph. Looking more closely at the load profile, it is clear that similar days behave alike; for example, five load profiles of three working days and a non-working day from five weeks in a row are shown in Fig. 15.7. The profile of loads in Fig. 15.7, panels (a), (b), and (d), are similar because these are working days, and they are completely different from Fig. 15.7(c) because this figure shows load's behavior on a non-working day.

Different weather parameters

The weather variables can also change customer's behavior. Important parameters that can change customer's behavior are the temperature, wind speed, daylight illumination, humidity, dew point, etc., which has been effectively utilized on different STLF researches, in different case studies all over the world. In the majority of them, the temperature is one of the most important parameters which can change the curve of the load. Different temperatures would have a casual impact on load profiles. For example, when the temperature is low, customers use heating devices; similarly, when the temperature is high, customers use cooling devices. Yet, the frequency of using electrical cooling devices is much more than electrical heating devices, which makes the impact of temperature on load profile more intricate and season dependent. The

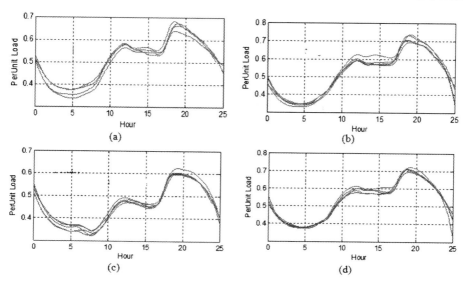

Figure 15.7 Load's profile for different days.

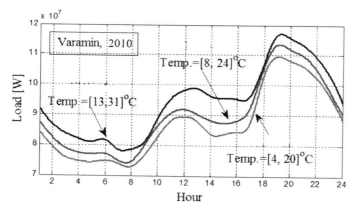

Figure 15.8 The profiles of load versus different temperature on three Wednesdays in July 2010.

changing load's behavior in a small area (Varamin) versus temperature during three consecutive Wednesdays in July 2010 are shown in Fig. 15.8.

Different methods for clustering analysis

So far, we mentioned several characteristics of the load pattern which could significantly assist the load prediction process. Yet, some patterns are far more hidden from the user's reach and could not be easily extracted and retrieved. Here, "clustering analysis" could come to the rescue since it could cluster different types of load profiles and introduce meaningful relations by putting similar data into a not pre-defined category.

Cluster analyses are used to describe a set of statistical data to discover classifications within complex data sets. The objective of these analyses is to group objects in different clusters so that objects in one cluster share more characteristics in common than members of other clusters. In other words, the similarity indexes between members of a cluster are minimized while it would be maximized between two members from two different clusters. Most of the cluster analyses have a similar process. Typically, an input sample is compared to the representative of a cluster (could be the average of the members in a cluster). If features of the input sample rather similar to the representative, it belongs to that specific cluster. Although a fairly comparable distance and similarity measures apply as an index of the similarity among data, selecting a clustering procedure is of great importance. The clustering methods provide a specific criterion for grouping data. There are some differences of opinions among researchers for choosing the most appropriate classification method for clustering. Yet, in general, cluster methods could be categorized into two distinct types: hierarchical clustering and k-means clustering. Hierarchical clustering algorithms seek to build a hierarchy of clusters. It starts with some initial clusters and gradually converges to the specific clusters. Hierarchical clustering has two categories, agglomerative and divisive. The agglomerative approach initially takes each data point as an individual cluster and then iteratively merges the clusters until no cluster contains all data points (agglomerative is also called a bottom-up approach). Divisive clustering techniques follow top-down route, which starts from a single cluster having all data points, and iteratively split the clusters into smaller ones until each cluster contains one data point (also called the top-down approach). Fig. 15.9 shows how hierarchical clustering cuts out K clusters.

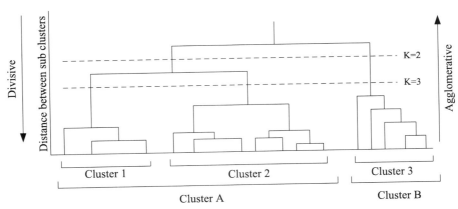

Figure 15.9 The clustering technique of the hierarchical clustering algorithm.

k-means clustering is an unsupervised learning algorithm. It is the simplest method used especially in data mining and statistics. Its goal is to form groups of data points

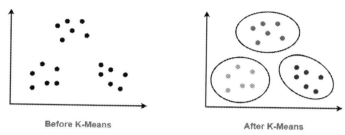

Figure 15.10 k-means algorithm.

based on the number of clusters, represented by the variable k. k-means uses an iterative refinement method to produce its final clustering based on the number of clusters defined by the user and the data set. Initially, k-means randomly chooses k points as the mean values of K clusters and finds the nearest data points of the chosen to form K clusters. In this procedure, the mean of each cluster would be dynamically changed. Algorithms would be usually stopped when there is no further change in means of each cluster; also called stabilization. Fig. 15.10 shows the clustering of 18 samples into three categories by k-means.

Different methods for STLF

Statistical and artificial intelligence techniques are two important categories for STLF. A large variety of statistical and artificial intelligence techniques have been developed for STLF.

One of the most widely used statistical methods is regression. Short-term load forecasting regression methods are usually used to model the relationship between load consumption and other factors (that were explained when discussing weather parameters and load behavior). Regression models incorporate deterministic influences such as those of weekdays and weekends, and stochastic influences such as those of weather conditions. References [10–13] describe different applications of regression models to STLF. Time-series techniques could also fall into the statistical methods category. Time-series techniques assume that there is a correlation between data history. Having extracted this correlation, the pattern of load data could be effectively predicted. Indeed, detecting such a correlation is the aim of time-series forecasting. Time series have been used for different fields and showed successful applications. In particular, linear and, ridge regression, ARMA,[1] ARIMA,[2] ARMAX,[3] and ARIMAX[4] are the most popular time-series techniques [14,15].

---

[1] Autoregressive moving average.
[2] Autoregressive integrated moving average.
[3] Autoregressive moving average with exogenous variables.
[4] Autoregressive integrated moving average with exogenous variables.

Among various forecasting methods, artificial intelligence techniques have received an important share of STLF [16–19]. Artificial neural networks (ANNs) are essentially nonlinear networks that have the capability of modeling interconnections and nonlinear patterns which are generally the two prime features of load profiles. An ANN consisting of three layers (input layer, hidden layer, and output layer) is shown in Fig. 15.11. The layers between the input node and the output layer are called the hidden layers. The hidden layers enable ANN to model nonlinear dynamics.

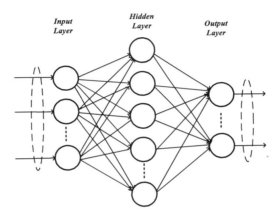

Figure 15.11 Structure of an ANN with a hidden layer.

Support vector machine (SVM) is another commonly used tool in machine learning, which is usually used for classification. Researchers have used support vector machine techniques for STLF [20,21]. Using a fuzzy-ARTMAP neural network (FANN) is another method for STLF. This method not only decreases processing time but also leads to results with faster and sometimes more accurate convergence compared to backpropagation (BP) [22]. The speed of convergence could be crucial in real-time load forecasts. Therefore, ANNs could not be readily applicable because of their immense computations, resulting in a lengthy process. Moreover, ANNs might suffer from local optima and overweight problems. An extreme learning machine (ELM) is a type of supervised learning that was introduced to tack the drawbacks of classical ANNs [23–25]. ELMs are a novel learning technology, which by ensemble method can increase the accuracy of each ELM. The prerequisite for achieving the proper performance of the ELM is the difference in the output error of each expert, therefore the number of middle layer neurons in a range $[M_1, M_2]$ is considered different. The structure of the ELM ensemble model is introduced in Fig. 15.12.

Fitness criteria

There are different ways to assess the accuracy of a forecast, depending on the definition of accuracy and goals of the forecaster [9,26]. Two important parameters for errors

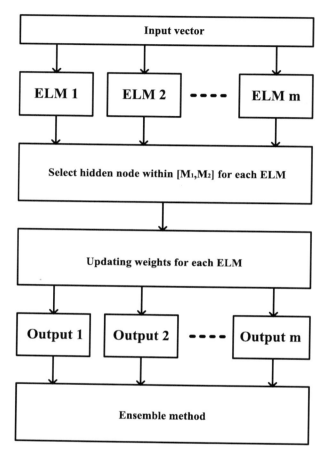

Figure 15.12 Ensemble model of ELM [25].

are the mean absolute percentage error (MAPE) and mean absolute error (MAE). Unlike MAE, MAPE is scale-independent; therefore, it can be used to compare different types of datasets. The MAPE and MAE are defined as:

$$MAPE = \frac{1}{N}\sum_{i=1}^{N}\left|\frac{p_i - r_i}{r_i}\right| \times 100, \quad (15.1)$$

$$MAE = \frac{1}{N}\sum_{i=1}^{N}|p_i - r_i|, \quad (15.2)$$

where $p = (p_1, p_2, ..., p_N)$ is the forecasted load and $r = (r_1, r_2, ..., r_N)$ is the actual load; N is the number of samples which would be determined by the nature of analysis (daily, weekly, and hourly).

Case study

In this part, we introduce an implementation of a powerful method based on the extreme learning machine. Short-term load forecast for Sydney, Australia, is simulated; the historical load demand data of the NSW region from July 1, 2010 to October 30, 2015 (about 5 years) is used [27]. As we stated earlier, different criteria affect the behavior of the load, including the type of load, environmental conditions, time factors, economic conditions, and accidental changes. The question that researchers face is which information to consider for load behavior? Some data may have a direct effect on load behavior in some areas and not in other areas. Therefore, selecting the appropriate parameters is essential since it will reduce the computational load and increase accuracy. An important characteristic of STLF that needs to be considered is the different patterns between Monday to Friday (working days) and Saturday, Sunday, and public holidays (non-working days).

Pearson correlation coefficient (PCC) has been employed to investigate the relationship between data and load behavior on the following day. PCC is calculated as follows:

$$\rho_{x,y} = \frac{\sum_{i=1}^{n}(x_i - \bar{x})(y_i - \bar{y})}{\sqrt{\sum_{i=1}^{n}(x_i - \bar{x})^2}\sqrt{\sum_{i=1}^{n}(y_i - \bar{y})^2}}, \quad (15.3)$$

where x and y are the two studied variables, $\bar{x}$ and $\bar{y}$ are the average values of the two data sets, respectively. The value of ρ ranges from -1 to 1, and a larger absolute value of ρ implies a stronger linear correlation between x and y. The type of correlation is introduced in Table 15.1.

Table 15.1 Interpretation of PCC.

The absolute value of PCC	Correlation
0–0.09	None
0.1–0.29	Small
0.3–0.49	Medium
0.5–1	Strong

The plot of load demand of day $k-1$ against day k for the hot season is shown in Fig. 15.13. The correlation between them is equal to 0.921; therefore, there is a strong correlation between electricity demand for today and the following day.

The correlation study results based on different days are given in Table 15.2, according to which there is a strong correlation between load on the following day and today's load. Loads of last days and last week are mostly related. So, we only selected these data as the load inputs.

The correlation study results for different weather variables (temperature, radiation, humidity, and wind speed) and electricity demand on the following day are introduced in Table 15.3, revealing that there are strong correlations between the environment temperature, radiation of the sun, and electricity demand of the following day. The

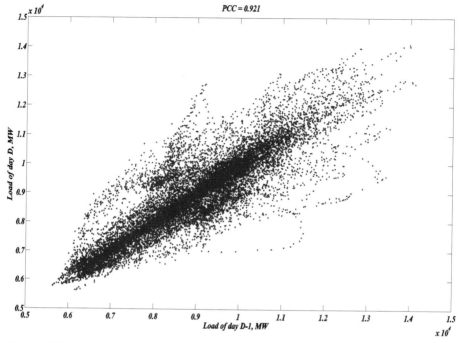

Figure 15.13 The load of day $k-1$ against the load of day k.

Table 15.2 Correlation between electricity demand the following day (k) and the previous day's ($k-d$).

Variable	PCC value
Load ($k-1$)	0.921
Load ($k-2$)	0.885
Load ($k-3$)	0.843
Load ($k-7$)	0.897

Table 15.3 Correlation between load and weather variables.

	Temperature	Radiation	Humidity	Wind speed
PCC value	0.723	0.501	-0.221	0.089

correlation between electricity demand for the following day and other weather variables (humidity and wind speed) is not significant.

One of the parameters that have a great impact on the performance of experts is the number of middle layer neurons. To find the appropriate range for the number of neurons in the middle layers, we calculated the MAPE value by incrementing the number of hidden layer neurons (Fig. 15.14). The interval for which MAPE value is minimized is considered as the appropriate interval. Because of the nature of the problem – the random selection of input weights and biases – the neurons in the middle layer of the plane were changed. The accuracy is calculated based on 20 repetitions, and its average is plotted in Fig. 15.14. The horizontal axis in Fig. 15.14 shows the number of neurons in the middle layer, and the vertical axis shows the MAPE value. In this figure, the minimum error for 34 neurons in the hidden layer is equal to 2.60%. The optimal range for the number of neurons in the hidden layer {20–50} has been selected.

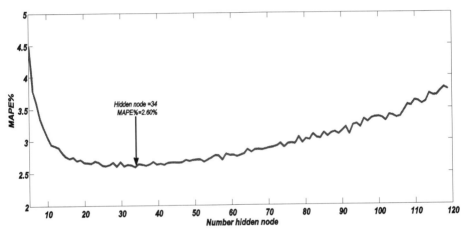

Figure 15.14 Hidden node number against validation MAPE (%).

After finding the optimal range for the number of middle layer neurons, the question arises as to how many EMLs must be used to achieve the optimal answer to increase the efficiency of the system. To determine the optimal number of experts, the numbers of neurons in the interval {20–50} were set, and then the number of ELMs was increased and the MAPE value was calculated. Similar to Fig. 15.14, the accuracy is calculated based on 20 repetitions, and its average is plotted in Fig. 15.15. The horizontal axis in Fig. 15.15 shows the number of ELMs, and the vertical axis shows the MAPE value of the error. The upper curve shows the error of an expert, and the lower curve shows the error of the expert community.

It is important to note that the combined experts must be different, which is why the number of hidden layer neurons in the range {20–50} has been chosen. As shown in Fig. 15.14, the lowest error is obtained for the accumulation of 96 experts and equal to 2.36%. In addition to using the ensemble expert method to solve the problem, common methods such as a backpropagation neural network (BPNN) and radial basis function neural network (RBFNN) have also been used for short-term load estimation, which shows the superiority of the ELM method. The results are presented in Table 15.4.

As can be seen in Table 15.4, the artificial neural network method based on the recursive method and the method of base radius functions have an error of 2.82% and

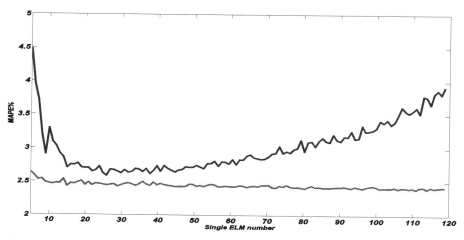

Figure 15.15 Single ELM number against validation MAPE (%).

Table 15.4 Generalizations and their performance

	Extreme Learning	RBFNN	BPNN
Min MAPE%	2.51	4.13	2.71
Ave MAPE%	2.63	4.13	2.82
Std MAPE%	0.09	-	0.12
Best Hidden node	34	-	20

4.13%, respectively. The coherent expert method has been able to achieve the desired answers with smaller errors than conventional methods.

Conclusion

The study of input data, more than any other analysis, is important for load forecasting. Moreover, those more rigorous standards and criteria should be adopted in the reporting of the experiments and the analysis of the results. Having adequate accuracy of electricity demand forecasting is very important for the system's operators. In this chapter, some statistical and artificial intelligence methods that have been applied for short-term load forecasting have been reviewed. Also, the factors that can affect the accuracy of forecasts such as weather conditions, time factors, customer classes, and economic factors of end-users are discussed comprehensively. Finally, a real case study has been simulated and discussed. Based on the results, ensemble learning methods have a better performance to solve short-term load forecasting problems compared to other methods.

References

[1] B. Wille-Haussmann, T. Erge, C. Wittwer, Decentralized optimization of cogeneration in virtual power plants, Sol. Energy 84 (4) (2010) 604–611.

[2] L. Hernandez, C. Baladron, J.M. Aguiar, B. Carro, A. Sanchez-Esguevillas, J. Lloret, et al., Multi-agent system architecture for smart grid management and forecasting of energy demand in virtual power plants, IEEE Commun. Mag. 51 (2013) 106–113.

[3] M.H. Shams, M. Shahabi, M. Kia, A. Heidari, M. Lotfi, M. Shafie-khah, J.P. Catalão, Optimal operation of electrical and thermal resources in microgrids with energy hubs considering uncertainties, Energy 187 (2019) 115949.

[4] M. Shafiekhani, A. Badri, M. Shafie-khah, J.P.S. Catalão, Strategic bidding of virtual power plant in energy markets: a bi-level multi-objective approach, Int. J. Electr. Power Energy Syst. (2019).

[5] M. Shafiekhani, A. Badri, F. Khavari, A bi-level model for strategic bidding of virtual power plant in day-ahead and balancing market, in: Smart Grid Conf., 2017.

[6] F. Khavari, A. Badri, A. Zangeneh, Energy management in multi-microgrids via an aggregator to override point of common coupling congestion, IET Gener. Transm. Distrib. 13 (5) (2018) 634–642.

[7] Y. Wang, Q. Xia, C. Kang, Secondary forecasting based on deviation analysis for short-term load forecasting, IEEE Trans. Power Syst. 26 (2) (2011) 500–507.

[8] Mayukha Pal, P. Madhusudana Rao, P. Manimaran, Multifractal detrended cross-correlation analysis of Indian electricity market, in: IEEE Xplore, ISBN 978-1-4673-9682-0, 2015, pp. 1–4.

[9] M. Jacob, C. Neves, D. Vukadinović Greetham, Forecasting and Assessing Risk of Individual Electricity Peaks, Springer, 2020.

[10] W. Charytoniuk, M.S. Chen, P. Van Olinda, Nonparametric regression based short-term load forecasting, IEEE Trans. Power Syst. 13 (1998) 725–730.

[11] T. Haida, S. Muto, Regression based peak load forecasting using a transformation technique, IEEE Trans. Power Syst. 9 (1994) 1788–1794.

[12] O. Hyde, P.F. Hodnett, An adaptable automated procedure for short-term electricity load forecasting, IEEE Trans. Power Syst. 12 (1997) 84–93.

[13] S. Ruzic, A. Vuckovic, N. Nikolic, Weather sensitive method for short-term load forecasting in electric power utility of Serbia, IEEE Trans. Power Syst. 18 (2003) 1581–1586.

[14] H.T. Yang, C.M. Huang, A new short-term load forecasting approach using self-organizing fuzzy ARMAX models, IEEE Trans. Power Syst. 13 (1998) 217–225.

[15] M.Y. Cho, J.C. Hwang, C.S. Chen, Customer short-term load forecasting by using ARIMA transfer function model, in: Proceedings of the International Conference on Energy Management and Power Delivery, vol. 1, 1995, pp. 317–322.

[16] A. Khotanzad, R. Afkhami-Rohani, T.L. Lu, A. Abaye, M. Davis, D.J. Maratukulam, ANNSTLF-a neural-network-based electric load forecasting system, IEEE Trans. Neural Netw. 8 (4) (1997) 835–846.

[17] A. Khotanzad, R. Afkhami-Rohani, D. Maratukulam, ANNSTLF-artificial neural network short-term load forecaster- generation three, IEEE Trans. Power Syst. 13 (4) (1998) 1413–1422.

[18] H.S. Hippert, C.E. Pedreira, R.C. Souza, Neural networks for short-term load forecasting: a review, evaluation, IEEE Trans. Power Syst. 16 (1) (2001) 44–55.

[19] D. Baczynski, M. Parol, Influence of artificial neural network structure on quality of short-term electric energy consumption forecast, IET Gener. Transm. Distrib. 151 (2) (2004) 241–245.

[20] Y. Li, T. Fang, Wavelet and support vector machines for short-term electrical load forecasting, in: Proceedings of International Conference on Wavelet Analysis and its Applications, vol. 1, 2003, pp. 399–404.
[21] M. Mohandes, Support vector machines for short-term electrical load forecasting, Int. J. Energy Res. 26 (2002) 335–345.
[22] T. Abreu, A.J. Amorim, C.R. Santos-Junior, A.D.P. Lotufo, C.R. Minussi, Multinodal load forecasting for distribution systems using a fuzzy-ARTMAP neural network, Appl. Soft Comput. 71 (Oct. 2018) 307–316.
[23] J. Ghasemi, J. Esmaily, R. Moradinezhad, Intrusion detection system using an optimized kernel extreme learning machine and efficient features, Sadhana 45 (1) (Dec 2019) 2.
[24] S. Fan, L. Chen, W.-J. Lee, Short-term load forecasting using comprehensive combination based on multi meteorological information, IEEE Trans. Ind. Appl. 45 (4) (2009) 1460–1466.
[25] R. Zhang, Z. Yang Dong, Y. Xu, Short-term load forecasting of Australian national electricity market by ensemble model of extreme learning machine, IET Gener. Transm. Distrib. 7 (2013) 391–397.
[26] E.A. Feinberg, D. Genethliou, Load forecasting, in: J.H. Chow, F.F. Wu, J.J. Momoh (Eds.), Applied Mathematics for Restructured Electric Power Systems: Optimization, Control and Computational Intelligence, Power Electronics and Power Systems, Springer, US, 2005, pp. 269–285.
[27] http://www.aemo.com.au/electricityops/3710.html.

Emission impacts on virtual power plant scheduling programs

Nazgol Khodadadi[a], Amin Mansour-Saatloo[a], Mohammad Amin Mirzaei[a], Behnam Mohammadi-Ivatloo[a,b], Kazem Zare[a], and Mousa Marzband[c]
[a]Faculty of Electrical and Computer Engineering, University of Tabriz, Tabriz, Iran,
[b]Department of Electrical and Electronics Engineering, Mugla Sitki Kocman University, Mugla, Turkey, [c]Department of Mathematics, Physics and Electrical Engineering, Northumbria University, Newcastle, England, United Kingdom

Nomenclature

Indices
I Index of CHP units
s Index of scenarios
T Index of time

Parameters
a_b, b_b, c_b Boiler cost function coefficients
$a_i, b_i, c_i, d_i, e_i, f_i$ CHP unit cost function coefficients
Em_{Daily} Daily permitted emission rate (kg/MWh)
λ_t^c Demand supply price
$EL_{t,s}$ Electrical load (MWh)
$\lambda_{t,s}$ Electricity market price
$NO_X^{eq}, CO_2^{eq}, SO_2^{eq}$ Emission factors associated with the purchased power from the grid (kg/MWh)
η_{ec}/η_{ed} ESS charging/discharging efficiency (%)
Pc^{max}/Pd^{max} ESS maximum charging/discharging power (MW)
η_{hc}/η_{hd} HSS charging/discharging efficiency (%)
Hc^{max}/Hd^{max} HSS maximum charging/discharging power (MWt)
Lf Load shifting factor
$H_{b,s}^{max}$ Maximum heat output of the boiler (MWt)
E^{max}/E^{min} Maximum/minimum level of ESS (MWh)
A^{max}/A^{min} Maximum/minimum level of HSS (MWh)
$PV_{t,s}$ Output power of photovoltaic panels (MW)
$PW_{t,s}$ Output power of wind turbines (MW)
π_s Probability of each scenario
NO_X^b, CO_2^b, SO_2^b The emission rate of boiler unit (kg/MWth)
$NO_X^{CHP}, CO_2^{CHP}, SO_2^{CHP}$ The emission rate of CHP units (kg/MWh)
$HL_{t,s}$ Thermal load (MWth)

Binary variables
$I_{i,t,s}^a, I_{i,t,s}^b$ Operation status of CHP type a and b
$V_{i,t,s}^1/V_{i,t,s}^2$ Commitment state of CHP unit type b in the first/second section

Scheduling and Operation of Virtual Power Plants. https://doi.org/10.1016/B978-0-32-385267-8.00021-4
Copyright © 2022 Elsevier Inc. All rights reserved.

Continuous variables

$C_{b,t,s}$ Boiler operation cost ($/h)
$C_{i,t,s}$ CHP unit operation cost ($/h)
$SUC_{i,t,s}/SDC_{i,t,s}$ CHP unit start-up/shut-down costs ($)
$L_{t,s}^{dr}$ Electrical load after DRP (MW)
$P_{i,t,s}$ Electrical output of the CHP unit (MW)
$Pc_{t,s}/Pd_{t,s}$ ESS charging/discharging power (MW)
$E_{t,s}$ ESS level (MWh)
$P_{t,s}^{eq}$ Exchanged power with the grid (MW)
$Hc_{t,s}/Hd_{t,s}$ HSS charging/discharging power (MWth)
$A_{t,s}$ HSS level (MWth)
$H_{b,t,s}$ Thermal output of the boiler (MWt)
$H_{i,t,s}$ Thermal output of the CHP unit (MWt)

16.1 Introduction

16.1.1 Motivation

The insertion of distributed energy resources (DERs) is growing globally fast due to environmental concerns, high energy generation costs, and fossil fuel shortage [1]. DER units are considered as clean and resilient small-scale manageable power sources which can alleviate the energy crisis [2]. Distributed generation (DG) units, energy storage systems, and renewable energy resources (RESs) are typical models of DERs that are dispersedly connected to the energy systems [3]. Due to the bulk power industry contribution in releasing a noticeable volume of carbon dioxide (CO_2) and in global warming, DER's role has become increasingly influential in providing a sustainable environment and handling the climate change issue [4]. The virtual power plant (VPP) concept is a proper alternative to realize the DERs integration into the energy systems [5]. A VPP is a collection of DG units, electrical storage systems (ESSs), RESs, flexible and inflexible loads that are aggregated together with an electrical boundary. In addition, the combined heat and power (CHP) unit can be integrated into a VPP to assist the energy system by generating power and heat simultaneously. CHP unit is a flexible source that can increase the efficiency of the system due to the recovery of the waste heat from a power generation procedure, so a CHP unit can deliver heat besides power for a respective feasible operation region (FOR) [6]. The operator of the VPP is assigned for optimal power dispatching of DERs, supplying the end-users' load, and interacting with the energy markets. Moreover, the operator of the VPP can control the flexible loads of consumers by employing demand response programs (DRPs) as a demand-side management tool [7]. In doing so, flexible loads of demand-side sectors can be curtailed at high energy price periods and shifted to low energy price periods. Hence, each of the DERs can access visibility in the energy markets with the interfacing of the VPP operator to maximize their profit. The VPP is able to control scheduling for resources and load efficiency. VPPs can handle cooperation in the energy and auxiliary markets simultaneously for a small-sized power system [4]. VPPs warrant the efficiency of the demand-side energy supply while minimizing costs [8]. CHP units are such critical technologies for providing electricity and heat since they are capable of raising the efficiency and lowering the pollution level of the plant. A CHP unit is the most efficient technology in a thermal unit, therefore applying a CHP unit is common and very popular in a VPP. It should be mentioned that a CHP is able to reduce systems' fluctuation and financial risks [9].

16.1.2 Literature review

Several approaches have already been investigated regarding the potential of the VPPs, their challenges, and opportunities in recent years. The studies can be classified from different points of view, such as integrating different sorts of DER technologies, the uncertainty factor, solving method, and formulation type. In [1], VPP's definition, concept, types, and components are identified. In [10], the trade-off between cost and emission has been demonstrated by applying particle swarm optimization (PSO), where smart programming reduces emissions and costs via maximizing the manipulation of RESs. In [11], utilizing a stochastic scheduling strategy, the contribution of the VPP in the day-ahead market, and the real-time market has been considered. The uncertainties included in the electricity price, generation of RESs, and consumption of loads were considered using cooperative game theory in [7]. In [12–14], a particular price-based unit involvement procedure has been performed as a proper solution for the entrance of VPPs in the energy market but without analyzing the attendance of renewable energy resources and DRP. In [15], a novel algorithm has been presented to optimize the electrical and thermal scheduling of a large-scale VPP. Amidst these methods, probabilistic planning is more suitable for the impact valuation of renewable generations, electricity prices, and load demand fluctuations. The output of renewable generators relies on their energy sources' characteristics, like wind speed, solar radiation, and environmental temperature [16]. These parameters' historical results are usually available, and they can be programmed by applying a probability density function (PDF) [12]. Regarding this issue, stochastic programming has been taken into consideration by many of the newly published research articles. For instance, in [17] the stochastic approach has been used for economic dispatch, and in [18] it has been used for unit commitment. Also, a short-term stochastic scheduling has been performed in [19] for VPPs, in which the potential of electric vehicles and smart buildings has been investigated for the participation in demand response. Owing to electric vehicle proliferation in recent years, an energy management model for VPPs was developed in [8], in which the model describes a real case study considering the development of hybrid vehicles. A scheduling approach was proposed in [20] to evaluate CHP systems' ability to decrease the PV installations imbalance errors in the VPP. In [21], a weekly self-scheduling model based on stochastic programming was presented, in which the electricity market, wind, and PV power generations are taken into account as uncertain parameters. In [22], a mathematical approach, namely mixed-integer linear programming (MILP), was used to formulate the optimization problem, where a medium-term coalition-forming-based approach is applied to solve the problem. In [23], an optimal mid-term stochastic scheduling of the VPP based on the maximization of profit was proposed, in which to cover the RESs intermittencies, a pumped hydro storage system was taken into account. In a similar context, mid-term scheduling of the commercial

VPP utilizing a two-stage framework was proposed in [24], where the first stage deals with the optimal dispatch of the DERs, bilateral contracts, and market contracts, while the second stage deals with day-ahead market clearing. Moreover, many studies have been dedicated to investigating the DRP, which are notable in economic problems and reliability in future power systems. The primary concepts of demand response, price-based demand response, and incentive-based demand response are defined in [25]. DR appears in load peak–valley shifting and decreases the renewable energy uncertainty [26], which is helpful in developing VPP power generation. A DRP can ensure the demanded comfort of customers while bringing financial benefits to both VPPs and consumers [27]. An economic approach based on applying DRP was studied in [28] to develop the load profile characteristics comfort of the customers. The authors of [29] described a stochastic profit-based model for a combined wind farm-cascade hydro system under a VPP in a day-ahead market, where a two-stage stochastic programming model is proposed to handle the wind power prediction errors. The authors of [30] presented a flexible demand response model managed by shiftable demand bids in the power market along with reserve bids in the reserve market. In [31], the authors are using day-ahead scheduling, choosing the best demand response programs on industrial technical VPPs, including loads and generations placed in an industrial grid in a short-term market. In [32], a mathematical approach for the energy bidding problem of a virtual power plant (VPP) has been represented, which is associated with the short-term market and the intraday demand response exchange market. In [33], a new structure for the development of distribution companies (discos) has been presented considering CO_2 emission costs and DERs which contain two phases, named day-ahead stage and hour-ahead stage. Reference [34] proposes a new structure for the VPP energy management problem studying correlated demand response, where the objective function is to minimize the system's cost while keeping the system's potential. In [35], the authors suggested the point estimate method (PEM) to model the uncertainties of VPP scheduling problem and lowering unexpected fluctuations risks of the system's uncertain parameters, in which power and reserve are scheduled altogether, considering renewable technologies' uncertainties, power demands, and market prices.

16.1.3 Contributions

To the best of the authors' knowledge, there is no focus on the evaluation of DRP and energy storage systems under a coordinated model in a CHP-based VPP considering emission limits. To this end, this chapter proposes an optimal energy management framework for a CHP-based VPP, which includes wind turbines, PV panels, an ESS, a heat storage system (HSS), a CHP unit, and a boiler. Fig. 16.1 shows the CHP-VPP schematically. The proposed CHP-based VPP aims to schedule its resources and supply the power and heat demand of the end-users considering carbon emission limitation. To achieve more benefits, the operator of the VPP, as an interface between DERs and electrical market operator, can exchange power with the electrical energy market. The DRP is applied as a demand-side management technology to ensure a viable operation of the VPP. In doing so, the operator of the VPP can control the

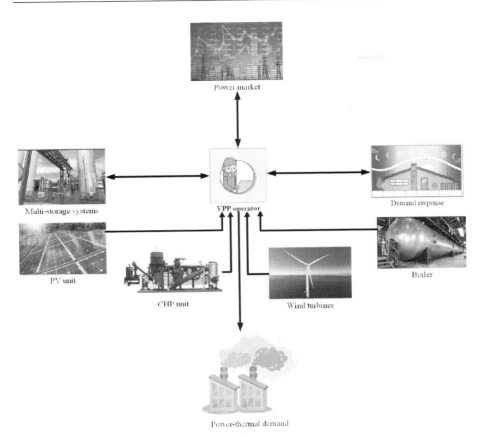

Figure 16.1 The proposed CHP-VPP schematic.

end-users' load profile to increase the benefit of the VPP and decrease the effect of carbon emission limitations on the VPP's profit. Finally, stochastic optimization is utilized to handle the uncertainty of renewable generations, electricity market, electrical load, and thermal load. The stochastic approach is one of the probabilistic-based optimization methods that can handle the uncertainties by considering different scenarios. The proposed model is formulated as an MILP that can be solved with a CPLEX solver. Obtained results show how carbon emission limitation influences the dispatch of DERs, exchanged power with the energy market, and end-users' strategies in participation in the DRP.

The rest of the chapter is organized as follows: Section 16.2 presents the objective function, mathematical models of utilized technologies, and associated constraints. Section 16.3 provides simulation and numerical results based on the proposed framework to verify the model. Finally, Section 16.4 concludes the chapter and discusses the outcomes.

16.2 Problem formulation

This section includes the mathematical model of the objective function, CHP units, multi-energy storage systems, i.e., ESS and HSS, boiler, wind turbines, PV panels, emission, DRP, and energy balance.

16.2.1 Objective function

The main objective of the suggested model is maximizing the day-ahead profit of CHP-VPP under a scenario-based stochastic technique, which is exposed in (16.1). The first term shows the power sold to consumers, and the second term demonstrates the cash flow between the CHP-VPP and the upstream power market. The third and fourth terms define the costs of fuel related to CHP units and boiler.

$$Profit = \max \sum_{s=1}^{Ns} \pi_s \times \sum_{t=1}^{Nt} \left[\lambda_t^c L_{t,s}^{dr} - \lambda_{t,s} P_{t,s}^{eq} - \sum_{i=1}^{Ni} C_{i,t,s} - \sum_{b=1}^{Nb} C_{b,t,s} \right]. \quad (16.1)$$

The cost functions of the CHP and boiler units are presented in (16.2)–(16.5). Also, the start-up and shut-down costs of the CHP units are shown in (16.6)–(16.7) [32,33]:

$$C_{i,t,s} = a_i + b_i \times P_{i,t,s} + c_i \times P_{i,t,s}^2 + d_i \times H_{i,t,s}^2 + e_i \times H_{i,t,s}$$
$$+ f_i \times H_{i,t,s} \times P_{i,t,s} + SUC_{i,t,s} + SDC_{i,t,s}, \quad (16.2)$$

$$C_{b,t,s} = a_b + b_b H_{b,t,s} + c_b H_{b,t,s}^2, \quad (16.3)$$

$$SUC_{i,t,s} = SU_{i,s}(I_{i,t,s} - I_{i,t-1,s}), \quad (16.4)$$

$$SUC_{i,t,s} \geq 0, \quad (16.5)$$

$$SDC_{i,t,s} = SD_{i,s}(I_{I,t-1,s} - I_{i,t,s}), \quad (16.6)$$

$$SDC_{i,t,s} \geq 0. \quad (16.7)$$

16.2.2 CHP unit

In CHP units, the thermal and electrical generated powers are dependent on each other. Therefore, the FOR is applied to model the dependency. Two types of CHP units are studied in this paper with convex and non-convex FORs, which are shown in Fig. 16.2. Eqs. (16.8)–(16.12) describe the restrictions of the first type of CHP unit [34]. Eqs. (16.13)–(16.19) express the constraints of the non-convex FOR of the second type of CHP:

$$P_{i,t,s}^a - P_{i,A}^a - \frac{P_{i,A}^a - P_{i,B}^a}{H_{i,A}^a - H_{i,B}^a}(H_{i,t,s}^a - H_{i,A}^a) \leq 0, \quad (16.8)$$

$$P_{i,t,s}^a - P_{i,B}^a - \frac{P_{i,B}^a - P_{i,C}^a}{H_{i,B}^a - H_{i,C}^a}(H_{i,t,s}^a - H_{i,B}^a) \geq -(1 - I_{i,t,s}) \times M, \quad (16.9)$$

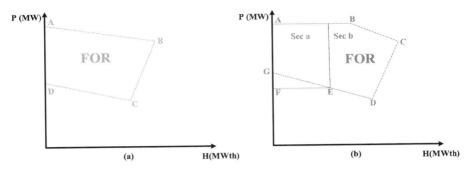

Figure 16.2 CHP units FOR characteristics: (a) CHP unit type a, (b) CHP unit type b.

$$P_{i,t}^a - P_{i,C}^a - \frac{P_{i,C}^a - P_{i,D}^a}{H_{i,C}^a - H_{i,D}^a}(H_{i,t}^a - H_{i,C}^a) \geq -(1 - I_{i,t}) \times M, \quad (16.10)$$

$$0 \leq P_{t,i,s}^a \leq P_{t,A}^a \times I_{i,t,s}, \quad (16.11)$$

$$0 \leq H_{t,i,s}^a \leq H_{t,A}^a \times I_{i,t,s}, \quad (16.12)$$

$$P_{i,t,s}^b - P_{i,B}^b - \frac{P_{i,B}^b - P_{i,C}^b}{H_{i,B}^b - H_{i,C}^b}\left[H_{i,t,s}^b - H_{i,B}^b\right] \leq 0, \quad (16.13)$$

$$P_{i,t,s}^b - P_{i,C}^b - \frac{P_{i,C}^b - P_{i,D}^b}{H_{i,C}^b - H_{i,D}^b}\left[H_{i,t,s}^b - H_{i,C}^b\right] \geq 0, \quad (16.14)$$

$$P_{i,t,s}^b - P_{i,E}^b - \frac{P_{i,E}^b - P_{i,F}^b}{H_{i,E}^b - H_{i,F}^b}\left[H_{i,t,s}^b - H_{i,E}^b\right] \geq -\left[1 - V_{i,t,s}^1\right] \times M, \quad (16.15)$$

$$P_{i,t,s}^b - P_{i,D}^b - \frac{P_{i,D}^b - P_{i,E}^b}{H_{i,D}^b - H_{i,E}^b}\left[H_{i,t,s}^b - H_{i,D}^b\right] \geq -\left[1 - V_{i,t,s}^2\right] \times M, \quad (16.16)$$

$$0 \leq P_{i,t,s}^b \leq P_{i,t,s}^b \times I_{i,t,s}^b, \quad (16.17)$$

$$0 \leq H_{t,i,s}^b \leq H_{i,C,s}^b \times I_{t,i,s}^b, \quad (16.18)$$

$$V_{i,t}^1 + V_{i,t}^2 = I_{i,t}^b. \quad (16.19)$$

16.2.3 Multi-energy storage systems

In this chapter, both electrical and heat energy storage systems are considered. Energy storage technologies are modeled between their minimum and maximum rate of energy. Eqs. (16.20)–(16.23) demonstrate the operational constraints of charging and discharging in their permitted range for electrical and heat storage systems, respectively. The state of charge (SoC) of electrical and heat storage systems in each hour is defined as in (16.24)–(16.25). The capacity limits of electrical and heat storages are stated in (16.26)–(16.27). The initial and final SoC of electrical and thermal storages are respectively displayed in (16.28)–(16.29) [35]. It is noteworthy that degradation costs of storage systems are not considered since the utilized storages are small-scale local storage systems and have very low degradation costs, which can be ignored in the objective function [40]:

$$0 \leq Pd_{t,s} \leq Pd^{\max}, \tag{16.20}$$

$$0 \leq Pc_{t,s} \leq Pc^{\max}, \tag{16.21}$$

$$0 \leq Hd_{t,s} \leq Hd^{\max}, \tag{16.22}$$

$$0 \leq Hc_{t,s} \leq Hc^{\max}, \tag{16.23}$$

$$E_{t,s} = E_{t-1,s} + \frac{Pc_{t,s}}{\eta_{ec}} - \eta_{ed} \times Pd_{t,s}, \tag{16.24}$$

$$A_{t,s} = A_{t-1,s} + \frac{H_c}{\eta_{hc}} + \eta_{hd} \times Hd_{t,s}, \tag{16.25}$$

$$E^{\min} \leq E_{t,s} \leq E^{\max}, \tag{16.26}$$

$$A^{\min} \leq A_{t,s} \leq A^{\max}, \tag{16.27}$$

$$E_0 = E_{Nt}, \tag{16.28}$$

$$A_0 = A_{Nt}. \tag{16.29}$$

16.2.4 Boiler

Utilizing only the CHP units to supply thermal demands is not cost-effective. So, a boiler can be a good alternative to supply thermal demand. The operating limits of the output of the boiler units are taken as in [36] to be

$$0 \leq H_{b,t,s} \leq H_{b,s}^{\max}. \tag{16.30}$$

16.2.5 Wind turbines and PV panels

The output power of the wind turbine is dependent on the wind speed, which is addressed in [2]. PV panel as a renewable distributed generation depends on the characteristics of the panel, the ambient temperature of the site, and the solar radiation that is modeled in this chapter according to [2].

$$PW_t(v) = \begin{cases} 0, & v \leq v^c \text{ or } v \geq v^c, \\ \frac{v-v^c}{v^r-v^c} PW^r, & v^c \leq v \leq v^r, \\ PW^r, & \text{otherwise,} \end{cases} \tag{16.31}$$

$$PV_t \leq S.\eta.R, \tag{16.32}$$

where v^c is the cut-in velocity, v^r is the rated velocity, and PW^r is the rated output power of the turbine. Also, the PV panel output depends on three parameters, namely size S, efficiency η, and solar radiation R.

16.2.6 Emission

Eqs. (16.33)–(16.35) calculate released emissions in each hour by CHP units, boilers units, and the power grid, respectively. The limitation of daily emissions is defined as in (16.36):

$$Em_{t,s}^{CHP} = \sum_{i=1}^{Ni}(NO_X)_i + (CO_2)_i + (SO_2)_i \times P_{i,t,s}, \quad (16.33)$$

$$Em_{t,s}^{Boiler} = \sum_{b=1}^{Nb}(NO_X)_b + (CO_2)_b + (SO_2)_b \times H_{b,t,s}, \quad (16.34)$$

$$Em_{t,s}^{eq} = (NO_X^{eq} + CO_2^{eq} + SO_2^{eq}) \times \max\{-P_{t,s}^{eq}, 0\}, \quad (16.35)$$

$$\sum_{t=1}^{Nt}(Em_{t,s}^{CHP} + Em_{t,s}^{Boiler} + Em_{t,s}^{eq}) \leq Em^{daily}. \quad (16.36)$$

16.2.7 Demand response program

In this chapter, DRP is implemented based on shiftable demands. The daily demand before and after the implementation of the DRP should be equal, as shown in (16.37). The demand shifting limit in each hour is also expressed as (16.38) [37,38]:

$$\sum_{t=1}^{Nt} EL_{t,s}^{dr} = \sum_{t+1}^{Nt} EL_{t,s}, \quad (16.37)$$

$$(1 - Lf)EL_{t,s} \leq El_{t,s}^{dr} \leq (1 + Lf)EL_{t,s}. \quad (16.38)$$

16.2.8 Power and heat balance

Eqs. (16.39)–(16.40) show relations regarding electrical and thermal power balance in each hour, respectively:

$$P_{t,s}^{eq} + \sum_{t=1}^{Nt} P_{i,t,s} + PW_{t,s} + PV_{t,s} + Pd_{t,s} - Pc_{t,s} = EL_{t,s}^{dr}, \quad (16.39)$$

$$\sum_{i=1}^{Ni} H_{i,t,s} + \sum_{b=1}^{Nb} H_{b,t,s} + Hd_{t,s} + Hc_{t,s} = HL_{t,s}. \quad (16.40)$$

16.3 Simulation and numerical results

The developed stochastic emission-constrained scheduling model of CHP-VPP is carried out on the system depicted in Fig. 16.1. The Monte Carlo simulation is used to

Table 16.1 Probability of reduced scenarios.

Scenario number	Probability
1	0.15
2	0.21
3	0.04
4	0.04
5	0.05
6	0.01
7	0.16
8	0.17
9	0.1
10	0.07

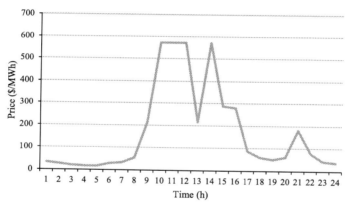

Figure 16.3 Hourly forecasted electricity market price.

generate 1000 scenarios for wind generation, PV generation, electricity price, electrical and thermal loads, then generated scenarios are reduced by the fast-backward selection algorithm using the SCENRED toolbox of GAMS software [39–41]. The probability of original scenarios after reduction has been added to the remaining scenarios based on the probability distance, where the probability of reduced scenarios is reported in Table 16.1 [42]. Moreover, optimization problems are also solved by GAMS employing CPLEX solver. All the simulations are performed on a PC with Intel Core i7-4710HQ CPU @ 2.5 GHz and 8 GB RAM. To evaluate the proposed day-ahead scheduling, two different case studies are considered: in case 1, the DRP is not implemented and the operator of the VPP cannot shift load; however, the impact of DRP is investigated in case 2. The forecasted electricity market price is shown in Fig. 16.3 and the rest of the techno-economic data for all embedded technologies are available in [43,44].

Emission impacts on virtual power plant scheduling programs

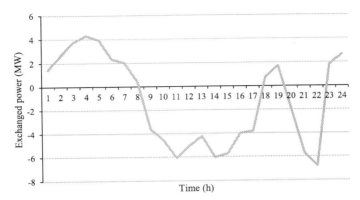

Figure 16.4 The expected exchanged power with the grid in case 1.

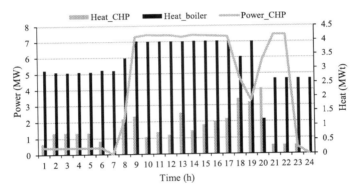

Figure 16.5 Expected dispatch of CHP and boiler units in case 1.

16.3.1 Case 1

In this case, the DRP is not considered and the operator of the VPP must meet the exact load of end-users for each time slot. The expected exchanged power of the VPP with the grid is depicted in Fig. 16.4. As can be seen, the operator of the VPP purchased power during low electricity price, i.e., $t = 1-8, 18-19, 23-24$, and sold the power at $t = 9-17, 20-22$, when the electricity price is high. Fig. 16.5 indicates the expected hourly dispatch of CHP and boiler units. The CHP units operate during peak hours to generate power; however, during off-peak hours, since the operation cost of the CHP units is higher than the market price, the operator prefers to supply the demand by purchasing power from the grid. Furthermore, owing to supplying thermal demands, the boiler is online during the whole time horizon. The contribution of the CHP units in supplying thermal demand is less than that of the boiler because of CHP units' FOR characteristics, according to which the generation of only heat is not possible for CHP units and, since the generation of power via CHP units is not affordable at off-peak hours, the operator runs the boiler to generate heat.

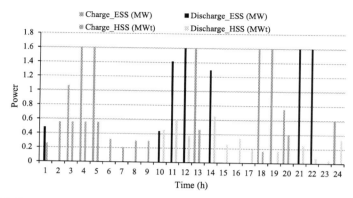

Figure 16.6 Expected scheduling of ESS and HSS in case 1.

Expected charging and discharging of energy storage systems are shown in Fig. 16.6. ESS charging and discharging mainly depend on the electricity market. During high electricity price hours, i.e., $t = 3–5, 13, 18–20, 24$, the ESS charges, and it discharges during high price hours, i.e., $t = 1, 10–12, 14, 21–22$. It is notable that the ESS charges at $t = 13$, while the electricity price is high. This is because of the electricity price at the next time slot that the ESS can charge during a lower electricity price hour, then discharge during a higher electricity price hour and bring economic benefits. Also, the ESS charging during the final hour is to satisfy the same level of power as during the initial hour. Despite the ESS, the HSS charging and discharging mainly depend on the amount of thermal load. As depicted in the figure, HSS enters the charge mode during hours $t = 1–9, 13, 18, 20$ when the thermal demand is lower compared to the hours $t = 10–12, 14–17, 19, 21–24$ when higher demand must be met. In total, the contribution of the boiler in supplying thermal demand is 77.73%, as well as 20.82% for CHP units, and 1.45% for HSS.

The profit of the CHP-VPP in this case is $33542.99, where 135386.64 kg of emissions are released to obtain this amount of profit without considering any limitation for emission releasing. However, a sensitivity analysis is conducted in this work to observe how emission limitation affects the profit. The expected result of the analysis is depicted in Fig. 16.7. The emission releasing was reduced through 10% steps, and in each step, the obtained profit is reported. Decreasing the emissions up to 50% leads to a reduction of nearly 39% in profit. So, it can be concluded that the more limited the emission releasing, the more reduction in the profit.

16.3.2 Case 2

In this case, the DRP is implemented by the operator and the effect of the end-user's participation in the DRP is evaluated. At first, the DRP impact on the electrical load is depicted in Fig. 16.8. According to the figure, the operator reduces the electrical load during high electricity price periods, i.e., $t = 9–17$ and 21, these amounts of load are shifted to the other electricity price periods that impose lower costs to the operator.

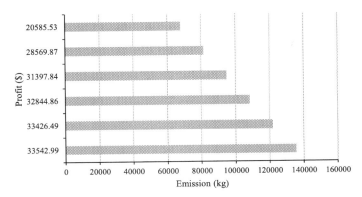

Figure 16.7 The expected profit of the CHP-VPP with respect to the emissions in case 1.

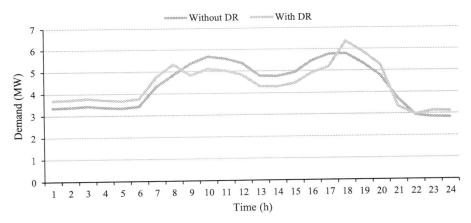

Figure 16.8 The expected electrical load with and without applying DRP.

This reduction of costs can be verified from Fig. 16.9 showing the expected exchanged power with and without applying DRP. As illustrated, the amount of purchased power is increased during those hours when the load shifts up due to DRP and, in contrast, the amount of sold power is also increased during load shift-down periods. In doing so, the CHP-VPP profit increases by shifting loads from peak to off-peak hours. The expected profit of the CHP-VPP in this case is $35185.87, which is increased by 4.9% with respect to case 1. The sensitivity analysis of the profit with respect to the emission is also performed in this case. (See Table 16.2.) Fig. 16.10 indicates the analysis results for case 2. According to the results, the profit decreases by 36.9% when the emission is limited to 50%, so that the reduction is less than in case 1. Further, a sensitivity analysis has been conducted on consumers' demand response participation rate with respect to emissions. Table 16.3 shows the results. According to this table, for each emissions rate, the amount of profit rises by increasing consumers' participation in the demand response program. At the end of this case study, it is noteworthy

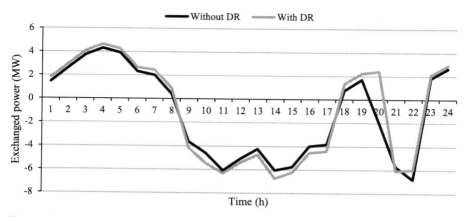

Figure 16.9 The expected exchanged power with the grid with and without applying DRP.

Table 16.2 Sensitivity analysis of DR.

Emission (kg)	DR Participation Rate	Profit ($)
135386.6	5	34206.19
	10	35185.87
	15	35741.83
121848	5	34206.19
	10	34979.77
	15	35741.83
108309.3	5	33580.76
	10	34297.55
	15	35007.82
94770.65	5	32035.03
	10	32642.38
	15	33220.27
81231.98	5	29195.34
	10	29787.46
	15	30306.90
67693.32	5	21467.64
	10	22173.17
	15	22732.62

to highlight stochastic programming. To do so, Table 16.3 is provided to report the profits for each scenario. By comparing the profit of each scenario, the impact of forecasting error can be seen. Also, Table 16.4 shows the expected profit with respect to the emission of 67693.32 kg for the different number of scenarios. As can be seen, increasing the number of scenarios has a minor impact on the expected profit, while causing computational burden.

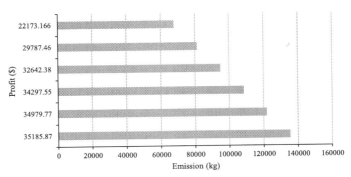

Figure 16.10 The expected profit of the CHP-VPP with respect to the emissions in case 2.

Table 16.3 Profit of each scenario with respect to the emission of 67693.32 kg.

Scenario number	Profit ($)
1	3325.976
2	4656.366
3	886.9268
4	886.9268
5	1108.659
6	221.7317
7	3547.707
8	3769.439
9	2217.317
10	1552.122

Table 16.4 Impact of the number of scenarios on the expected profit and execution time with respect to the emission of 67693.32 kg.

Scenario number	Expected profit ($)	Execution time (ms)
10	22173.17	97
20	22176.02	122
30	22174.97	271
40	22170.55	453
50	22171.89	702

16.4 Conclusions

In this work, a scheduling framework for optimal operation of the virtual power plant (VPP) was introduced. The combined heat and power (CHP) unit is also integrated into the VPP to assist the system in supplying heat and power demand. The proposed framework was modeled as a scenario-based problem to tackle uncertain parameters associated with wind generation, photovoltaic generation, electrical and thermal demands. A sensitivity analysis was performed to see the impact of the emission limitations on the profit of the CHP-VPP. Obtained results illustrate that a 50% decrease in emissions causes a 39% reduction in the expected profit. Indeed, considering environmental issues in the power sector reduces energy systems' profit. In addition, the demand response program (DRP) was embedded into the energy system, and its benefits were verified. Applying the DRP causes a 4.9% increase in the expected profit. Moreover, the DRP alleviates financial losses due to emission releasing limitations up to 2.1%, based on the obtained results.

References

[1] H. Saboori, M. Mohammadi, R. Taghe, Virtual power plant (VPP), definition, concept, components and types, https://doi.org/10.1109/APPEEC.2011.5749026, 2011.

[2] S. Hadayeghparast, A. SoltaniNejad Farsangi, H. Shayanfar, Day-ahead stochastic multi-objective economic/emission operational scheduling of a large scale virtual power plant, Energy 172 (2019), https://doi.org/10.1016/j.energy.2019.01.143.

[3] M.A. Mirzaei, et al., Evaluating the impact of multi-carrier energy storage systems in optimal operation of integrated electricity, gas and district heating networks, Appl. Therm. Eng. (2020), https://doi.org/10.1016/j.applthermaleng.2020.115413.

[4] A.G. Zamani, A. Zakariazadeh, S. Jadid, A. Kazemi, Stochastic operational scheduling of distributed energy resources in a large scale virtual power plant, Int. J. Electr. Power Energy Syst. (2016), https://doi.org/10.1016/j.ijepes.2016.04.024.

[5] Z. Tan, et al., Dispatching optimization model of gas-electricity virtual power plant considering uncertainty based on robust stochastic optimization theory, J. Clean. Prod. (2020), https://doi.org/10.1016/j.jclepro.2019.119106.

[6] M.A. Mirzaei, M. Nazari-Heris, B. Mohammadi-Ivatloo, K. Zare, M. Marzband, S.A. Pourmousavi, Robust flexible unit commitment in network-constrained multicarrier energy systems, IEEE Syst. J. (2020).

[7] M.A. Mirzaei, et al., A novel hybrid two-stage framework for flexible bidding strategy of reconfigurable micro-grid in day-ahead and real-time markets, Int. J. Electr. Power Energy Syst. 123 (2020) 106293.

[8] O. Arslan, O.E. Karasan, Cost and emission impacts of virtual power plant formation in plug-in hybrid electric vehicle penetrated networks, Energy (2013), https://doi.org/10.1016/j.energy.2013.08.039.

[9] S. Ghavidel, L. Li, J. Aghaei, T. Yu, J. Zhu, A review on the virtual power plant: components and operation systems, https://doi.org/10.1109/POWERCON.2016.7754037, 2016.

[10] A.Y. Saber, G.K. Venayagamoorthy, Intelligent unit commitment with vehicle-to-grid — a cost-emission optimization, J. Power Sources 195 (3) (2010), https://doi.org/10.1016/j.jpowsour.2009.08.035.

[11] S.R. Dabbagh, M.K. Sheikh-El-Eslami, Risk-based profit allocation to DERs integrated with a virtual power plant using cooperative Game theory, Electr. Power Syst. Res. (2015), https://doi.org/10.1016/j.epsr.2014.11.025.

[12] E. Mashhour, S.M. Moghaddas-Tafreshi, Bidding strategy of virtual power plant for participating in energy and spinning reserve markets—part II: numerical analysis, IEEE Trans. Power Syst. (2011), https://doi.org/10.1109/TPWRS.2010.2070883.

[13] E. Mashhour, S.M. Moghaddas-Tafreshi, Bidding strategy of virtual power plant for participating in energy and spinning reserve markets—part I: problem formulation, IEEE Trans. Power Syst. (2011), https://doi.org/10.1109/TPWRS.2010.2070884.

[14] E. Mashhour, S.M. Moghaddas-Tafreshi, Mathematical modeling of electrochemical storage for incorporation in methods to optimize the operational planning of an interconnected micro grid, J. Zhejiang Univ. Sci. C (2010), https://doi.org/10.1631/jzus.C0910721.

[15] M. Giuntoli, D. Poli, Optimized thermal and electrical scheduling of a large scale virtual power plant in the presence of energy storages, IEEE Trans. Smart Grid (2013), https://doi.org/10.1109/TSG.2012.2227513.

[16] A. Soroudi, Taxonomy of uncertainty modeling techniques in renewable energy system studies, Green Energy Technol. (2014), https://doi.org/10.1007/978-981-4585-30-9_1.

[17] V. Suresh, et al., Stochastic economic dispatch incorporating commercial electric vehicles and fluctuating energy sources, IEEE Access 8 (2020), https://doi.org/10.1109/ACCESS.2020.3041309.

[18] H. Liu, et al., Application of modified progressive hedging to stochastic unit commitment in electricity-gas coupled systems, CSEE J. Power Energy Syst. (2020), https://doi.org/10.17775/cseejpes.2020.04420.

[19] H. Rashidizadeh-Kermani, M. Vahedipour-Dahraie, M. Shafie-khah, P. Siano, A stochastic short-term scheduling of virtual power plants with electric vehicles under competitive markets, Int. J. Electr. Power Energy Syst. 124 (2021), https://doi.org/10.1016/j.ijepes.2020.106343.

[20] J. Zapata, J. Vandewalle, W. D'Haeseleer, A comparative study of imbalance reduction strategies for virtual power plant operation, Appl. Therm. Eng. (2014), https://doi.org/10.1016/j.applthermaleng.2013.12.026.

[21] H. Pandžić, I. Kuzle, T. Capuder, Virtual power plant mid-term dispatch optimization, Appl. Energy (2013), https://doi.org/10.1016/j.apenergy.2012.05.039.

[22] M. Shabanzadeh, M.K. Sheikh-El-Eslami, M.R. Haghifam, A medium-term coalition-forming model of heterogeneous DERs for a commercial virtual power plant, Appl. Energy (2016), https://doi.org/10.1016/j.apenergy.2016.02.058.

[23] S. Fan, Q. Ai, L. Piao, Fuzzy day-ahead scheduling of virtual power plant with optimal confidence level, IET Gener. Transm. Distrib. (2016), https://doi.org/10.1049/iet-gtd.2015.0651.

[24] S.V. Papaefthymiou, S.A. Papathanassiou, Optimum sizing of wind-pumped-storage hybrid power stations in island systems, Renew. Energy (2014), https://doi.org/10.1016/j.renene.2013.10.047.

[25] J. Wang, C. Liu, D. Ton, Y. Zhou, J. Kim, A. Vyas, Impact of plug-in hybrid electric vehicles on power systems with demand response and wind power, Energy Policy (2011), https://doi.org/10.1016/j.enpol.2011.01.042.

[26] E. Heydarian-Forushani, M.E.H. Golshan, M.P. Moghaddam, M. Shafie-Khah, J.P.S. Catalão, Robust scheduling of variable wind generation by coordination of bulk energy storages and demand response, Energy Convers. Manag. (2015), https://doi.org/10.1016/j.enconman.2015.09.074.

[27] M. Pasetti, S. Rinaldi, D. Manerba, A virtual power plant architecture for the demand-side management of smart prosumers, Appl. Sci. (2018), https://doi.org/10.3390/app8030432.

[28] M.P. Moghaddam, A. Abdollahi, M. Rashidinejad, Flexible demand response programs modeling in competitive electricity markets, Appl. Energy (2011), https://doi.org/10.1016/j.apenergy.2011.02.039.

[29] I.G. Moghaddam, M. Nick, F. Fallahi, M. Sanei, S. Mortazavi, Risk-averse profit-based optimal operation strategy of a combined wind farm-cascade hydro system in an electricity market, Renew. Energy 55 (2013), https://doi.org/10.1016/j.renene.2012.12.023.

[30] G. Liu, K. Tomsovic, A full demand response model in co-optimized energy and reserve market, Electr. Power Syst. Res. 111 (2014), https://doi.org/10.1016/j.epsr.2014.02.006.
[31] R.A. Hooshmand, S.M. Nosratabadi, E. Gholipour, Event-based scheduling of industrial technical virtual power plant considering wind and market prices stochastic behaviors – a case study in Iran, J. Clean. Prod. 172 (2018), https://doi.org/10.1016/j.jclepro.2017.12.017.
[32] H.T. Nguyen, L.B. Le, Z. Wang, A bidding strategy for virtual power plants with the intraday demand response exchange market using the stochastic programming, IEEE Trans. Ind. Appl. 54 (4) (2018), https://doi.org/10.1109/TIA.2018.2828379.
[33] H.M. Ghadikolaei, E. Tajik, J. Aghaei, M. Charwand, Integrated day-ahead and hour-ahead operation model of discos in retail electricity markets considering DGs and CO_2 emission penalty cost, Appl. Energy 95 (2012), https://doi.org/10.1016/j.apenergy.2012.02.034.
[34] Z. Liang, Q. Alsafasfeh, T. Jin, H. Pourbabak, W. Su, Risk-constrained optimal energy management for virtual power plants considering correlated demand response, IEEE Trans. Smart Grid 10 (2) (2019), https://doi.org/10.1109/TSG.2017.2773039.
[35] A.G. Zamani, A. Zakariazadeh, S. Jadid, Day-ahead resource scheduling of a renewable energy based virtual power plant, Appl. Energy 169 (2016), https://doi.org/10.1016/j.apenergy.2016.02.011.
[36] M.A. Mirzaei, et al., Two-stage robust-stochastic electricity market clearing considering mobile energy storage in rail transportation, IEEE Access 8 (2020) 121780–121794.
[37] M. Hemmati, B. Mohammadi-Ivatloo, M. Abapour, A. Anvari-Moghaddam, Optimal chance-constrained scheduling of reconfigurable microgrids considering islanding operation constraints, IEEE Syst. J. (2020).
[38] M.A. Mirzaei, et al., Network-constrained joint energy and flexible ramping reserve market clearing of power and heat-based energy systems: a two-stage hybrid IGDT-stochastic framework, IEEE Syst. J. (2020) 1–10.
[39] A. Mansour-Saatloo, M. Agabalaye-Rahvar, M.A. Mirzaei, B. Mohammadi-Ivatloo, K. Zare, Robust scheduling of hydrogen based smart micro energy hub with integrated demand response, J. Clean. Prod. (2020) 122041.
[40] M.H. Shams, M. Shahabi, M.E. Khodayar, Stochastic day-ahead scheduling of multiple energy Carrier microgrids with demand response, Energy 155 (2018), https://doi.org/10.1016/j.energy.2018.04.190.
[41] A. Mansour-Saatloo, M.A. Mirzaei, B. Mohammadi-Ivatloo, K. Zare, A risk-averse hybrid approach for optimal participation of power-to-hydrogen technology-based multi energy microgrid in multi-energy markets, Sustain. Cities Soc. (2020) 102421.
[42] M. Ahrabi, M. Abedi, H. Nafisi, M.A. Mirzaei, B. Mohammadi-Ivatloo, M. Marzband, Evaluating the effect of electric vehicle parking lots in transmission-constrained AC unit commitment under a hybrid IGDT-stochastic approach, Int. J. Electr. Power Energy Syst. 125 (2020) 106546.
[43] M.A. Mirzaei, et al., Integrated energy hub system based on power-to-gas and compressed air energy storage technologies in the presence of multiple shiftable loads, IET Gener. Transm. Distrib. 14 (13) (2020) 2510–2519.
[44] N. Nasiri, et al., A bi-level market-clearing for coordinated regional-local multi-carrier systems in presence of energy storage technologies, Sustain. Cities Soc. 63 (2020) 102439.

Multi-objective scheduling of a virtual power plant considering emissions

17

Morteza Shafiekhani[a] and Ali Hashemizadeh[b]

[a]Department of Electrical Engineering, Faculty of Engineering, Pardis Branch, Islamic Azad University, Pardis, Tehran, Iran, [b]School of Management and Economics, Beijing Institute of Technology, Beijing, China

17.1 Introduction

The use of renewable energy sources causes various problems in the performance and operation of power systems because these sources are unpredictable and there are many changes in their primary fuel (wind speed and solar radiation) that have caused to call these resources undispatchable [1–3]. The main purpose of the virtual power plant is to provide access to distributed energy resources (renewable and nonrenewable) with low power range to the energy market. Due to random changes in the energy produced by renewable sources such as wind and solar, the risk of participation of a single unit in the energy market is very high; indeed, if the power is not provided, this unit has to buy power at a high price from the balancing markets. Another problem with distributed energy sources is their low production capacity to participate in the energy market since participating in energy or ancillary services' markets requires having a minimum exchange capacity. To solve these problems, the virtual power plant combines several distributed energy units into one cluster. Therefore, the existence of several production units together allows the production to be balanced or, in other words, to reduce the existing uncertainties related to each distributed production unit.

According to the FENIX project [4], a virtual power plant aggregates the capacities of several distributed energy resources that can provide a separate performance characteristic of the parameters related to each distributed energy resource and apply the effect of the network on the output of distributed energy resources. The objective of FENIX is to boost DER (distributed energy resources) by maximizing their contribution to the electric power system, through aggregation into large-scale virtual power plants (LSVPP) and decentralized management. This project was tested in real networks, namely with Iberdrola in Spain and EDF Energy in the UK [5–12]. One of the important features of the virtual power plant is the ability to participate in the energy market, so the owners of these power plants can move in the direction of maximizing their profits by determining the optimal bidding strategy to the network.

There is a lot of research to determine the optimal bidding models in the day-ahead market for traditional power plants. In virtual power plants, some references have also discussed this issue [13,14]. Relatively few references have examined economic issues along with emissions in the virtual power plant. Among them, reference [15] addresses the optimal operation problem for VPP. In this reference two conflicting objective functions are considered and a multi-objective problem is formulated mathematically. The first objective function accounts for costs. The second objective function considers pollutants. Particle swarm optimization (MOPSO) in the form of multi-objective optimization is applied to extract the Pareto frontier which is a set of non-inferior solutions. Reference [16], in order to make full use of distributed energy resources and decrease the abandoned clean energy, proposes a multi-objective optimization model in VPP with three objectives, namely maximizing operation revenue, minimizing operation risk, and minimizing carbon emissions. Reference [17] represents a bi-level model for finding the strategic bidding equilibrium of a virtual power plant in a joint energy and regulation market in the presence of rivals. One of goals of this paper is to solve the bi-level problem in a bi-objective way using the augmented epsilon constraint method, which maximizes the profit and minimizes the emissions of virtual power plant units.

Some references in the virtual power plant literature have dealt with the issue of carbon trading, among which in the reference [18] the bidding strategy of the VPP by considering the carbon-electricity integration trading in an auxiliary service (AS) market is studied. In [19] a virtual power plant (VPP) including demand response resources, gas turbines, wind power, and photovoltaics with participation in carbon emission trading is examined, and an optimal dispatching model of the VPP is presented. First, the carbon emission trading mechanism is briefly described, and then the framework of optimal dispatching in the VPP is discussed.

Similar to the Fenix project, virtual power plants have many operational applications. The following are some examples of practical projects of virtual power plants.

✓ Global CCS Institute (Large-scale virtual power plant integration (DERINT)) [20]

Main findings:

- The virtual power plant (power hub) enables reliable delivery of ancillary services, like voltage control and reserves, by intelligent control of distributed generation, including wind farms and industrial consumption.
- Power hub was able to optimize the output from the available resources in different units across different markets, deciding when the highest value would be generated.
- It is economically attractive for all stakeholders to participate in virtual power plants. Power hub can be replicated across Europe, although challenges include attracting and integrating industrial units to participate in a virtual power plant.
- Another challenge was scaling up the virtual power plant on commercial terms in Denmark due to the Danish regulatory regime and market design. Similar challenges have been identified in Germany and Spain.

Results in detail:

The project integrated 47 units into the VPP representing 15 different unit types. One of the conclusions from the project is that it is a challenging task to mobilize industrial units to participate in a virtual power plant; the task involves complexity unit flexibility assessment and unit owner education in the complex issues of VPPs, power markets, and the future energy system.

✓ **Kansai Project [21]**

Fourteen companies, including Sumitomo Electric, have launched the virtual power plant project, subsidized by the Agency for Natural Resources and Energy of the Ministry of Economy, Trade and Industry. Amid ongoing electricity deregulation and electric power system reforms, this project aims to realize an unprecedented new energy management toward constructing an energy infrastructure for efficient energy use by the entire society.

✓ **Simply Energy Virtual Power Plant (AGL project, Adelaide) [22,23]**

AGL's virtual power plant project is being rolled out in three phases over about 18 months. By the next year, 1,000 batteries are expected to be deployed across metropolitan Adelaide. AGL is one of Australia's leading integrated energy companies. It is taking action to responsibly reduce its greenhouse gas emissions while providing secure and affordable energy to its customers.

✓ **Next-Kraftwerke [24,25]**

In this virtual power plant, they connect thousands of power producers such as biogas, wind or solar power plants and power consumers. The production of polluting units is reduced by the same amount and the amount of emissions is also reduced. By intelligently controlling their feed-in and consumption, they are able to valorize their power and flexibility in different markets. With a total capacity of more than 4,000 megawatts, they also help balancing frequency fluctuations of the power grid. Together with all participants they are shaping the energy landscape of the future – with a reliable power production from 100% sustainable energy sources.

Japan [26]

According to the Paris Agreement, to pay more attention to environmental issues, the Tokyo Metropolitan Government is expanding the local production and consumption of renewable energy, and Yokohama City in Kanagawa Prefecture is establishing VPPs.

ABB [27]

Conventional multi-unit power plants can improve their flexibility, reduce fuel consumption, and lower their carbon dioxide emissions by operating the plant units as a virtual power pool.

Therefore, in this chapter, the economic scheduling of a virtual power plant considering emissions has been discussed.

17.2 Problem formulation

In this model the VPP is a price taker and it can't affect the market prices. It is assumed that the market price and load amounts are known. The objective function of the virtual power plant, which refers to profit maximization, is stated as follows. Eq. (17.1) shows the profit of the virtual power plant obtained from the difference between revenue and cost. Eq. (17.2) refers to revenue. The revenue of the virtual power plant comes from the sale of power to internal loads and the sale of power to the upstream network. These are shown in Eq. (17.2). If the virtual power plant buys power from the upstream network, the term P_t^{GSP} becomes negative and shows its cost. Eq. (17.3) shows the cost of the virtual power plant. It includes the cost of operation, the cost of starting up, and the cost of shutting down the thermal units in the virtual power plant. The cost of operation is linear and will be explained further. The mentioned equations are as follows:

$$Profit = \sum_t Revenue_t - Cost_t, \tag{17.1}$$

$$Revenue_t = P_t^{Demand} \lambda_t + P_t^{GSP} \lambda_t^{GSP}, \tag{17.2}$$

$$Cost_t = \sum_g FC_{g,t} + SUC_{g,t} + SDC_{g,t}. \tag{17.3}$$

17.2.1 Linear equivalent of the cost function

The linear equivalent of the cost function is given in Eqs. (17.4)–(17.12) [28]. (See Fig. 17.1.) For this purpose, the cost function, which is a second-order expression, is approximated using lines. The equations related to the linearization of the cost function are Eqs. (17.4)–(17.12). The higher the number of lines, the better the approximation. Variable z represents the number of cost function intervals, indicating the number of these lines, too. Each interval has an initial and final power, which is represented by $P_{g,in}^z$ and $P_{g,fi}^z$ respectively. Power $P_{g,t}^z$ represents the generated power in each of these intervals. The expression ΔP_g^z represents the ratio of the power variations of each unit to the number of desired intervals. Variable H_g^z represents the ratio of cost to power variations in each interval. The considered equations are:

$$0 \leq P_{g,t}^z \leq \Delta P_g^z \alpha_{g,t}, \quad \forall z = 1:n, \tag{17.4}$$

$$\Delta P_g^z = \frac{P_g^{Max} - P_g^{Min}}{n}, \tag{17.5}$$

$$P_{g,in}^z = (z-1)\Delta P_g^z + P_g^{Min}, \tag{17.6}$$

$$P_{g,fi}^z = \Delta P_g^z + P_{g,in}^k, \tag{17.7}$$

$$P_{g,t} = P_g^{Min} \alpha_{g,t} + \sum_z P_{g,t}^z, \tag{17.8}$$

$$Cost_{g,in}^z = a_g (P_{g,in}^z)^2 + b_g P_{g,in}^z + c_g, \tag{17.9}$$

$$Cost_{g,fi}^z = a_g (P_{g,fi}^z)^2 + b_g P_{g,fi}^z + c_g, \tag{17.10}$$

$$H_g^z = \frac{Cost_{g,fi}^z - Cost_{g,in}^z}{\Delta P_g^z}, \tag{17.11}$$

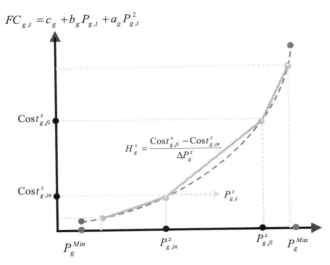

Figure 17.1 Cost function linearization [28].

$$FC_{g,t} = a_g(P_g^{Min})^2 + b_g P_g^{Min} + c_g \alpha_{g,t} + \sum_z H_g^z P_{g,t}^z. \quad (17.12)$$

The start-up and shut-down cost of unit g at period t are as given in Eqs. (17.13) and (17.14), respectively. The values of CS_g and SD_g are parameters which are given in Table 17.1:

$$SUC_{g,t} = CS_g \beta_{g,t}, \quad (17.13)$$
$$SDC_{g,t} = SD_g \gamma_{g,t}. \quad (17.14)$$

17.2.2 Problem constraints

There are various constraints in this problem, such as technical and economic constraints, which are described as follows.

✓ Ramp rate constraints

Eqs. (17.15)–(17.19) represent the ramp-up and ramp-down constraints of thermal units. The production capacity of thermal units has a maximum and a minimum value at any time interval; $\underline{P}_{g,t}$ and $\overline{P}_{g,t}$ represent minimum and maximum, respectively. These values are not necessarily equal to P_g^{Min} and P_g^{Max}:

$$\underline{P}_{g,t} \leq P_{g,t} \leq \overline{P}_{g,t}, \quad (17.15)$$
$$\overline{P}_{g,t} \leq P_g^{Max}[\alpha_{g,t} - \gamma_{g,t+1}] + SD_g \gamma_{g,t+1}, \quad (17.16)$$
$$\overline{P}_{g,t} \leq P_{g,t-1} + RU_g \alpha_{g,t-1} + SU_g \beta_{g,t}, \quad (17.17)$$

$$\underline{P}_{g,t} \geq P_g^{Min} \alpha_{g,t}, \tag{17.18}$$

$$\underline{P}_{g,t} \geq P_{g,t-1} - RD_g \alpha_{g,t} - SD_g \gamma_{g,t}. \tag{17.19}$$

✓ Minimum up and down time constraints

Constraints related to the minimum up and down times of the units are given in Eqs. (17.20)–(17.21). Eqs. (17.22)–(17.24) also show the logical relations between the binary variables $\alpha_{g,t}$, $\beta_{g,t}$, and $\gamma_{g,t}$:

$$\sum_{t=1}^{MUP} \alpha_{g,t+1} - 1 \geq MUT_g, \quad \forall \beta_{g,t} = 1, \tag{17.20}$$

$$\sum_{t=1}^{MDN} 1 - \alpha_{g,t+1} \geq MDN_g, \quad \forall \gamma_{g,t} = 1, \tag{17.21}$$

$$\beta_{g,t} - \gamma_{g,t} = \alpha_{g,t} - \alpha_{g,t-1}, \tag{17.22}$$

$$\beta_{g,t} + \gamma_{g,t} \leq 1, \tag{17.23}$$

$$\beta_{g,t}, \gamma_{g,t}, \alpha_{g,t} \in \{0, 1\}. \tag{17.24}$$

✓ Power balance constraint

The power balance equation in a virtual power plant is provided by Eq. (17.25). In this regard, the total power generated by thermal units, wind unit, discharge of batteries, and power purchased from the upstream network must be able to provide the internal load of the virtual power plant. On the other hand, the total power of the virtual power plant should be able to charge the batteries and sell power to the upstream network:

$$\sum_g P_{g,t} - P_t^{GSP} + P_t^{wind} + P_t^{dch} - P_t^{ch} \geq P_t^{Demand}. \tag{17.25}$$

✓ Modeling of energy storages

As mentioned earlier, the virtual power plant also has storage. Storage equations are given in Eqs. (17.26)–(17.30). Eq. (17.26) shows the charge status of the storage. This amount depends on the charge level of the previous hour and the amount of charge and discharge in this hour. Eqs. (17.27) and (17.28) show the maximum amount of charge and discharge per hour, respectively. Eq. (17.29) shows the minimum and maximum charge levels, and finally, Eq. (17.30) indicates that charging and discharging will not occur simultaneously in one hour:

$$SOC_t^{ESS} = SOC_{t-1}^{ESS} + P_t^{ch} \mu^{ch} - P_t^{dch}/\mu^{dch}, \tag{17.26}$$

$$P_t^{ch} \leq P_t^{ch,Max} \alpha_t^{ch}, \tag{17.27}$$

$$P_t^{dch} \leq P_t^{dch,Max} \alpha_t^{dch}, \tag{17.28}$$

$$SOC^{ESS,Min} \leq SOC_t^{ESS} \leq SOC^{ESS,Max}, \tag{17.29}$$

$$\alpha_t^{ch} + \alpha_t^{dch} \leq 1. \tag{17.30}$$

✓ Emission objective function

The amount of emissions of the virtual power plant units is modeled according to Eq. (17.31). In order not to get involved in the complexities of nonlinear problems, a linear model has been used. In this equation the sum of the productions of all the blocks of the virtual power plant units is multiplied by the emission coefficients of the same units, yielding the emissions of those units [17]:

$$f_2 = \sum_t \sum_g (u_{sg} P_{g,t} + m_{sg}). \tag{17.31}$$

17.2.3 Multi-objective programming

To solve the problem of economic scheduling of the virtual power plant by considering emissions, one of the best models is to solve this problem in a bi-objective manner. By introducing the profit and emission objective functions for the virtual power plant, this problem can be solved in a bi-objective manner using the existing methods. Two methods will be used to solve this problem, namely the epsilon constraint method and weighting coefficient method. Each will be explained in detail as follows:

✓ Multi-objective scheduling of profit and emissions using the epsilon constraint method

In the epsilon constraint method, one of the objective functions is considered as the main objective function and the other objective functions are considered as constraints in the optimization problem. In the two-objective problem of profit and emissions for virtual power plant units, f_1 is considered as the main objective function and f_2 is considered as a constraint.

The format of the multi-objective optimization problem by the epsilon method is as follows [29,30]:

$$\text{Min} f_1(x) \text{ subject to } f_2(x) \leq e_2. \tag{17.32}$$

The advantage of the epsilon constraint method over the weighting coefficient method is that there is no need to balance the functions. In this method there is no scale problem for the objective functions.

✓ Multi-objective scheduling of profit and emissions using weighting coefficient method

Another method of solving multi-objective problems is the weight coefficient method. In this method, using Eq. (17.33), the problem is solved in a bi-objective manner. In this regard, the values of each objective function are divided to their maximum value to solve the scale problem to some extent. Here n represents the number of Pareto responses, which in this case is considered to be 11; w is a constant coefficient that changes from zero to $n - 1$ to extract the Pareto response values. As mentioned before, f_1 is here for profit and f_2 is for emissions. If f_2 is maximized, its sign will be positive, and if it is minimized, it will be negative. The negative sign is used here:

$$WF = \frac{((n-1-w)\frac{f_1}{f_1^{Max}} \pm w\frac{f_2}{f_2^{Max}})}{n-1}. \tag{17.33}$$

17.2.4 Fuzzy decision making method

After solving all the sub-problems and obtaining all the Pareto solutions to the problem, the decision maker must choose one of the Pareto solutions as the final answer to the desired problem among the Pareto solutions obtained according to different priorities, as well as different uses. Therefore, the proposed method for selecting the best answer is to use a fuzzy approach with a linear membership function for the decision maker.

One of the advantages of this method is good flexibility with the operational parameters [31]. The fuzzy modeling approach has several advantages when compared to other nonlinear modeling techniques, such as neural networks. In general, fuzzy systems can provide a more transparent representation of the system under study and can also give a linguistic interpretation in the form of rules [32].

The membership function of the proposed fuzzy method is defined in Eqs. (17.34) and (17.35), which are used for maximization and minimization, respectively. In these equations, f_n^r is the value of the objective function f_n in response to the rth Pareto solution, and μ_n^r is the membership function f_n in response to the rth Pareto solution, μ_n^r actually represents the degree of optimization of the objective function in the rth Pareto solution [33].

Maximization:

$$\mu_n^r\Big|_{n=1} = \begin{cases} 0, & f_n^r \leq f_n^{SN}, \\ \frac{f_n^r - f_n^{SN}}{f_n^U - f_n^{SN}}, & f_n^{SN} \leq f_n^r \leq f_n^U, \\ 1, & f_n^r \geq f_n^U. \end{cases} \tag{17.34}$$

Minimization:

$$\mu_n^r\Big|_{n=2} = \begin{cases} 1, & f_n^r \leq f_n^U, \\ \frac{f_n^{SN} - f_n^r}{f_n^{SN} - f_n^U}, & f_n^U \leq f_n^r \leq f_n^{SN}, \\ 0, & f_n^r \geq f_n^{SN}. \end{cases} \tag{17.35}$$

The general membership function of the rth Pareto solution is called μ^r, which is calculated according to Eq. (17.36), where w_n is the weight coefficient of the nth objective function. The values of the weighting coefficients are determined by the decision maker. For example, if economic issues are a priority, the decision maker should assign a large weight factor to f_1. But if the emissions are considered, it assigns a small weight factor to f_1. The final solution to the problem will be the Pareto solution with the maximum μ^r value:

$$\mu^r = \frac{\sum_{n=1}^{p} w_n \mu_n^r}{\sum_{n=1}^{p} w_n}. \tag{17.36}$$

17.3 Case studies

The virtual power plant under study has thermal units, wind unit, storage and internal loads, and its structure is shown in Fig. 17.2.

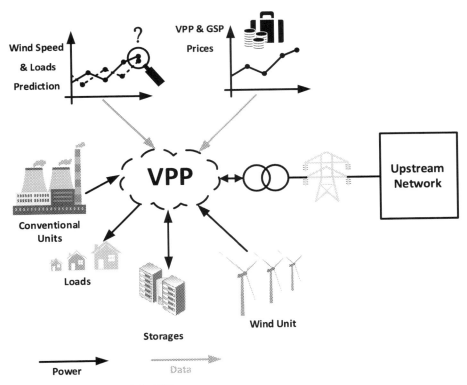

Figure 17.2 Structure of the VPP.

In order to maximize its profit, the virtual power plant exchanges power with the upstream network and has 5 thermal units, the specifications of which are given in Table 17.1. It should be noted that the power exchange limit with the upstream network in the virtual power plant is 20 MW.

The market price of the virtual power plant, as well as the upstream market price, is shown in Fig. 17.3. As can be seen from this figure, the price of the upstream network

Table 17.1 Characteristics of VPP units.

Unit	a_g ($/MW^2)	b_g ($/MW)	c_g ($)	P_g^{Min} (MW)	P_g^{Max} (MW)	RU_g (MWh^{-1})	RD_g (MWh^{-1})	MUT_g (h)	MDT_g (h)	SD_g ($)	CS_g ($)	u_{sg} (lb/MW)	m_{sg} (lb)
g1	0.0148	12.1	82	8	20	4	4	3	2	9	11	1.05	1.1
g2	0.0289	12.6	49	12	32	6.4	6.4	4	2	13	14	1.74	2.58
g3	0.0135	13.2	100	5	15	3	3	3	2	7	8	0.8	1.25
g4	0.0127	13.9	105	25	52	10.4	10.4	5	3	24	25	1.05	1.87
g5	0.0261	13.5	72	8	28	5.6	5.6	4	2	11	13	2.16	2.19

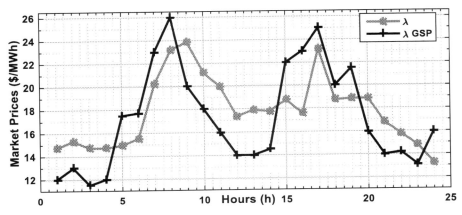

Figure 17.3 Day-ahead market price in VPP and GSP.

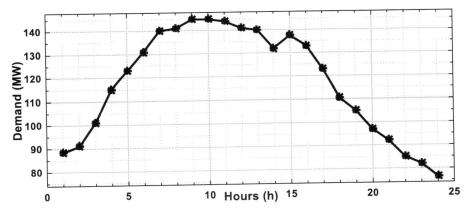

Figure 17.4 VPP demand.

is higher during some hours than the market price inside the virtual power plant and lower during others.

As mentioned earlier, the virtual power plant also has storage. The storage information available in the virtual power plant is in accordance with Table 17.2.

The amount of load in the virtual power plant is shown in Fig. 17.4.

Optimal economic scheduling of the virtual power plant considering emissions is considered as a two-objective problem in this chapter. The objective function of the economic problem was expressed together with its constraints, as well as the objective function of emissions. The outputs of the problem will be stated as follows. The studies in this chapter are implemented in three cases. First, the single-objective problem of economic scheduling of the virtual power plant is investigated, then the two-objective problem of profit and emissions is solved using the epsilon constraint method, and finally, the same problem will be solved using the weighting coefficient method.

Table 17.2 Characteristics of ESS.

Parameters	μ^{ch}	μ^{dch}	$P_t^{ch,Max}$ (MW/h)	$P_t^{dch,Max}$ (MW/h)	SOC_0^{ESS} (MW)	$SOC^{ESS,Max}$ (MW)	$SOC^{ESS,Min}$ (MW)
Value	0.95	0.9	3	3	2	10	2

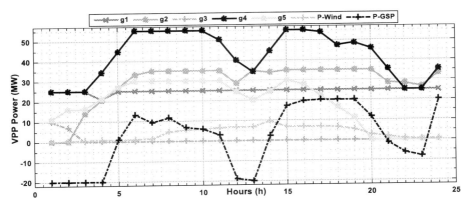

Figure 17.5 Power produced in VPP component.

17.3.1 Case 1: Solving economic scheduling of VPP in form of a single-objective problem

The goal in this problem is to maximize the profit of the virtual power plant, and for this purpose Eqs. (17.1)–(17.30) have been solved. It should be noted that in this problem, no attention has been paid to emissions and only their amount has been calculated. The generated power of the wind and thermal units, and the power exchanged with the upstream network are shown in Fig. 17.5. As it is clear from this figure, in the early hours of the day, the virtual power plant has purchased power from the upstream network, the reason is that the price of the upstream network is lower than price of the virtual power plant. On the other hand, the wind unit has not produced almost any power during these hours. In the middle of the day, for exactly this reason, the virtual power plant has purchased power from the upstream network. From 15:00 to 20:00, the opposite process occurred. During these hours, the price of the upstream network is higher than the price of the virtual power plant, so the virtual power plant has been a seller of power. During these hours, most thermal and wind units have high production.

The amounts of charge, discharge, and charge level of the storage are shown in Table 17.3. As can be seen from this table and Fig. 17.5, the storage devices are in the charge mode during the hours when the price of the upstream network is lower than the price of the virtual power plant. The reason is that during these hours, a lot of power is purchased by the virtual power plant. For example, this has happened during hours 1–2, 4, and 12–14. In the hours when the price of the upstream network is higher than the virtual power plant, the opposite has happened. For example, during hours 7–8 and 15–16, the virtual power plant has become a power seller, so the storage devices have been discharged and have given their power to the virtual power plant. In general, looking at Table 17.3 and Fig. 17.5, it is clear that the charge level rises in the case where the virtual power plant acts as a purchaser, and vice versa.

Table 17.3 Values of charge, discharge and SOC of ESS in single objective problem.

Time (h)	1	2	3	4	5	6	7	8	9	10	11	12
SOC (MW)	4.8	7.2	7.2	10	10	10	6.7	3.3	2	2	2	4.8
p^{ch} (MW)	3	2.2	0	3	0	0	0	0	0	0	0	3
p^{dch} (MW)	0	0	0	0	0	0	3	3	1.2	0	0	0
Time (h)	13	14	15	16	17	18	19	20	21	22	23	24
SOC (MW)	5.8	8.7	5.3	2	2	2	2	2	2	2	4.8	2
p^{ch} (MW)	1	3	0	0	0	0	0	0	0	0	3	0
p^{dch} (MW)	0	0	3	3	0	0	0	0	0	0	0	2.6

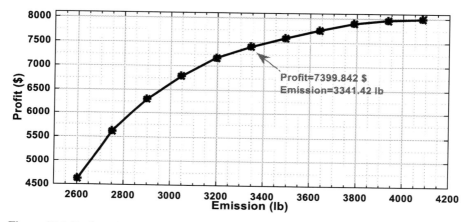

Figure 17.6 Profit and emissions in the epsilon constraint method.

17.3.2 Case 2: Solving two-objective problem of profit and emissions in VPP using the epsilon constraint method

The purpose of this problem is to maximize the profit of the virtual power plant and minimize its emissions, and for this purpose, Eqs. (17.1)–(17.32) and (17.34)–(17.36) have been solved. As described, in the epsilon constraint method we seek to extract a number of Pareto solutions. In this problem, 11 Pareto solutions have been extracted. Fig. 17.6 shows the amounts of profit and emissions in these solutions. As it is clear from this figure, with the increase of profit, the amount of emissions has also increased. The reason is that the increase in profit is mainly accompanied by an increase in production so this increase also causes higher emissions (it should be noted that this increase is limited due to the epsilon constraint method).

Now, one of the Pareto solutions obtained in the figure above must be selected as the best solution. The fuzzy decision making method is used for this purpose. The importance of the profit function is considered to be 3 to 2 compared to the emission function. Therefore, according to the formulation of the fuzzy decision-making method and according to Fig. 17.7, the sixth Pareto solution has been selected as the best solution. The following are the outputs for this answer.

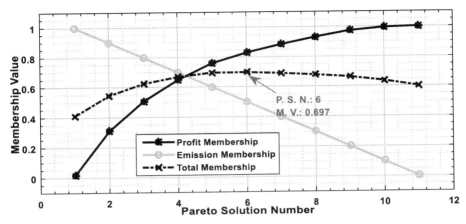

Figure 17.7 Membership values in the epsilon constraint method.

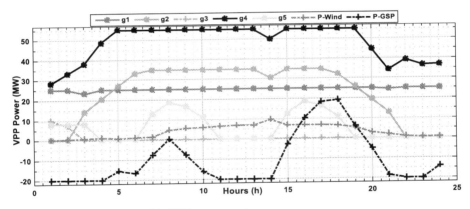

Figure 17.8 Power produced in VPP component.

In the sixth Pareto response, the generated power of the wind and thermal units and the power exchanged with the upstream network are shown in Fig. 17.8. In this case, the amount of emissions of the units is limited, so, as it is clear from this figure, the amount of unit production and power exchanged with the upstream network is different from Fig. 17.5. In this case, the production of more polluting units such as the fifth unit is much less. The second and fourth units have reduced their production in the same proportion. The exchange of virtual power plants with the upstream network has changed a lot. The virtual power plant purchases power from the upstream network most of the time, because purchasing power from the upstream network is more economical than generating emissions by its units. During hours 16–19, because the price of the upstream network is higher than the price of the virtual power plant and also because the amount of load power is decreasing, the virtual power plant has become a seller of power to the upstream network.

Table 17.4 Values of charge, discharge, and SOC of ESS in multi-objective problem (epsilon constraint method).

Time (h)	1	2	3	4	5	6	7	8	9	10	11	12
SOC (MW)	4.8	7.2	10	10	10	10	6.7	3.3	3.3	6.2	3.3	4.2
p^{ch} (MW)	3	2.4	3	0	0	0	0	0	0	3	0	0.9
p^{dch} (MW)	0	0	0	0	0	0	3	3	0	0	2.5	0
Time (h)	13	14	15	16	17	18	19	20	21	22	23	24
SOC (MW)	6	8.9	8.7	5.3	2	2	2	2	2	2	2	2
p^{ch} (MW)	1.9	3	0	0	0	0	0	0	0	0	0	0
p^{dch} (MW)	0	0	0.2	3	3	0	0	0	0	0	0	0

Charge amount, discharge, and storage charge level are shown in Table 17.4. As can be seen from this table, the storage devices are in the charge mode during the hours when the price of the upstream network is lower than the price of the virtual power plant. The reason is that during these hours, a lot of power is purchased by the virtual power plant.

17.3.3 Case 3: Solving two-objective problem of profit and emissions in VPP using the weighting coefficient method

The purpose of this problem is to maximize the profit of the virtual power plant and minimize its emissions, and for this purpose, Eqs. (17.1)–(17.31) and (17.33)–(17.36) have been solved. As previously explained, in this method we also seek to extract a number of Pareto solutions. In this problem, 11 Pareto solutions have been extracted. Fig. 17.9 shows the amounts of profit and emissions in these solutions. As it is clear from this figure, with the increase of profit, the amount of emissions has also increased. As can be seen from this figure, the solutions are less uniform in comparison with Fig. 17.6. In this case, the amount of profit and emissions has increased compared to the previous case.

Now one of the Pareto solutions obtained in the figure above should be selected as the best solution. Here, the fuzzy decision method is chosen for this purpose. The importance of the profit function is considered to be 3 to 2 compared to the emission function. Therefore, according to the formulation of the fuzzy decision-making method and according to Fig. 17.10, the fifth Pareto solution has been selected as the best solution. The following are the outputs for this answer.

The generated power of the wind and thermal units and the power exchanged with the upstream network are shown in Fig. 17.11. As is clear from this figure. In this case, the amount of emissions of the units is limited. So, as it is clear from this figure, the amount of production of units and power exchanged with the upstream network is very different from Fig. 17.5 and slightly different from Fig. 17.8. In this case, the amount of profit and, of course, the amount of emissions has increased slightly compared to the epsilon constraint method, so the amount of production of thermal units is also higher. Therefore, the amount of sales of virtual power plants has also increased, which is

Multi-objective scheduling of a virtual power plant considering emissions

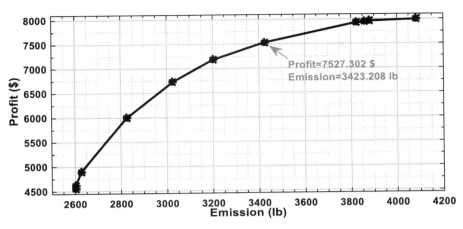

Figure 17.9 Profit and emissions using the weighting coefficient method.

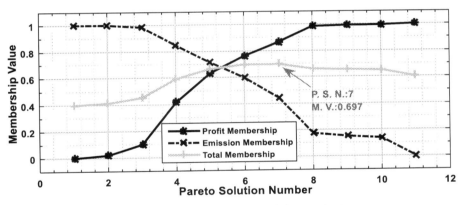

Figure 17.10 Membership values in the weighting coefficient method.

clear from the figure. In this case, the production of more polluting units, such as the fifth unit, is much less. The second and fourth units have reduced their production in the same proportion.

The amount of charge, discharge, and charge level of the storage are shown in Table 17.5. As can be seen from this table, the storage unit is in the discharge mode during the hours when the virtual power plant is selling power.

The amounts of profit and emissions in the above three cases are shown in Table 17.6. In the single objective case (case 1), the amount of profit is at its maximum. In this case, the amount of emissions is at its highest, because in the first case there is no limit to the emissions of the units. In the second and third cases, emissions have had their effect. As the amount of emissions is limited, the amount of profit is also affected.

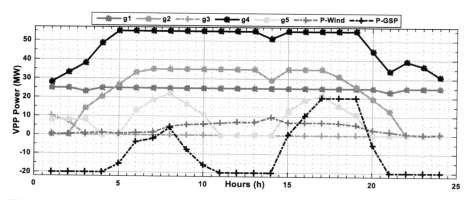

Figure 17.11 Power produced in VPP component.

Table 17.5 Values of charge, discharge and SOC of ESS in multi-objective problem (weighting coefficient method).

Time (h)	1	2	3	4	5	6	7	8	9	10	11	12
SOC (MW)	4.8	7.2	10	10	10	10	6.7	3.3	3.3	6.2	3.3	4.2
p^{ch} (MW)	3	2.4	3	0	0	0	0	0	0	3	0	0.9
p^{dch} (MW)	0	0	0	0	0	0	3	3	0	0	2.5	0
Time (h)	13	14	15	16	17	18	19	20	21	22	23	24
SOC (MW)	6	8.8	5.5	2.2	2.3	5.1	2	2	2	2	2	2
p^{ch} (MW)	1.9	3	0	0	0.1	3	0	0	0	0	0	0
p^{dch} (MW)	0	0	3	3	0	0	2.8	0	0	0	0	0

Table 17.6 Profits and emissions of all cases.

	First case	Second case	Third case
Profit ($)	7992.82	7399.842	7527.302
Emissions (lb)	4081.117	3341.42	3423.298

17.4 Conclusion

In this chapter, the economic scheduling of the virtual power plant with regard to emissions was discussed. The virtual power plant includes thermal units, wind unit, storage and load. First, the economic scheduling of the virtual power plant without considering emissions was studied. It was found that in this case, the maximum profit was obtained by the virtual power plant. On the other hand, the largest amount of emissions was created. Then, the problem was solved in a two-objective manner, considering economic problems along with emissions. For this purpose, two methods were used, the epsilon constraint method and the weighting coefficient method.

In the two-objective method, several Pareto solutions were extracted. Then, using the fuzzy decision making method and considering the importance of each objective functions, the best solution was obtained. The obtained solution showed that considering emissions reduces the amount of profit of the virtual power plant, because it causes more limitations on the generation of thermal units.

Nomenclature

Indices
t Index for time period
g Index for generator
ESS Index for energy storage

Parameters
P_t^{wind} Wind power generation in period t
λ_t Day-ahead market price in period t
λ_t^{GSP} Market price in GSP in period t
μ^{dch} Discharge efficiency of ESS
μ^{ch} Charge efficiency of ESS
MUT_g Minimum up time of unit g
$P_t^{ch,Max}$ Maximum charge limitation of ESS in period t
$P_t^{dch,Max}$ Maximum discharge limitation of ESS in period t
SOC_t^{ESS} State of charge of ESS in period t
MDN_g Minimum down time of unit g
$SUC_{g,t}$ Start-up cost of unit g in period t
$SDC_{g,t}$ Shut-down cost of unit g in period t
P_g^{Min} Minimum limit of power generation of unit g
P_g^{Max} Maximum limit of power generation of unit g
$\underline{P}_{g,t}$ Minimum time-dependent operating limit of unit g in period t
$\overline{P}_{g,t}$ Maximum time-dependent operating limit of unit g in period t
P_t^{Demand} Demand power in period t
u_{sg}, m_{sg} Emission coefficients of unit g
a_g, b_g, c_g Cost coefficients of unit g

Variables
P_t^{GSP} Power exchanged by upstream network in period t
$P_{g,t}$ Power produced by unit g in day-ahead market in period t
P_t^{ch} Power charge of ESS in period t
P_t^{dch} Power discharge of ESS in period t
$FC_{g,t}$ Operation cost of unit g in period t
$Cost_t$ Total cost of VPP in period t
$Revenue_t$ Total revenue of VPP in period t
$Profit$ Total profit of VPP
α_t^{ch} Binary variable for ESS that indicate charge status in period t
α_t^{dch} Binary variable for ESS that indicate discharge status in period t
$\alpha_{g,t}$ Binary variable related to on/off status of unit g, 1 if unit g is on, 0 otherwise
$\beta_{g,t}$ Binary variable related to start up status of unit g, 1 if unit g starts up
$\gamma_{g,t}$ Binary variable related to shut down status of unit g, 1 if unit g shuts down

References

[1] M. Manohar, E. Koley, S. Ghosh, Stochastic weather modeling-based protection scheme for hybrid PV–wind system with immunity against solar irradiance and wind speed, IEEE Systems Journal 14 (2020) 3430–3439.
[2] W. Li, Y. Liu, H. Liang, Y. Shen, A new distributed energy management strategy for smart grid with stochastic wind power, IEEE Transactions on Industrial Electronics 1 (1) (2020).
[3] A. Hashemizadeh, Y. Ju, S.M. Hosseini Bamakan, H. Phong Le, Renewable energy investment risk assessment in belt and road initiative countries under uncertainty conditions, Energy 214 (2021) 118923.
[4] D. Pudjianto, C. Ramsay, G. Strbac, The FENIX vision: The Virtual Power Plant and system integration of distributed energy resources, 2008.
[5] J.M. Corera, Integrated project FENIX – what is it all about?, FENIX Bulletin 1 (Nov 2007).
[6] C. Ramsay, The virtual power plant: enabling integration of distributed generation and demand, FENIX Bulletin 2 (Jan 2008).
[7] C. Andrieu, M. Fontela, B. Enacheanu, H. Pham, B. Raison, Y. Besanger, M. Randrup, U.B. Nilsson, R. Kamphuis, G.J. Schaeffer, Distributed Network Architectures. European project CRISP (ENK5-CT-2002-00673), Deliverable D1.7, 30 August 2005.
[8] D. Pudjianto, C. Ramsay, G. Strbac, The FENIX vision: The Virtual Power Plant and system integration of distributed energy resources, FENIX Deliverable 1.4.0, 21 Dec 2006.
[9] M. Sebastian, Towards virtual power plant integration into the transmission and distribution grid tools, FENIX Bulletin 5 (Nov 2008).
[10] P. Lang, C. Ramsay, The FENIX scenarios – part 1, the northern scenario, FENIX Bulletin 3 (April 2008).
[11] J. Martí, The FENIX scenarios – outlines of the southern demonstration, FENIX Bulletin 6 (Jan 2009).
[12] J. Maire, A. Ilo, A. Heher, M. Muscholl, J.M. Yarza, K. Kok, M. Sanduleac, M.B. Rivas, New features developed by manufactures within FENIX, FENIX Bulletin 7 (April 2009).
[13] M. Shafiekhani, A. Badri, A risk-based gaming framework for VPP bidding strategy in a joint energy and regulation market, Iranian Journal of Science and Technology, Transactions of Electrical Engineering 43 (2019) 545–558.
[14] M. Shafiekhani, A. Badri, F. Khavari, A bi-level model for strategic bidding of virtual power plant in day-ahead and balancing market, in: Smart Grid Conference (SGC), 2017.
[15] J. Zhang, Z. Xu, W. Xu, F. Zhu, X. Lyu, M. Fu, Bi-objective dispatch of multi-energy virtual power plant: deep-learning-based prediction and particle swarm optimization, Applied Sciences 9 (2019).
[16] L. Ju, Q. Tan, Y. Lu, Z. Tan, Y. Zhang, Q. Tan, A CVaR-robust-based multi-objective optimization model and three-stage solution algorithm for a virtual power plant considering uncertainties and carbon emission allowances, International Journal of Electrical Power & Energy Systems 107 (2019) 628–643.
[17] M. Shafiekhani, A. Badri, M. Shafie-Khah, J.P. Catalão, Strategic bidding of virtual power plant in energy markets: a bi-level multi-objective approach, International Journal of Electrical Power & Energy Systems 113 (2019) 208–219.
[18] D. Yang, H. Shaowen, C. Qiuyue, L. Dingqian, P. Hrvoje, Bidding strategy of a virtual power plant considering carbon-electricity trading, CSEE Journal of Power and Energy Systems 5 (2019) 306–314.
[19] Z. Liu, Z. Weimin, Q. Feng, W. Lei, Z. Bo, W. Fushuan, X. You, Optimal dispatch of a virtual power plant considering demand response and carbon trading, Energies 11 (2018) 1488.

[20] [Online] Available: https://cordis.europa.eu/project/id/249812.
[21] [Online] Available: https://global-sei.com/company/sei-world/2016/09/feature.html.
[22] [Online] Available: https://www.agl.com.au/about-agl/media-centre/asx-and-media-releases/2017/march/agl-virtual-power-plant-goes-live.
[23] [Online] Available: https://arena.gov.au/assets/2018/03/ARENA-Media-Release_Simply-Energy-Virtual-Power-Plant-in-Adelaide-280318-v15-FINAL.pdf.
[24] [Online] Available: https://www.next-kraftwerke.com/.
[25] [Online] Available: https://www.next-kraftwerke.com/vpp/virtual-power-plant/virtual-power-plant-design.
[26] [Online] Available: https://www.eujapan.eu.
[27] [Online] Available: https://library.e.abb.com.
[28] M. Shafiekhani, A. Zangeneh, Integration of electric vehicles and wind energy in power systems, in: Electric Vehicles in Energy Systems, Springer, Cham, 2020, pp. 165–181.
[29] M. Shafiekhani, M. Hesami, O. Homaee, A new bi-level multi-objective model for market clearing in the presence of strategic producers using augmented ε-constraint method, Journal of Iranian Association of Electrical and Electronics Engineers 17 (2020) 161–173.
[30] M. Kia, M. Shafiekhani, H. Arasteh, S.M. Hashemi, M. Shafie-khah, J.P.S. Catalão, Short-term operation of microgrids with thermal and electrical loads under different uncertainties using information gap decision theory, Energy 208 (2020) 118418.
[31] J. Chen, Jun Ye, Some single-valued neutrosophic Dombi weighted aggregation operators for multiple attribute decision-making, Symmetry 9 (6) (2017) 82.
[32] J. Sousa, M. Costa, Uzay Kaymak, Fuzzy Decision Making in Modeling and Control, vol. 27, World Scientific, 2002.
[33] A. Hashemizadeh, Y. Ju, P. Dong, A combined geographical information system and Best–Worst Method approach for site selection for photovoltaic power plant projects, International Journal of Environmental Science and Technology 17 (2020) 2027–2042.

Author index

Note: Page numbers followed by "*f*" indicate figures, and "*t*" indicate tables.

A

Aalami, H.A., 179
Abapour, M., 234, 235, 367
Abaye, A., 351
Abdollahi, A., 362
Abedi, M., 325, 368
Abedinia, O., 339
Abreu, T., 351
Abrudean, M., 3*t*, 8*t*
Achterberg, T., 239
Afkhami-Rohani, R., 351
Afonso, João Luiz, 303
Aftab, Suhail, 193
Agabalaye-Rahvar, M., 368
Aghaei, J., 218–220, 222, 229, 361, 362, 364
Aghaeic, J., 222
Aguiar, J.M., 343
Ahangar, R.A., 133, 139*t*, 141*t*, 143*t*–145*t*
Ahlstrom, M., 326
Ahmadi, A., 218, 219, 222
Ahmadian, A., 40
Ahmadian, Ali, 257, 259, 265
Ahmed, Manzar, 193
Ahmed, Zaki, 193
Ahrabi, M., 368
Ai, Q., 41, 163, 361
Ai, Qian, 198
Ai, X., 39, 111, 113
Akçay, H., 326, 328, 329, 331, 333, 336, 339
Akçay, Hüseyin, 325–339
Akkaş, Ö.P., 230
Al-Awami, A.T., 216, 219*t*–221*t*
Alahyari, A., 38, 112, 113, 239
Albadi, M.H., 163
Alexander, M., 316
Alexiadis, M., 325
Alhamali, Abdulwahab, 269, 270*f*
Alharbi, W., 220

Alhelou, H.H., 137, 139*t*, 141*t*, 143*t*–145*t*, 146
Aliakbar-Golkar, Masoud, 257, 259, 265
Aliasghari, P., 234
Alipour, H., 163
Alipour, M., 234
Alizadeh, B., 217, 219*t*–221*t*
Alizadeh Pahlavani, Mohammad Reza, 199
Alkaran, D.S., 4
Almabsout, Emad Ali, 320
Alsafasfeh, Q., 362, 364
Amin, Uzma, 193
Amini, M. Hadi, 264–266
Aminifar, Farrokh, 260
Amjady, N., 220
Amjady, Nima, 75
Amleh, N.A., 216, 219*t*–221*t*
Amorim, A.J., 351
An, Osman Nuri Uç, 320
Anaya-Lara, Olimpo, 194
Andersen, P.B., 19
Andersson, Göran, 82
Andrieu, C., 377
Anvari-Moghaddam, A., 113, 231, 367
Arasteh, H., 383
Ardakani, A.J., 21*t*
Ardakani, F.J., 21*t*
Arefi, A., 230
Aristidou, P., 158
Arrigo, F., 132
Arroyo, J.M., 21*t*
Arroyo, Jose M., 59, 60, 76, 77, 85, 101, 101*f*, 103*f*, 104*f*
Arslan, O., 360, 361
Arslan, Okan, 316*f*, 317*f*
Asmus, P., 5, 10, 11
Asmus, Peter, 300
Athari, M.H., 220
Athari, Mir Hadi, 193

Atieh, Ahmad, 196
Australian Energy Market Operator, 131
Ausubel, L.M., 6
Awerbuch, S., 5

B
Babaei, S., 216, 219t–221t
Baczynski, D., 351
Badar, A.Q., 7
Badar, Altaf Q.H., 1–22
Badri, A., 21t, 38, 40, 41, 111, 113, 179, 180f, 182f, 184, 184f, 187f–189f, 189t, 215, 219t–221t, 231, 343, 378, 383
Bagchi, A., 13, 217, 219t–221t
Bagheri, M., 339
Bai, X., 339
Bakirtzis, A.G., 40, 41, 167, 217, 219, 219t, 220, 220t, 221t
Bakirtzis, Anastasios G., 59, 67, 257, 258, 274
Baladron, C., 343
Baniasadi, A., 230
Baringo, A., 21t
Baringo, Ana, 59, 60, 76, 77, 85, 101, 101f, 103f, 104f
Baringo, L., 14, 21t
Baringo, Luis, 59, 60, 75–77, 85, 101, 101f, 103f, 104f
Barman, J., 138, 139t, 141t–145t, 146
Baskaran, S., 4
Basu, Mayur, 30t
Bayat, Oguz, 320
Bazaraa, M., 40
Baziar, A., 218, 220
Beagon, P., 151, 152f, 153f, 155
Behi, B., 230
Bell, K.R.W., 326, 338
Belmans, R., 6
Benbouzid, M., 229
Berizzi, A., 131
Berseneff, B., 5
Besanger, Y., 5, 377
Bessa, Ricardo, 82
Bevrani, H., 136, 137, 142t, 143t, 145t, 146
Bezergianni, S., 179
Bhattacharya, K., 147, 147f, 165
Bică, D., 3t, 8t
Bignucolo, F., 5, 179
Binding, C., 138, 139t, 141t, 142t, 144t, 145t

Bindner, H.W., 217, 219t–221t
Bindner, Hendrik, 195
Bliek, F.W., 16
Bo, Z., 378
Bocquet, A., 325
Boisvert, R.N., 164
Bollen, M.H., 135, 136, 139t, 141t, 143t–145t
Bompard, E., 132
Bose, A., 179
Bostan, A., 115, 116
Botterud, A., 222
Boyd, S., 330
Bremer, J., 5
Brownsword, R., 325
Bruckstein, A.M., 330
Bruninx, K., 216, 219t–221t
Burlando, M., 325
Burt, Graeme, 194
Bustamante, M.D., 151, 152f, 153f, 155

C
Caire, Raphaël, 194
Caldon, R., 5, 179
Callaway, D.S., 158
Calvert, S., 326
Çam, E., 230
Candés, E.J., 330
Canha, L.N., 9–11
Cañizares, C., 147, 147f
Cañizares, C.A., 219
Cantoni, M., 112, 113
Cao, Y., 111, 113
Cappers, P.A., 164
Capuder, T., 14, 230, 361
Cardoso, G., 218
Cardoso, M., 5
Carr, Stephen, 273
Carrion, M., 183
Carrión, M., 238
Carro, B., 343
Casasola-Aignesberger, L., 138, 139t, 141t, 143t, 145t
Cassola, F., 325
Catalão, J.P., 21t, 343, 378, 383
Catalão, J.P.S., 40, 41, 325, 339, 343, 362, 383
Cavalcante, R.L., 13, 138, 139t, 141t, 142t, 144t, 145t
Çelik, A.N., 326

Chan, K.W., 111, 113
Changsong, Chen, 67
Chartouni, D., 137, 143t, 145t
Charwand, M., 362, 364
Charytoniuk, W., 350
Chau, K.T., 257, 258, 261
Chen, B., 302
Chen, C.S., 350
Chen, H., 339
Chen, J., 132f, 339, 384
Chen, Ji, 203
Chen, K., 112, 113, 325
Chen, L., 351
Chen, M.S., 350
Chen, R., 302
Chen, Z., 339
Chen, Zonghai, 272
Cheng, M., 302
Cho, M.Y., 350
Chow, J.H., 351
Chtourou, Zied, 196
Chu, F., 339
Chung, C.Y., 111, 113
Cobelo, I., 135, 139t, 141t, 144t, 145t, 179
Cocco, V.M., 339
Cochran, J.J., 239
Conejo, A.J., 1, 2, 6, 37, 40, 41, 49, 53, 183, 218, 219t–221t, 238
Conejo, Antonio J., 59, 72, 74, 194
Coonick, A.H., 147, 147f
Corchero, C., 271
Corera, J.M., 377
Costa, M., 384
Cousins, Terry, 63
Cramton, P., 6
Cruz-Zambrano, M., 271
Cuccia, P., 132
Cui, K., 21t, 111, 113
Cui, M., 112, 113, 339

D
Dabbagh, S.R., 215, 219t–221t, 361
Dai, C., 13
Dai, T., 41
Dalala, Z.M., 302
Damiano, Alfonso, 258, 261, 265, 266t, 267
D'Arco, S., 12
Das, Choton K., 267
Das, D.C., 138, 139t, 141t–145t, 146

Dash, P.P., 134, 145t
Dasouki, S., 135, 139t, 141t, 144t, 145t
Davis, M., 351
De Grève, Z., 179
De Haan, S.W., 133, 143t, 145t, 147, 148f
De Paola, A., 219
De Vos, K., 6
de Wit, J., 16
Deconinck, Geert, 195
Delille, G., 137, 143t, 145t
Delnooz, A., 6, 14
Dempe, S., 167
Denhard, M., 325
Denholm, Paul, 28f, 316
Devine, M.T., 135, 139t, 141t–145t, 151, 152f, 153f, 155, 156
D'haeseleer, W., 216, 219t–221t
D'Haeseleer, W., 361
Di Silvestre, M.L., 219
Díaz-González, Francisco, 194
Dielmann, K., 229
Dierich, F., 325
Ding, H., 7, 11
Dingqian, L., 378
Djapic, Predrag, 300
Dokopoulos, P., 325
Dominguez, X., 137, 139t, 141t, 143t–145t
Domínguez-García, A.D., 265
Dong, F., 111, 113, 215, 219t, 221t
Dong, P., 384
Dong, Z., 218, 219, 219t–221t
Dong, Z.Y., 39, 181, 230, 231
Donoho, D.L., 330
Dörfler, F., 131
Dowell, J., 338
Draxl, C., 325
Du, P., 339
Duan, X., 339
Dulău, L.I., 3t, 8t
Duvall, M., 316
Dykeman, D., 19

E
Eghbal, M., 165
Ehsan, M., 38, 112, 113, 239
Ehsan, Mehdi, 194
Eijgelaar, M., 16
Ekanayake, J., 133, 143t, 145t
El Bakari, K., 3, 4, 13
El-Khattam, Walid, 193

El-Khattan, W., 165
El-Saadany, E.F., 163
El-Sehiemy, Ragab A., 320
El-Zobaidi, H., 147, 147f
Elad, M., 330
Eldoromi, Mojtaba, 193–206, 299–321
Elkamel, A., 40
Eller, Alex, 258, 268f
Emarati, M., 231
Emhemed, Abdullah S., 194
Enacheanu, B., 377
EnergyVille, 14
ENTSO-E, 132
Erdem, E., 325
Erdinc, O., 217, 218, 219t, 220t, 221, 221t
Erdinc, Ozan, 193
Erge, T., 343
Erge, Thomas, 194
Esmaily, J., 351
Esmaily, Jamal, 343–356
Espassandim, Helena M.D., 257, 263
Etherden, N., 135, 136, 139t, 141t, 143t–145t
European Commission's Directorates-General for Research and Innovation, Energy and the Joint Research Centre, 3
European Virtual Fuel Cell Power Plant, 16, 17t

F

Faisal, Mohammad, 258, 267, 272
Fallahi, F., 362
Fan, L., 216, 219t–221t, 325
Fan, M., 13
Fan, S., 112, 113, 351, 361
Fan, Songli, 198
Fang, F., 16, 214, 219t
Fang, Fang, 197f
Fang, T., 351
Fang, X., 167
Fang, Xi, 300
Farham, H., 163
Faria, P., 13
Faria, Pedro, 195
Fateh, Davood, 193–206, 299–321
Fathi, A., 136, 143t, 145t
Fayaz-Heidari, A., 113
Feblowitz, J., 14
Federal Energy Regulatory Commission, 193

Fei, S.W., 325
Feinberg, E.A., 351
Felsenstein, Simon, 266
Feng, C., 339
Feng, Q., 378
Fennell, S., 151, 152f, 153f, 155
Fernando, T., 229
Ferreira, J., 133, 143t, 145t, 147, 148f
Filik, T., 326, 328, 329, 336, 339
Filik, Tansu, 325–339
Fontela, M., 377
Foroughi, M., 113
Fotuhi-Firuzabad, Mahmud, 59–107, 260
Francois, B., 4, 137, 143t, 145t
Fu, M., 378
Fu, Y., 136, 143t, 145t
Fuller, J.D., 53
Fushuan, W., 378

G

Gabriel, S.A., 53
Galanis, G., 325
Gallardo, C., 137, 139t, 141t, 143t–145t
Galván, E., 218, 220
Gan, D., 13, 134, 139t, 141t–143t, 145t
Gandomkar, Majid, 200, 204
Gantenbein, D., 19, 138, 139t, 141t, 142t, 144t, 145t
Gao, C., 135, 139t, 141t, 143t–145t
Gao, Shuang, 257, 258, 261
Gauntlett, Dexter, 258, 268f
Genethliou, D., 351
Genton, M.G., 325–327, 338
Gevorgian, V., 30t, 32, 136, 139t, 141t, 143t–145t, 146
Ghadikolaei, H.M., 362, 364
Ghafouri, A., 136, 139t, 143t, 145t
Ghahgharaee Zamani, A., 112, 113
Gharehpetian, G.B., 4, 136, 139t, 143t, 145t
Ghasemi, H., 217, 219t–221t
Ghasemi, J., 351
Ghavidel, S., 229, 361
Ghazi, R., 302
Gholipour, E., 6, 39, 112, 113, 229, 362
Gholipour, Eskandar, 197, 257
Ghorbankhani, E., 179
Ghosh, S., 377
Giebel, G., 325
Giuntoli, M., 13, 40, 361
Giuntoli, Marco, 59

Gkatzikis, L., 164
Goel, L., 13, 217, 219t–221t
Gogoi, N., 138, 139t, 141t–145t, 146
Golestaneh, F., 239
Golkar, M.A., 40
Golshan, M.E.H., 362
Gomes, R.D.S., 339
Gomis-Bellmunt, Oriol, 194
Gonçalves, Henrique, 303
Gong, Xun, 318
González-Cabrera, N., 218, 220
Gorinevsky, D., 330
Gorjy, A., 230
Grant-Peters, J., 151, 152f, 153f, 155
Granville, S., 222
Gu, Y., 325–327
Guerrero, J.M., 136, 137, 139t, 141t, 143t, 145t, 146
Guo, Hongxia, 302
Guo, Q., 229
Guo, W., 21t, 113
Guo, X., 135, 139t, 141t, 143t–145t
Gutiérrez-Alcaraz, G., 218, 220

H

Hadayeghparast, S., 112, 113, 360, 366
Hadjsaid, N., 5
Hadjsaid, Nouredine, 194
Haes Alhelou, H., 132
Haghifam, M.-R., 215, 219t
Haghifam, M.R., 6, 182, 230, 231, 237, 361
Haida, T., 350
Hall, C., 151, 152f, 153f, 155
Hamedani-Golshan, M.E., 132
Han, Hua, 302
Han, Q., 339
Han, S., 138, 143t, 145t
Han, Wang, 196
Hasanien, H.M., 229
Hashemi, S.M., 383
Hashemizadeh, A., 377, 384
Hashemizadeh, Ali, 377–395
Hashim, H., 218, 220
Hatziargyriou, N., 134, 184, 217, 219, 219t, 220t, 325
Hatziargyriou, N.D., 134, 139t, 141t–145t
Hatziargyriou, Nikos, 300
Hauksson, E.B., 19
He, G., 112, 113

He, Y., 325
Hegazy, Y., 165
Heher, A., 377
Heidari, A., 222, 343
Hemmati, M., 367
Hennebel, M., 220, 222
Henríquez, Rodrigo, 67, 71
Heredia, F., 271
Hering, A.S., 326, 338
Hernandez, L., 343
Hesami, M., 383
Heydari-Doostabad, H., 302
Heydarian-Forushani, E., 362
Heydt, G.T., 265
Hill, D., 338
Hill, D.C., 326, 338
Hill, D.J., 131
Hill, R., 151, 152f, 153f, 155
Hippert, H.S., 351
Ho, W.S., 218, 220
Hodge, B.M., 339
Hodnett, P.F., 350
Holladay, J.D., 4
Homaee, O., 383
Hommelberg, M., 179
Hommelberg, M.P.F., 14
Hong, J., 30t, 32
Hooshmand, R., 39, 112, 113
Hooshmand, R.A., 6, 229, 362
Hooshmand, Rahmat-Allah, 197, 257
Horchler, S., 20
Hosseini, S.H., 302
Hosseini Bamakan, S.M., 377
Hovsapian, R., 30t, 32, 136, 139t, 141t, 143t–145t, 146
Hrvoje, P., 378
Hu, Q., 167
Hu, T., 339
Hu, X., 136, 143t, 145t
Hu, Z., 138, 143t, 145t
Huang, C.M., 350
Huang, N.E., 330
Hug, G., 131, 158
Hwang, J.C., 350
Hyde, O., 350

I

Iervolino, R., 137, 139t, 141t, 143t–145t, 146
Igualada, L., 271

Illinois Center for a Smarter Electric Grid (ICSEG), 156
Ilo, A., 377
Imhoff, C.H., 4
Infield, D., 326, 338
INTERRFACE Consortium, 151
Ippolito, M.G., 219
Ipsakis, D., 179
Iravani, R., 136, 143*t*, 145*t*
Iria, Jose P., 60, 83, 84
Islam, Arif, 264–266
Iu, H.H., 135, 139*t*, 141*t*, 144*t*, 145*t*

J

Jacob, M., 345, 351
Jacobsen, J.H., 179
Jadid, S., 112, 113, 163, 164, 184, 218–220, 219*t*–221*t*, 360, 362, 366
Jadid, Shahram, 59, 76, 79, 195
Jadidbonab, M., 231
Jaffe, S., 14
Jahanbani Ardakani, A., 40
Jahanbani Ardakani, F., 40
Jahangir, H., 40
Jaimoukha, I.M., 147, 147*f*
Jain, Manu, 303
Jampeethong, P., 136, 139*t*, 141*t*, 143*t*
Jannesar, Mohammad Rasol, 271
Jansen, B., 19, 138, 139*t*, 141*t*, 142*t*, 144*t*, 145*t*
Jansen, Bernhard, 261, 275
Jansen, J., 19
Jasiński, M., 21*t*
Javadi, M.S., 218, 220
Jayasekara, Nadeeshani, 258, 272, 273
Jenkins, N., 133, 143*t*, 145*t*, 300
Jennings, N.R., 13, 138, 139*t*, 141*t*, 142*t*, 144*t*, 145*t*
Jennings, P., 230
Ji, Feng, 302
Ji, T., 339
Jia, H., 13
Jia, J., 339
Jiang, C., 21*t*
Jiang, X., 15
Jiang, Y., 113, 339
Jin, Q., 21*t*, 111, 113
Jin, T., 362, 364
Jirutitijaroen, Panida, 61
Jofré, Alejandro, 88
Ju, J., 111, 113
Ju, L., 112, 113, 215, 219*t*, 221*t*, 378
Ju, Liwei, 75
Ju, Y., 377, 384

K

Kabirifar, Milad, 59–107, 257–293
Kaczorowska, D., 21*t*
Kádár, P., 5
Kadar, Peter, 195
Kádár, Péter, 195
Kær, S.K., 137, 139*t*, 141*t*, 143*t*–145*t*
Kallos, G., 325
Kaltschmitt, M., 180*f*, 182*f*, 184, 184*f*, 187*f*–189*f*, 189*t*, 231
Kamphuis, I., 179
Kamphuis, I.G., 16
Kamphuis, R., 16, 377
Kamwa, I., 135, 139*t*, 141*t*–145*t*, 229
Kang, C., 343
Kang, Y.C., 30*t*, 32
Karakas, A., 217, 218, 219*t*, 220*t*, 221, 221*t*
Karami, H., 7
Karami, M., 218
Karasan, O.E., 360, 361
Karasan, Oya Ekin, 316*f*, 317*f*
Kardakos, E.G., 40, 41, 167, 217, 219, 219*t*, 220, 220*t*, 221*t*
Kardakos, Evaggelos G., 59, 67, 257, 258, 274
Karimi, I.A., 220
Karimyan, Peyman, 59, 60, 78, 81
Kariniotakis, G., 325
Kasaei, Mohammad Javad, 200, 204
Kashani, Navid Zare, 299–321
Katsafados, P., 325
Kauhaniemi, K., 137, 139*t*, 141*t*, 143*t*, 145*t*, 146
Kavasseri, R.G., 325
Kavousi-Fard, A., 218, 220
Kaymak, Uzay, 384
Kazemi, A., 112, 113, 360
Kazempour, S.J., 182
Kazerani, M., 219
Ke, D., 113
Keane, A., 149
Kërçi, T., 135, 139*t*, 141*t*–145*t*, 151, 152*f*, 153*f*, 155, 156

Kërçi, Taulant, 131–158
Kessels, K., 14
Keynia, F., 231
Khan, H., 135, 139t, 141t, 144t, 145t
Khan, R., 138, 139t, 141t–145t, 146
Khavari, F., 40, 41, 343, 378
Khavari, Farshad, 37–55, 343–356
Khezri, R., 230, 234, 237
Khodadadi, Nazgol, 359–374
Khodayar, M.E., 366, 368
Khodr, H., 5
Khodr, H.M., 195
Khomfoi, S., 136, 139t, 141t, 143t
Khotanzad, A., 351
Kia, M., 343, 383
Kieny, C., 5
Kieny, Christophe, 68
Kim, J., 30t, 32, 136, 139t, 141t, 143t–145t, 146, 362
Kim, Jinho, 27–33, 30t
Kim, S.J., 330
Kirar, Mukesh K., 257, 258
Kirschen, D.S., 164
Kirschen, Daniel, 86
Kirtley, Jim L., 30t
Kjær, P.C., 137, 139t, 141t, 143t–145t
Kling, W.L., 3, 4, 13, 133, 143t, 145t, 147, 148f
Kocvara, M., 40
Koeppel, G., 179
Koh, K., 330
Kok, J., 179
Kok, J.K., 16
Kok, K., 377
Koley, E., 377
Kolhe, M., 326
Kong, D., 21t, 111, 113
Kong, X., 21t, 111, 113
Koritarov, V., 30t, 32, 136, 139t, 141t, 143t–145t, 146
Kosek, Anna Magdalena, 195
Kostyla, P., 21t
Kota, R., 13, 138, 139t, 141t, 142t, 144t, 145t
Koutsopoulos, I., 164
Kurrat, Michael, 194
Kuzle, I., 13, 14, 40, 41, 218, 219, 219t–221t, 230, 361
Kuzle, Igor, 194

L
Lærke, R., 137, 139t, 141t, 143t–145t
Lai, J., 302
Lang, C., 179
Lang, P., 377
Latif, A., 138, 139t, 141t–145t, 146
Le, L., 339
Le, L.B., 362, 364
Leduc, Guillaume, 265
Lee, D.-Y., 220
Lee, M.H., 325
Lee, W.-J., 351
Lee, W.J., 325
Lehn, P.W., 302
Lehnhoff, S., 13
Lehtonen, Matti, 260
Lei, W., 378
Li, C., 111, 113
Li, D.F., 325
Li, F., 136, 143t, 145t, 167
Li, Fucun, 302
Li, G., 325
Li, H., 13, 111–113, 215, 219t, 221t
Li, Hao, 302
Li, Hua, 302
Li, L., 229, 361
Li, M., 21t, 215, 219t–221t, 339
Li, N., 13
Li, P., 15
Li, Shuyu, 258
Li, W., 377
Li, Xiangjun, 258, 259, 273, 278
Li, Y., 339, 351
Li, Y.-F., 220, 222
Li, Y.F., 325, 329, 333
Li, Z., 41, 163
Li, Zuyi, 86
Lian, H., 21t
Liang, H., 377
Liang, Z., 362, 364
Liao, J., 113
Liao, X., 339
Lim, J.S., 218, 220
Lin, J., 138, 143t, 145t
Liu, C., 21t, 362
Liu, Chang, 272
Liu, Chunhua, 257, 261, 262f
Liu, D., 325
Liu, G., 362

Liu, H., 138, 143*t*, 145*t*, 325, 329, 333, 361
Liu, H.H., 330
Liu, J., 4, 16, 112, 113, 214, 219*t*
Liu, Jizhen, 197*f*
Liu, L., 339
Liu, M., 132*f*, 134, 139*t*, 141*t*, 143*t*–145*t*, 147, 148, 148*f*
Liu, P., 21*t*, 113
Liu, S., 39, 111, 113, 339
Liu, W.H., 218, 220
Liu, X., 111, 113
Liu, Y., 15, 16, 21*t*, 134, 139*t*, 141*t*–143*t*, 145*t*, 214, 219*t*, 325, 339, 377
Liu, Yajuan, 197*f*
Liu, Yonglu, 302
Liu, Z., 7, 11, 13, 378
Lloret, J., 343
Lombardi, P., 5, 258, 261
Long, S.R., 330
Lopes, J.A. Pecas, 300
López, Miguel A., 276
Lotfi, M., 339, 343
Lotfi, Mohamed, 257, 258, 263, 274
Lotufo, A.D.P., 351
Louie, Henry, 300
Louka, P., 325
Lu, Chan-Nan, 193
Lu, J., 339
Lu, N., 13
Lu, Ning, 63
Lu, T., 41, 163
Lu, T.L., 351
Lu, Y., 378
Lundkvist, J., 135, 139*t*, 141*t*, 143*t*–145*t*
Lünsdorf, O., 5
Luo, F., 39, 218, 219, 219*t*–221*t*, 230
Luo, Y., 30*t*, 32, 136, 139*t*, 141*t*, 143*t*–145*t*, 146, 222
Lynch, P., 325
Lyu, X., 136, 143*t*, 145*t*, 378

M
Ma, P., 339
Ma, Z., 339
Macchietto, S., 218, 220
MacDougall, Pamela, 195
Madawala, Udaya K., 302
Madhusudana Rao, P., 344

Madsen, H., 1, 2, 6, 37, 325, 326
Mahfoud, F., 137, 139*t*, 141*t*, 143*t*–145*t*, 146
Mahindara, Vincentius Raki, 30*t*
Mahmoudi, N., 165
Mai, Trieu, 28*f*
Maire, J., 5, 377
Malarange, G., 137, 143*t*, 145*t*
Malekpour, A., 239
Malinowski, M., 4
Mancarella, P., 21*t*, 112, 113, 221
Mancarella, Pierluigi, 196
Manerba, D., 362
Manimaran, P., 344
Manohar, M., 377
Mansour-Saatloo, A., 368
Mansour-Saatloo, Amin, 359–374
Maratukulam, D., 351
Maratukulam, D.J., 351
Markovic, U., 158
Marquez, F.P.G., 235
Marquis, M., 326
Marra, F., 19
Marti, I., 325
Martí, J., 17, 377
Martin, Richard, 16
Martinez, S., 138, 139*t*, 141*t*, 143*t*, 145*t*
Marzband, M., 231, 360, 368
Marzband, Mousa, 359–374
Mashhour, E., 11, 229, 237, 361
Mashhour, Elaheh, 59, 65, 76, 77, 85, 87, 89, 90, 92, 92*f*, 93*f*, 95*f*–99*f*, 193
Masoum, Mohammad A.S., 258, 272, 273
Matos, Manuel, 82
Matos, Manuel A., 60, 83, 84
McMillan, D., 326, 338
Mehta, K.C., 325
Mena, R., 220, 222
Mendes, V.M.F., 325
Meng, K., 39, 181, 230, 231
Mercado, Pedro Enrique, 30*t*, 32
Mi, X., 339
Miao, W., 13
Milano, F., 131, 132*f*, 134, 135, 139*t*, 141*t*–145*t*, 147, 148, 148*f*, 150, 151, 156
Milano, Federico, 131–158
Milewski, J., 4
Milewski, Jarosław, 300
Milimonfared, J., 136, 139*t*, 143*t*, 145*t*

Ming Kwok, J.J., 220
Mingke, Z., 136, 143t, 145t
Minniti, Simone, 59, 82, 257
Minussi, C.R., 351
Mirzaei, M.A., 235, 360, 361, 366–368
Mirzaei, Mohammad Amin, 359–374
Mishra, S., 230
Mishra, Y., 135, 139t, 141t, 144t, 145t
Misra, Satyajayant, 300
Mitton, Nathalie, 196
Mo, O., 12
Moeini-Aghtaie, M., 113
Moeini-Aghtaie, Moein, 59–107, 257–293
Moghaddam, Amjad Anvari, 199
Moghaddam, I.G., 362
Moghaddam, M.P., 362
Moghaddam, Mohsen Parsa, 194
Moghaddas-Tafreshi, S.M., 11, 87, 229, 237, 361
Moghaddas-Tafreshi, Seyed Masoud, 59, 65, 76, 77, 85, 89, 90, 92, 92f, 93f, 95f–99f, 193, 257, 259, 265
Moghassemi, Ali, 131–158, 209–223
Moghimi, H., 219, 222
Moghimi, M., 239
Mohammadi, M., 4, 8, 9, 59, 360, 361
Mohammadi-Ivatloo, B., 230, 231, 234, 235, 238, 360, 367, 368
Mohammadi-Ivatloo, Behnam, 227–253, 359–374
Mohammadian, L., 163
Mohammadpour Shotorbani, A., 238
Mohammadpour Shotorbani, Amin, 227–253
Mohandes, M., 351
Mohanpurkar, M., 30t, 32, 136, 139t, 141t, 143t–145t, 146
Mohsenian-Rad, H., 179
Molina, Marcelo Gustavo, 30t, 32
Molzahn, Daniel K., 277
Momoh, J.J., 351
Monache, L.D., 325
Monteiro, Vítor, 303
Monticelli, A., 222
Moradinezhad, R., 351
Moradzadeh, A., 235
Morais, H., 5, 13, 195
Morais, Hugo, 195
Morales, J.M., 1, 2, 6, 37, 40, 41, 183, 218, 219t–221t, 238

Morales, Juan M., 194
Moreno, S.M., 339
Morenoff, D.L., 164
Morren, J., 133, 143t, 145t, 147, 148f
Mortazavi, S., 362
Mosca, C., 132
Moslehi, K., 179
Moti Birjandi, Ali Akbar, 193–206, 299–321
Mousavizadeh, M., 38, 112, 113
Moutis, P., 134, 139t, 141t–145t, 217, 219, 219t, 220t
Moutis, Panayiotis, 131–158, 209–223
Muis, Z.A., 218, 220
Muljadi, E., 30t, 32, 136, 139t, 141t, 143t–145t, 146, 229
Muljadi, Eduard, 30t
Muñoz, Almudena, 265
Munuera, Luis, 270f
Muqbel, A.M., 216, 219t–221t
Murad, M.A.A., 134, 135, 139t, 141t–145t, 147, 148, 148f, 151, 156
Murphy, C., 149
Muscholl, M., 377
Musio, Maura, 258, 261, 265, 266t, 267
Mutale, Joseph, 300
Muto, S., 350
Muttaqi, Kashem M., 258, 274
Muyeen, S., 135, 139t, 141t–145t
Muyeen, S.M., 229, 230, 234, 237

N
Nafisi, H., 368
Nair, N.K., 7
Nasiri, N., 368
Naughton, J., 112, 113
Naval, N., 6, 21t, 40, 112, 113, 230
Nazar, M.S., 115, 116
Nazar, Mehrdad Setayesh, 59, 76, 79
Nazari, F., 167, 231
Nazari-Heris, M., 234, 235, 360
Neenan, B., 164
Nelms, Robert M., 30t
Nemhauser, George L., 316
Nemry, Françoise, 265
Neves, C., 345, 351
Nezamabadi, H., 112, 113
Nezamabadi, Hossein, 59, 76, 79
Nguyen, H.T., 362, 364
Nguyen-Duc, Huy, 59, 60, 258, 274

Nguyen-Hong, Nhung, 59, 60, 258, 274
Nicholson, R., 14
Nick, M., 362
Nielsen, K.S., 134, 139*t*, 141*t*, 143*t*–145*t*
Nielsen, T.S., 325
Niknam, T., 239
Niknam, Taher, 199
Nikolic, N., 350
Nikonowicz, Ł.B., 4
Nikonowicz, Łukasz Bartosz, 300
Nikoukar, Javad, 200, 204
Nilsson, U.B., 377
Nipen, T., 325
Niu, D., 325
Niu, T., 339
Njenda, T.C., 132
Nojavan, S., 179
Nolan, Sh., 179
Nosratabadi, S.M., 6, 39, 112, 113, 229, 362
Nosratabadi, Seyyed Mostafa, 197, 257

O

Ohler, C., 137, 143*t*, 145*t*
O'Keefe, G., 151, 152*f*, 153*f*, 155
Okwelum, E., 179
Olivares, D.E., 219
O'Malley, M., 179
Önen, A., 135, 139*t*, 141*t*–145*t*, 229
Onen, A., 230
Ortega, Á., 131, 134, 139*t*, 141*t*, 143*t*–145*t*, 147, 148, 150
Ortega, L., 137, 139*t*, 141*t*, 143*t*–145*t*
Oshnoei, A., 230, 234, 237
Ossowski, S., 13, 138, 139*t*, 141*t*, 142*t*, 144*t*, 145*t*
Oudalov, A., 137, 143*t*, 145*t*
Outrata, J., 40
Oyarzabal, J., 135, 139*t*, 141*t*, 144*t*, 145*t*, 179

P

Pai, M.A., 148
Pal, B.C., 147, 147*f*
Pal, Mayukha, 344
Paliwal, Priyanka, 257, 258
Palizban, O., 137, 139*t*, 141*t*, 143*t*, 145*t*, 146
Palma-Behnke, Rodrigo, 88
Pan, D.F., 325
Pandžić, H., 13, 14, 40, 41, 218, 219, 219*t*–221*t*, 230, 361
Pandžić, Hrvoje, 194
Pantoš, Miloš, 194
Papadopoulou, S., 179
Papaefthymiou, S.V., 362
Papathanassiou, S., 184
Papathanassiou, S.A., 362
Parazdeh, Moein Aldin, 299–321
Park, J.W., 30*t*, 32
Parol, M., 351
Parsa Moghaddam, M., 182
Pasban, A., 113
Pasetti, M., 362
Pasupuleti, Jagadeesh, 59
Patil, P., 2, 4
Patil, Piyush, 1–22
Patria, A.R., 5
Peart, Mervyn R., 203
Pedram, Massoud, 272
Pedreira, C.E., 351
Peik-Herfeh, M., 40, 163, 214, 219*t*–221*t*, 230, 247
Peik-Herfeh, Malahat, 59, 194
Peng, T., 339
Pereira, M., 222
Pham, H., 377
Phong Le, H., 377
Piao, L., 361
Piao, Longjian, 198
Pinson, P., 1, 2, 6, 37, 325, 326
Pinto, J.G., 303
Pinto, T., 13
Pisarenko, Zhanna, 258
Pivrikas, A., 230
Plancke, G., 6
Poli, D., 13, 40, 361
Poli, Davide, 59
Poncelet, K., 216, 219*t*–221*t*
Poolla, K., 325, 330, 339
Pouladi, J., 163
Pourbabak, H., 362, 364
Pourghaderi, Niloofar, 59–107, 257–293
Pourmousavi, S.A., 360
Pousinho, H.M.I., 325
Powalko, M., 5
Pozo, M., 137, 139*t*, 141*t*, 143*t*–145*t*
Praça, I., 13
Prandoni, V., 5, 179

Preston, A., 5
Priestley, M.B., 331
Primadianto, Anggoro, 193
Pudjianto, D., 5, 10, 10f, 11, 59, 68, 135, 139t, 141t, 144t, 229, 257, 274, 377
Pudjianto, Danny, 300
Pytharoulis, I., 325

Q
Qiao, W., 41
Qiu, J., 39, 181, 230, 231
Qiuyue, C., 378

R
Raahemifar, K., 220
Rabiee, A., 219, 222
Rahimi, M., 21t, 40
Rahimi-Eichi, Habiballah, 272
Rahimi-Kian, A., 14, 215, 217, 219t–221t
Rahimiyan, M., 14
Rahimiyan, Morteza, 75
Rahmani, K., 113, 115
Rahmani, Kiumars, 111–129
Raison, B., 377
Ralon, Pablo, 268f, 270f
Ramachandaramurthy, Vigna K., 261, 275
Ramdas, A., 330
Ramos, L.F., 9–11
Ramsay, C., 5, 10, 10f, 11, 135, 139t, 141t, 144t, 229, 377
Ramsay, Charlotte, 300
Ramteke, M.R., 2, 4
Ramtharan, G., 133, 143t, 145t
Randrup, M., 377
Rangaraju, Jayanth, 318
Rashidinejad, M., 231, 362
Rashidizadeh-Kermani, H., 113, 231, 361
Rastegar, Mohammad, 260
Re, G.L., 219
Rebours, Yann, 86
Reddy, S.S., 164
Rehtanz, C., 13
Rekik, Mouna, 196
Riahy, G.H., 325
Riaz, S., 21t, 112, 113
Riaz, Shariq, 196
Rinaldi, S., 362
Rivas, M.B., 377

Riveros, J.Z., 216, 219t–221t
Roder, Gerold, 194
Romero-Cadaval, E., 4
Rong, Peng, 272
Roossien, B., 16
Ropuszyńska-Surma, E., 21t
Rosenthal, R.E., 239
Rossi, A., 179
Roux, G., 325
Rudion, K., 5
Ruiz, B.F.H.C., 53
Ruiz, C., 49, 220, 222
Ruiz, Carlos, 59, 72, 74
Ruiz, N., 135, 139t, 141t, 144t, 145t, 179
Ruthe, S., 13
Ruzic, S., 350

S
Saber, A.Y., 361
Saboori, H., 4, 8, 9, 360, 361
Saboori, Hedayat, 59
Sadeghi, M., 222
Sadeghi, S., 40
Sadeghi Yazdankhah, A., 235
Sadeghian, O., 230, 231, 234, 235, 237, 238
Sadeghian, Omid, 227–253
Saha, T.K., 165
Sahsamanoglou, H., 325
Saint-Pierre, A., 221
Salama, M.M.A., 165
Salama, Magdy M.A., 193
Salonidis, T., 164
Sampani, A., 134
Sanandaji, B.M., 325, 330, 339
Sánchez, R., 6, 21t, 40
Sanchez, R., 112, 113
Sánchez, R., 230
Sanchez-Esguevillas, A., 343
Sandmeier, Patricia, 266
Sanduleac, M., 377
Sanei, M., 362
Sanjari, M.J., 1–22
Sanseverino, E.R., 219
Santos, L.C.D., 339
Santos-Junior, C.R., 351
Santoso, S., 30t
Saranchimeg, S., 7
Sauer, P.W., 148
Schaeffer, G.J., 377

Schulz, Christian, 194
Schwaegerl, C., 136
Sebastian, M., 377
Sedghi, Mahdi, 257, 259, 265
Seetharaman, K., 325
Seferlis, P., 179
Seifi, Alireza, 199
Seifi, H., 40, 59, 163, 194, 214, 219t–221t, 230, 247
Semsar, S., 302
Senroy, N., 133, 143t, 145t
Setayesh Nazar, M., 112, 113, 115
Setayesh Nazar, Mehrdad, 111–129
Setiadi, Herlambang, 258
Setiawan, E.A., 5, 135, 139t
Seven, S., 230
Sezaki, K., 138, 143t, 145t
Sfetsos, A., 325, 338
Shabanzadeh, M., 6, 215, 219t, 230, 231, 237, 361
Shafie-Khah, M., 21t, 40, 41, 362, 378, 383
Shafie-khah, M., 339, 343, 361, 383
Shafie-Khah, Miadreza, 194
Shafiee, Q., 136, 143t, 145t
Shafiekhani, M., 21t, 38, 40, 41, 44, 343, 378, 380, 381f, 383
Shafiekhani, Morteza, 37–55, 343–356, 377–395
Shah, Zeal, 209–223
Shahabi, M., 343, 366, 368
Shahani, D.T., 303
Shahidehpour, Mohammad, 86
Shams, M.H., 343, 366, 368
Shanxu, Duan, 67
Shaowen, H., 378
Sharma, Angshuman, 308
Sharma, H., 230
Sharma, Santanu, 308
Sharp, J., 326
Shawky El Moursi, Mohamed, 30t
Shayanfar, H., 112, 113, 218, 220, 360, 366
Shayegan Rad, A., 231
Shayegan Rad, Ali, 257
Shayegan-Rad, A., 111, 113, 167, 167f, 169, 171, 171f, 172t, 173f–175f, 175t, 176t, 180f, 182f, 184, 184t, 187f–189f, 189t, 215, 219t–221t, 231
Shayegan-Rad, Ali, 163–177, 179–191
Sheblé, Gerald B., 66, 277

Sheikh-El-Eslami, M., 214, 219t–221t
Sheikh-El-Eslami, M.-K., 215, 219t
Sheikh-El-Eslami, M.K., 6, 40, 59, 163, 194, 215, 219t–221t, 230, 231, 237, 247, 361
Sheikh-El-Eslami, Mohamad Kazem, 194
Shen, C., 136, 143t, 145t
Shen, Y., 215, 219t–221t, 377
Shen, Z., 330, 339
Sherali, H., 40
Shetty, C.M., 40
Sheykholeslami, A., 133, 139t, 141t, 143t–145t
Shi, D., 339
Shi, Haiyun, 203
Shi, J., 7, 11, 325
Shi, L., 222
Shih, H.H., 330
Shokoohi, S., 137, 142t, 143t, 145t, 146
Shotorbani, A.M., 230, 231
Shu, X., 21t, 113
Shu, Zhen, 61
Shukur, O.B., 325
Siano, P., 113, 132, 137, 139t, 141t, 143t–145t, 146, 163, 164, 184, 220, 231, 361
Sideratos, G., 325
Siebert, N., 325
Sikorski, T., 21t, 137, 139t, 141t, 144t
Silva, Marco, 195
Simoglou, C.K., 40, 41, 167, 217, 219, 219t, 220, 220t, 221t
Simoglou, Christos K., 59, 67, 257, 258, 274
Singh, Bhim, 303
Singh, J., 302
Singh, M., 30t
Sioshansi, Ramteen, 316
Sivakumar, Bellie, 203
Six, D., 14
Skovmark, R., 179
Smith, D.A., 325
Smith, J.C., 326
Soares, Filipe Joel, 60, 83, 84
Sobhani, B., 339
Soltaninejad, A., 112, 113
SoltaniNejad Farsangi, A., 360, 366
Song, S.H., 30t, 32
Song, Y., 138, 143t, 145t
Sonnenschein, M., 5
Soong, T., 302

Soran, A., 230
Soroudi, A., 14, 215, 219*t*, 221*t*, 361
Soroudi, Alireza, 194
Sousa, J., 384
Souza, R.C., 351
Spelta, S., 5
Sreeram, V., 135, 139*t*, 141*t*, 144*t*, 145*t*
Stan, A.I., 135, 137, 139*t*, 141*t*, 143*t*–145*t*
Stanojev, O., 158
Stern, A., 326
Strbac, G., 5, 10, 10*f*, 11, 135, 139*t*, 141*t*, 144*t*, 229, 377
Strbac, Goran, 300
Stroe, D.I., 135, 137, 139*t*, 141*t*, 143*t*–145*t*
Strunz, K., 184
Strunz, Kai, 300
Stull, R., 325
Su, Mei, 302
Su, W., 362, 364
Sumper, Andreas, 194
Sun, G., 216, 219*t*–221*t*
Sun, H., 230
Sun, Y., 113
Sun, Yao, 302
Sun, Yinong, 28*f*
Sundstrom, O., 138, 139*t*, 141*t*, 142*t*, 144*t*, 145*t*
Suresh, V., 361
Suryanarayanan, S., 265
Sutanto, Danny, 258, 274
Suul, J.A., 12
Suvire, Gastón Orlando, 30*t*, 32
Swierczynski, M., 135, 139*t*, 141*t*, 143*t*–145*t*
Świerczyński, M., 137, 139*t*, 141*t*, 143*t*–145*t*

T

Tabatabaei, S.M., 200
Taghe, R., 4, 8, 9, 59, 360, 361
Taheri, H., 217, 219*t*–221*t*
Taheri, S.S., 234, 235
Tahmasebi, Mehrdad, 59
Tajeddini, M.A., 14, 215, 219*t*, 221*t*
Tajik, E., 362, 364
Tambke, J., 325
Tamimi, B., 147, 147*f*
Tan, J., 111, 113
Tan, Kang Miao, 261, 275
Tan, Q., 111–113, 215, 219*t*, 221*t*, 378

Tan, Z., 39, 111–113, 215, 219*t*, 221*t*, 360, 378
Tang, Wenjun, 61, 258, 274, 278, 289, 289*f*, 291*f*
Tang, X., 21*t*
Tao, Cai, 67
Tao, L., 136
Tao, T., 330
Tascikaraoglu, A., 217, 218, 219*t*, 220*t*, 221, 221*t*
Taşçıkaraoğlu, A., 325, 330, 339
Tastu, J., 325, 326
Teodorescu, R., 135, 137, 139*t*, 141*t*, 143*t*–145*t*
Thavlov, A., 217, 219*t*–221*t*
Thrimawithana, Duleepa J., 302
Tian, H.Q., 325, 329, 333
Tibshirani, R.J., 330
Tipaldi, M., 137, 139*t*, 141*t*, 143*t*–145*t*, 146
Tiwari, R., 302
Tomsovic, K., 362
Ton, D., 362
Torchia, M., 14
Toubeau, J-F., 179
Trombe, P.J., 325, 326
Tröschel, M., 5
Trotignon, Marc, 86
Tu, G., 222
Tung, C.C., 330
Türkay, S., 333, 339
Turri, R., 5, 179

U

U.S. Federal Energy Regulatory Commission, 163
Ulbig, Andreas, 82
Usaola, J., 325
Uzunoglu, M., 217, 218, 219*t*, 220*t*, 221, 221*t*
Uzunoğlu, M., 325, 339
Uzunoglu, Mehmet, 193

V

Vahedipour-Dahraie, M., 113, 231, 361
Vahidi, B., 200
Vale, Z., 13
Vale, Z.A., 5, 195
Vale, Zita A., 195

Vallée, F., 179
van den Noort, A., 16
van der Kam, Mart, 265, 266t
van der Velde, J., 16
Van Der Velden, A., 229
Van Haaren, Rob, 266
Van Olinda, P., 350
van Sark, Wilfried, 265, 266t
Vandewalle, J., 361
Varaiya, P., 325, 330, 339
Vargas, Luis S., 88
Vasirani, M., 13, 138, 139t, 141t, 142t, 144t, 145t
Vasirani, Matteo, 59
Vasquez, J.C., 136, 143t, 145t
Vassilakis, A., 134
Vatandoust, B., 40
Vatani, M., 4
Venayagamoorthy, G.K., 361
Verbič, G., 131
Verdult, V., 334
Verhaegen, M., 334
Verma, Arun Kumar, 303
Vezzola, M., 5
Vickers, Neil J., 201
Vidyanandan, K., 133, 143t, 145t
Villafáfila-Robles, Roberto, 194
Villar, José, 82
Virlot, S., 325
Voutetakis, S., 179
Vrettos, E., 158
Vuckovic, A., 350
Vukadinović Greetham, D., 345, 351
Vyas, A., 362
Vyatkin, V., 136, 139t, 141t, 144t, 145t

W

Waldl, I.H-P., 325
Wamer, C., 179
Wang, B., 167
Wang, C., 13, 21t, 111, 113, 339
Wang, D., 13
Wang, G., 112, 113
Wang, H., 21t, 112, 113, 325
Wang, Hui, 302
Wang, J., 113, 218, 339, 362
Wang, P., 13, 217, 219t–221t
Wang, Qiang, 258
Wang, Shangxing, 258, 259, 273, 278
Wang, X., 7, 11
Wang, Y., 39, 111, 113, 343
Wang, Yujie, 272
Wang, Z., 134, 139t, 141t–143t, 145t, 220, 302, 362, 364
Wang, Zhenhua, 302
Wang, Zhifang, 193
Wankhede, Sumeet Kumar, 257, 258
Warmer, C.J., 16
Weearsinghe, Saranga, 302
Węglarz, M., 21t
Wei, C., 113
Weimin, Z., 378
Weiss, S., 338
Wellinghoff, H.J., 164
Wilczak, J., 326
Wille-Haussmann, B., 343
Wille-Haussmann, Bernhard, 194
Wittwer, C., 343
Wittwer, Christof, 194
Wolfs, Peter J., 258, 272, 273
Wollenberg, Bruce F., 66, 277
Wolsey, Laurence A., 316
Wong, K.P., 39, 230
Wood, Allen J., 66, 277
Wu, B., 339
Wu, Diyun, 257, 258, 261
Wu, F.F., 179, 351
Wu, H., 339
Wu, M.C., 330
Wu, Q., 112, 113, 339
Wu, W., 230
Wu, X., 136, 143t, 145t

X

Xia, Q., 343
Xiao, G., 339
Xiao, H., 302
Xiao, J., 21t, 111, 113
Xiao, Weidong, 30t
Xie, L., 325–327
Xin, H., 13, 15, 134, 139t, 141t–143t, 145t
Xin, J., 339
Xiong, Shang-fei, 198
Xiong, Wenjing, 302
Xu, J., 113, 339
Xu, W., 339, 378
Xu, Y., 230, 351, 352f
Xu, Z., 378
Xue, Guoliang, 300

Y

Yamin, Hatim, 86
Yan, L., 39, 111, 113
Yan, R., 339
Yang, D., 30*t*, 32, 378
Yang, Dechang, 257–259, 278–280, 279*f*, 282*f*, 285*f*–289*f*
Yang, Dejun, 300
Yang, H., 218, 219, 219*t*–221*t*
Yang, H.T., 350
Yang, Hong-Tzer, 61, 258, 274, 278, 289, 289*f*, 291*f*
Yang, Hongming, 277
Yang, J., 39, 135, 139*t*, 141*t*, 143*t*–145*t*, 230
Yang, W., 112, 113
Yang, X., 339
Yang, Y., 325
Yang, Z., 4
Yang Dong, Z., 351, 352*f*
Yao, G., 230
Yarza, J.M., 377
Yavuz, L., 135, 139*t*, 141*t*–145*t*, 229
Yazdani, A., 134, 145*t*
Yazdani-Damavandi, Maziar, 267
Ye, Jun, 384
Yen, N.-C., 330
Yi, D., 218, 219, 219*t*–221*t*
Yi, Z., 230
Yin, S., 41, 163
Yin, Shuangrui, 257
Yong, Jia Ying, 261, 275
You, S., 302
You, X., 378
Yousefi, G.R., 182
Yu, H., 136, 143*t*, 145*t*
Yu, J., 325
Yu, N., 220
Yu, S., 16, 214, 219*t*
Yu, Songyuan, 197*f*
Yu, T., 229, 361
Yuan, J., 111, 113, 215, 219*t*, 221*t*
Yusta, J.M., 6, 21*t*, 40, 112, 113, 230

Z

Zahedmanesh, Arian, 258, 274
Zahid, Z.U., 302
Zakariazadeh, A., 112, 113, 163, 164, 184, 218–220, 219*t*–221*t*, 360, 362, 366
Zakariazadeh, Alireza, 59, 76, 79, 195
Zamani, A.G., 218, 219, 219*t*–221*t*, 360, 362, 366
Zamani, Ali Ghahgharaee, 59, 76, 79, 195
Zanganeh, A., 111, 113
Zangeneh, A., 44, 167, 167*f*, 169, 171, 171*f*, 172*t*, 173*f*–175*f*, 175*t*, 176*t*, 180*f*, 182*f*, 184, 184*f*, 187*f*–189*f*, 189*t*, 215, 219*t*–221*t*, 231, 343, 380, 381*f*
Zangeneh, Ali, 37–55, 163–177, 179–191
Zapata, J., 361
Zare, K., 234, 235, 360, 368
Zare, Kazem, 359–374
Zdrilić, M., 13, 219
Zhang, C., 339
Zhang, H., 7, 11, 339
Zhang, J., 339, 378
Zhang, Jingjing, 65
Zhang, K., 339
Zhang, N., 30*t*, 32
Zhang, R., 351, 352*f*
Zhang, Y., 13, 41, 163, 302, 339, 378
Zhang, Z., 113
Zhao, C., 216, 219*t*–221*t*
Zhao, J., 112, 113, 135, 136, 139*t*, 141*t*, 143*t*–145*t*, 218, 219, 219*t*–221*t*
Zhao, N., 339
Zhao, Q., 215, 219*t*–221*t*
Zhao, S., 339
Zhao, Y., 7, 11
Zhao-Xia, X., 136, 143*t*, 145*t*
Zheng, F., 339
Zheng, Q., 135, 139*t*, 141*t*, 143*t*–145*t*, 330
Zheng, T., 30*t*, 32
Zheng, Y., 181, 231, 339
Zhong, Q.C., 4
Zhong, W., 134, 139*t*, 141*t*, 143*t*–145*t*, 147, 148, 148*f*
Zhong, Weilin, 131–158
Zhou, B., 111, 113
Zhou, H., 112, 113
Zhou, J., 230, 339
Zhou, Y., 362
Zhou, Yizhou, 59, 60, 62, 65, 75, 78, 79, 258
Zhu, F., 378
Zhu, J., 229, 361
Zhu, L., 339
Zhu, Q., 339
Zhu, X., 325–327, 338

Zio, E., 220, 222
Ziogou, C., 179
Zou, Xiao-yan, 198

Zowe, J., 40
Zubov, D., 14
Zugno, M., 1, 2, 6, 37

Subject index

A

Abandoned clean energy, 378
Active distribution network (ADN), 9
Advanced metering infrastructure (AMI), 13
Aggregated
    DERs, 72, 179, 274, 276, 289
    energy, 182
    power, 144
    power balance, 81
    units, 5
Aggregating DERs, 85, 272
Allocated reserve energy, 124
Alternating direction method of multipliers (ADMM), 277, 278
Ancillary service, 8, 11, 13–15, 27–29, 61, 69, 113, 135, 136, 141, 196, 212, 221, 258, 274, 378
    for frequency regulation, 143
    from VPPs, 136
    market, 38, 151, 200, 377
Ant colony optimization (ACO), 142
Artificial bee colony (ABC), 142
Artificial intelligence (AI), 214
Artificial neural network (ANN), 142, 195, 214, 339, 351
Automatic generation control (AGC), 132
Autoregressive moving average, 214, 222
Auxiliary service (AS), 378

B

Backup power, 267
Balanced market (BM), 200
Balancing
    energy markets, 215
    energy services, 217
    market, 38, 41–43, 111, 194, 212, 215, 217, 274, 280, 287, 377
    market prices, 41, 42, 49

Battery
    capacity, 260, 263–265, 272
    charging, 318
    cycle aging, 272
    damage, 308
    degradation, 301, 316
    degradation cost, 290, 299, 308, 315–317
    depreciation cost, 230
    deterioration cost, 316
    storage, 139, 140
    storage systems, 133
    types, 265
    units, 136
    voltages, 319
Battery energy storage system (BESS), 2, 194, 267, 272, 273, 277
Bidirectional AC–DC converter (BADC), 302–304
Bidirectional DC–DC converter (BDC), 302, 303, 308, 309
Bidirectional power
    conversion stages, 302
    interface, 261, 301
Bilateral contract (BC), 200
Bulk power industry contribution, 360

C

Centralized
    controlled VPP, 8, 9
    scheduling scheme, 274
Charging
    battery, 318
    capacities, 280
    costs, 276, 288
    energy, 263, 276
    ESSs, 370
    EVs, 280
    facilities, 280
    loads, 257

locations, 280
mode, 259, 260, 319
model, 259
power, 70, 78, 96, 279, 280
schedule, 223
smart, 20
stations, 259, 260, 263, 265, 280, 319, 321
time, 279, 280, 284
Circulating power, 318, 319
Cleared power, 40, 45, 51, 53
Clustering analysis, 348, 349
Combined heat and power (CHP) units, 5, 194, 203, 275, 360, 364, 366, 367, 369
Commercial virtual power plant (CVPP), 8–10, 19, 38, 67, 68, 72, 81, 135, 140, 141, 257, 277, 362
Compressed air energy storage (CAES), 140, 194
Conditional value-at-risk (CVaR)
  model, 112
  profit, 112
Conflicting operation, 181
Constituent units, 38
Consumed power, 63, 65, 78
Consumption bidding powers, 74
Continuous regulation reserve (CRR), 13
Controllable
  loads, 1, 4–6, 11, 40, 134, 139, 140, 163, 194, 197, 200, 209, 212, 259
  power, 194
  units, 38
Conventional
  energy sources, 21
  generation units, 42, 195
  power systems, 132
  units, 38, 40, 196
Conventional power plant (CPP), 2, 11, 133
Cost
  analysis, 193
  considerations, 213
  curves, 41, 61
  function, 60, 77, 86, 87, 232, 380
  market, 212, 215
  minimization, 212, 230, 232, 235, 314
  operational, 87, 180, 220, 232, 251
Cumulative distribution function (CDF), 238
Curtailable loads, 230
Custom energy conversion technologies, 111

Customers
  electricity, 171, 172, 184, 186, 190, 217
  industrial, 184, 186
  interrupted, 72, 80, 90
  interrupted power, 80
  loads, 213
  profits, 212, 223
Cyber-attack, 15

D

Daily charging, 272
Day-ahead (DA)
  energy, 274
  energy market, 38, 184, 215, 222, 275
Day-ahead market (DAM), 85, 200, 275, 280, 283, 290
Decentralized
  DERs, 277
  energy transformation, 4
Decision tree (DT), 134
Demand response aggregator (DRA), 164
Demand response (DR), 8, 10, 11, 111, 113, 117, 143, 172, 176, 180, 194, 212–215, 217, 220, 257, 361, 362, 371, 378
Demand response program (DRP), 124–127, 163, 230, 360, 374
Depth of discharge (DoD), 315, 316
Deregulated electric grids, 15
Discharge
  cycles, 276
  efficiency, 202
  energy, 262, 300
  mode, 393
  power, 259, 261, 279, 290
  rate, 202
  time, 267
Discharging
  capacities, 78
  cycles, 257, 259, 271, 272
  efficiencies, 62, 63, 271, 315
  ESSs, 70
  power, 7, 61, 62, 71, 72, 77, 241, 261, 263, 271, 273, 276
  schedules, 220, 314
  statuses, 61, 62
Dispatchable
  DERs, 32
  units, 231

Subject index

Distributed controlled VPP, 8, 9
Distributed energy resource (DER), 1, 5, 27–29, 31, 59, 60, 133, 135, 179, 193, 209, 211, 257, 258, 300, 360, 362
  active power output, 148
  aggregated, 72, 179, 274, 276, 289
  commitment strategies, 111
  constraints, 72, 74, 81
  energy scheduling, 59
  frequency, 150, 157
  frequency control, 150
  integration, 360
  operation, 274, 277
  optimal
    energy scheduling, 95
    power dispatching, 360
    scheduling, 85
  reactive powers, 70
  renewable, 28
  scheduling, 195
  technical constraints, 274
Distributed generation (DG), 1–3, 111, 115, 116, 138–142, 301, 366, 378
  energy extraction, 142
  facilities, 112
  operational nature, 139
  units, 37, 119, 123, 127, 135, 139, 141, 229, 230, 237, 360
Distributed generator (DG), 59, 60, 257
  units, 87, 96, 113, 115
Distributed scheduling scheme, 277
Distribution
  network, 37, 59, 68, 90, 279
  power losses, 172
Distribution management system (DMS), 19
Distribution system operator (DSO), 8, 168, 193
Distribution system state estimation (DSSE), 193
Doubly fed induction generators (DFIG), 136
Droop controller, 133, 138
Dynamic programming (DP), 142

E

Electric
  grid, 15, 144
  power, 29, 222, 377, 379

Electric vehicle (EV), 29, 59, 62, 69, 112, 123, 133, 136, 137, 140, 209, 211, 230, 234, 257, 279, 300, 301, 320
  charging, 280
  charging schedule, 211
  discharge, 115, 320
Electrical
  energy, 5, 301
    market, 362
    resources, 112
    storage systems, 267, 301
  grid, 210, 222
  loads, 13
  market operator, 362
  power, 1
Electricity
  customers, 171, 172, 184, 186, 190, 217
  customers participation, 181
  distribution networks, 259, 272
  grid, 20, 257, 301, 321
  market, 39, 40, 59, 65, 101, 112–115, 139, 141, 142, 165, 172, 173, 179, 183, 193, 194, 199, 213, 219, 221, 229, 230, 232, 274, 279, 289, 361, 363
  market prices, 163, 171, 186
  networks, 267
  price, 230–232, 237, 239, 247, 251, 361, 368–370
Energy
  aggregated, 182
  balance, 364
  balance equation, 315
  charging, 263, 276
  consumption, 123–127, 301
  conversion, 111
  costs, 67, 112, 186, 209
  demand, 1, 218
  discharge, 262, 300
  distribution, 300
  electrical, 5, 301
  generation cost, 315
  interface, 301
  management, 6, 20, 61, 301
  management framework, 362
  management problem, 318
  models, 60
  operation, 13
  prices, 15, 184
  procurement, 190

production, 123, 125, 126
renewable, 1, 16, 20, 22, 221, 301, 362, 379
reserve, 123
resources, 99–101, 112, 179, 180, 212
scheduling, 71–73, 83, 95
scheduling framework, 67
scheduling problem, 62, 72, 84, 85
sources, 1, 2, 22, 38, 215, 300, 361, 377
storage, 44, 111, 215, 382
  equipment, 301
  technologies, 7, 365
units, 38, 315, 377
virtual power plant, 300, 379
wind, 4, 200, 326, 337
Energy consumption profile (ECP), 65
Energy management model (EMM), 41, 62, 81, 220, 301, 316, 361
Energy management system (EMS), 137, 196, 263, 301
Energy market (EM), 7, 9–12, 37–41, 59, 69, 71, 76, 182, 186, 194, 209, 211, 220, 360, 361, 363, 377
  operation, 8
  participation, 82
  price, 95, 99, 186, 211
  pricing, 211
Energy storage device (ESD), 217
Energy storage element (ESE), 139
Energy storage system (ESS), 7, 27–29, 59–61, 71, 77, 133, 135, 137, 140, 193, 209, 211, 230, 232, 234, 257–259, 301, 360, 362
  battery, 267, 277
  characteristics, 267
  charging, 370
  charging power, 72
  charging schedules, 80
  constraints, 87
  discharge power, 271
  discharging, 70, 370
  operation, 271, 290
  powers, 87, 89
  scheduling, 273, 278
  types, 259
  utilization, 272
Epsilon constraint method, 378, 383, 387, 390–392, 394

Excess
  energy, 43
  power, 151
Exchanged
  energy, 81
  power, 80, 89, 96, 264, 285, 287, 363, 369, 371
  power TVPP, 96
Extreme learning machine (ELM), 351–353, 355, 356

F
Feasible operation region (FOR), 360
Firefly algorithm (FA), 138, 142
Flexi-renewable virtual power plant (FRVPP), 41
Flexibility
  market, 83, 84
  power, 83
  scheduling, 60, 83
Flexible loads, 2, 63, 72, 76, 78, 79, 83, 85, 99, 100, 103, 135, 212, 280, 360
Flywheel energy storage (FES), 140
Forward markets, 221
Frequency
  control, 27–29, 32, 133, 134, 136–138, 147, 150
  regulation, 11, 133–135, 137, 138, 143–145, 267, 272, 319
  regulation services, 135, 144
  regulation signals, 134
  regulation VPPs, 133, 157
Frequency containment reserve (FCR), 143
Frequency regulation market (FRM), 40
Frequency restoration reserve (FRR), 143
Fuel cell (FC), 139, 195, 203
Fuel costs, 60, 184, 200, 212, 216
Fuzzy chance constraint programming (FCCP), 199
Fuzzy decision making method, 384, 390, 392, 394
Fuzzy logic controller (FLC), 142

G
General algebraic modeling system (GAMS), 203, 206, 232, 239, 252
Genetic algorithm (GA), 89, 142

Subject index

Grid
  assets, 218, 219
  electric, 15, 144
  electrical, 222
  electricity, 20, 257, 301, 321
  infrastructure, 210
  local, 141
  operation, 135, 222, 261
  operators, 194
  performance, 301
  power, 195, 214, 261, 300, 318, 367, 379
  security, 222
  services, 134, 222
  stability, 158
  stakeholder, 19
  supports, 267
  system, 193
  topology, 220, 222
  visibility, 158
  voltage, 318

H

Harmony search (HS), 142
Heat storage system (HSS), 362, 370
Heating, ventilating, and air-conditioning (HVAC) units, 13
Hierarchical control strategy (HCS), 27, 29
High temperature (HT), 16
Hourly
  energy prices, 186
  interruptible loads, 220
  profits, 53, 188
Hydraulic pumped energy storage (HPES), 140
Hydroelectric
  power plants, 230
  units, 132
Hydropower plants, 20

I

Imbalance markets, 216, 221
Independent system operator (ISO), 72, 179
Industrial
  customers, 184, 186
  technical VPPs, 362
  units, 378
Inflexible loads, 360
Information and communication technology (ICT), 6, 14, 20, 135, 136, 139

Information gap decision theory (IGDT), 199, 231, 265
Injected power, 45, 264
Interconnected
  power system, 132
  units, 194
Intermittent renewable energy output, 41
Interrupted
  customers, 72, 80, 90
  loads, 65, 72, 90
  power, 65, 88
Interruptible
  customers interrupted loads, 89
  loads, 60, 63, 65, 85, 86, 88, 89, 99, 193, 220
  loads scheduling, 98
Interruption cost, 85, 88, 93, 98
Interval analysis (IA), 200
Islanded
  operation, 4
  power systems, 137

K

Karush–Kuhn–Tucker (KKT)
  conditions, 40, 277
  optimality conditions, 101, 167, 169, 176

L

Large-scale virtual power plant (LSVPP), 377
Latin hypercube sampling (LHS), 210
Linear programming (LP), 142
Linear regression, 214, 222
Linearization, 48, 53, 116, 316, 380, 381
Load curtailment (LC), 163
Load frequency control (LFC), 136
Load shifting (LS), 163
Loads
  charging, 257
  controllable, 1, 4–6, 11, 40, 134, 139, 140, 163, 194, 197, 200, 209, 212, 259
  customers, 213
  electrical, 13
  forecasting, 140
  interrupted, 65, 72, 90
  interruptible, 60, 63, 65, 85, 86, 88, 89, 99, 193, 220
  powers, 69

Local
  grid, 141
  markets, 197
  network, 85, 88, 92, 95
  network energy price, 93
  power systems, 8
Locational marginal price (LMP), 229

M

Maintenance
  costs, 87, 204, 212
  scheduling, 230
Marginal costs, 45, 49, 52, 210
Market
  cleared price, 52, 53
  clearing, 40, 41, 45, 46, 59, 67, 72, 74
  clearing price, 40, 53, 67, 73
  conditions, 42
  contracts, 362
  cost, 212, 215
  data, 263
  design, 378
  dynamics, 210
  electrical energy, 362
  electricity, 39, 40, 59, 65, 101, 112–115, 139, 141, 142, 165, 172, 173, 179, 183, 193, 194, 199, 213, 219, 221, 229, 230, 232, 274, 279, 289, 361, 363
  energy, 7, 9–12, 37–41, 59, 69, 71, 76, 182, 186, 194, 209, 211, 220, 360, 361, 363, 377
  flexibility, 83, 84
  maturity, 221
  model, 67
  operator, 45
  operator electrical, 362
  participants, 71, 196
  participation, 221
  power, 33, 40, 67, 194, 362, 364, 379
  price, 40, 41, 49, 63, 67, 71, 72, 92, 98, 112, 113, 197, 198, 206, 210–213, 222, 237, 280, 362, 369, 379, 385
  price uncertainty, 197, 199, 211, 214
  structures, 221
  trading, 76, 104
  uncertainty, 237
  wholesale, 38, 60, 67, 72, 79, 80, 85, 164, 165, 168, 171, 229

Mathematical problem with equilibrium constraints (MPEC), 41, 46
Maximum
  available power, 319
  charging power, 87
  discharging power, 87
  power, 147, 319
  power transfer, 135
  profit, 212, 215, 394
Maximum power point tracking (MPPT), 147
Mean absolute error (MAE), 352
Mean absolute percentage error (MAPE), 352
Mean-variance portfolio (MVP), 213
Microgrid (MG), 133, 136, 140, 195, 229
  concept, 229
  exchange power, 229
Micronetwork, 300
Microturbine (MT), 202, 203, 263, 275, 279, 285
Mixed-integer linear programming (MILP), 133, 361
Mixed-integer program (MIP), 134
Model
  charging, 259
  market, 67
  network operational constraints, 277
  uncertainty, 219
Monte Carlo simulation (MCS), 75, 198, 210, 280, 367
Multi-objective programming, 383

N

National grid domain, 316
National household travel survey (NHTS), 260
Negotiation power, 164
Net
  energy, 315
  profit, 175
Network
  characteristics, 69
  codes, 143
  congestion management, 69
  constraints, 40, 72, 81, 84, 88
  electricity losses, 174
  load, 45, 278–280, 287
  local, 85, 88, 92, 95
  management, 37
  modeling, 277

nodes, 135
operating constraints, 67, 70
operation, 69–71, 88, 257, 259
operational constraints, 60, 72, 74, 81, 85, 86, 104, 257, 274
operator, 59, 301
reliability, 163
reserve margin, 89
security, 40
services, 59, 69, 76, 301
structure, 40
Nonlinear programming (NLP), 142

O

Operation
cost, 180, 212, 283, 380
DERs, 274, 277
efficiency, 271
energy, 13
energy market, 8
ESSs, 271, 290
grid, 135, 222, 261
network, 69–71, 88, 257, 259
optimal, 280, 289, 290, 378
power, 377
power system, 12, 210, 214
units, 237
VPPs, 14, 142, 144, 158, 163, 218, 220, 222, 232, 274, 280, 285
Operational
constraints, 72, 84, 184, 277
cost, 87, 180, 220, 232, 251
data, 31
dispatchable nature, 139
functionality, 222
nature, 139
parameters, 384
state optimum, 274
Optimal
bidding strategy, 39–43, 48, 53, 111, 216, 258, 261, 274, 377
dispatch, 71, 85, 116, 231, 287, 362
energy management, 67
energy scheduling, 13, 72
operation, 280, 289, 290, 378
operation strategy, 274, 277, 278
power, 71
scheduling, 68, 85, 111, 112, 195, 197, 222, 230–232, 274, 290

TVPP, 98
VPP operation, 289
VPP scheduling, 218

P

Pareto solutions, 384, 390, 392, 394
Participation
energy market, 82
market, 221
percentage, 246, 247, 250
profitable, 211
rate, 371
schedule, 103
TVPP, 90, 97
VPPs, 103, 280, 289
Particle swarm optimization (PSO), 138, 142, 361
Pearson correlation coefficient (PCC), 353
Penalty cost, 180, 184, 186, 188
Permissible participation, 186
Photovoltaic (PV)
inverter operation, 134
power, 134, 215, 361
power plant, 29
units, 111, 263, 273, 279
Plugged-in hybrid electric vehicle (PHEV), 124, 265
Point estimate method (PEM), 40, 111, 214, 215, 218, 219, 230, 362
Point estimation method (PEM), 200, 214
Point of common coupling (PCC), 29
Polluting units, 379, 391, 393
Pool market, 42
Power
actuation, 221
aggregated, 144
balance, 43, 45, 73, 74, 138, 165, 201, 277, 283, 367
balance constraint, 382
balance equation, 45, 277, 382
capacity, 12
charging, 70, 78, 96, 279, 280
compression, 303
consumers, 194
consumption, 2, 131, 134, 197, 265
control, 135
controllable, 194
conversion, 302

converters, 134, 137
demand, 21, 65, 164, 362, 373
density, 303, 308
discharge, 259, 261, 279, 290
discharging, 7, 61, 62, 71, 72, 77, 241, 261, 263, 271, 273, 276
electric, 29, 222, 377, 379
electrical, 1
electricity market, 164
electronics converters, 300
exchange, 101, 103, 385
exchange limits, 81
factor, 303
flexibility, 83
flow, 3, 4, 70, 81, 88, 89, 140, 165, 193, 194, 203, 220, 318
flow analysis, 164, 165
flow directions, 300
fluctuation, 273
generating elements, 229
generation, 1, 16, 42, 71, 76, 155, 163, 194, 196, 199, 223, 235, 237
grid, 195, 214, 261, 300, 318, 367, 379
hub, 378
imbalances, 131
industry, 196
interrupted, 65, 88
limits, 45, 204
loss, 89, 164, 217
loss minimization, 142
management, 267, 320
market, 33, 40, 67, 194, 362, 364, 379
maximum, 147, 319
networks, 267
operation, 377
optimal, 71
output, 134, 144, 150, 151, 153, 156, 194, 272
plant, 1, 6, 7, 10, 27, 31, 32, 37–40, 134, 147, 153, 156, 163, 164, 179, 193, 209, 215, 300, 343, 377–379
producers, 194, 379
production, 12, 139
quality, 3, 267
rating, 267
reserve, 132
sector, 22, 374
signal, 147, 195
sources, 2, 5, 200, 360
system
  blackouts, 131, 132
  engineering, 209
  operation, 12, 210, 214
  planning, 214
  stability, 32
unbalance, 146
units, 40, 203
variations, 380
wind, 19, 28, 29, 101, 136, 210, 212, 215, 362, 378
Power factor correction (PFC), 302
Power Matcher (PM), 16
Price based unit commitment (PBUC), 214
Price uncertainty, 199, 217
Price uncertainty market, 197, 199, 211, 214
Price-maker agent, 67, 71–73
Price-taker agent, 67, 71, 85, 280, 283
Primary frequency control (PFC), 132
Probability density function (PDF), 75, 361
Profit
  function, 167, 390, 392
  maximization, 40, 41, 43, 212, 380
  maximum, 212, 215, 394
  optimization, 189
  uncertainty, 231
  VPPs, 127
Profitable
  for TVPP, 99
  participation, 211
  trading, 212, 213
Proportional-integral (PI) controller, 133, 136, 146–148, 157
Pumped hydro energy storage (PHES), 194
Pumped storage hydropower, 29
Purchased power, 95, 371, 389
PWM control, 302

R
Radial basis function neural network (RBFNN), 355
Ramping costs, 212, 216
Rate of change of frequency (RoCoF) control, 147
Rated output power, 66, 366
Reactive power, 31, 32, 69, 70, 136, 149, 221
  balance, 81
  control, 12, 33, 135
  distribution, 33
  transfer, 135

Subject index

Real time (RT) market, 272, 280
Regulating reserve, 111
Regulation
  market, 40, 179, 189, 215, 378
  market prices, 182–184, 190
  reserve markets, 215
Regulation reserve (RR) market, 180, 182, 186
Regulatory, 15
  barriers, 15, 229
  environment, 14
  framework, 14, 19, 211
  reasons, 15
  standards, 318
  structures, 229
Renewable
  DERs, 28
  energy, 1, 16, 20, 22, 221, 301, 362, 379
    output, 216
    resources, 111, 361
    uncertainty, 197
  power generation, 206
  power uncertainty, 198, 214
  units, 42, 43
Renewable energy source (RES), 1, 59, 66, 67, 136, 193–195, 209, 210, 213, 229, 230, 257, 258, 261, 301, 321, 360, 361, 377
  energy uncertainties, 211
  generations, 231
  integration, 66
  intermittencies, 361
  uncertainty modeling, 193
  units, 194
Replacement reserve (RR), 143
Reserve
  energy, 123
  energy prices, 218
  market, 41, 80, 85, 86, 90, 92, 97, 98, 112, 113, 122, 190, 215, 362
  market prices, 93, 99
  power, 100, 132
  scheduling, 79–81, 85, 88, 90, 98, 104
Residential
  customers, 184, 258
  loads, 300

Responsive
  customers, 63, 67
  loads, 13, 60, 69, 71, 72, 163, 165, 179, 181, 182, 188, 190, 231, 276, 277, 279
Risk-averse, 231, 246
  bidding strategies, 213
  conditions, 120
  decision, 231, 238, 246
  decision parameter, 113
  model, 112
  parameters, 118–121
  participation, 215
  scheduling, 231, 246
  strategy, 115, 119
Risk-seeking, 231, 238, 246
Robust optimization (RO), 198
Rolling horizon approach (RHA), 198, 199
Roulette wheel mechanism (RWM), 210

S

Scenario tree, 75, 183, 190, 210
Scheduling
  algorithm, 217
  charging/discharging cycles, 265
  constraints, 234
  DERs, 195
  DERs optimal, 85
  energy, 71–73, 83, 95
  ESSs, 273, 278
  flexibility, 60, 83
  framework, 83
  horizon, 114
  models, 67
  optimal, 68, 85, 111, 112, 195, 197, 222, 230–232, 274, 290
  problem, 75, 80, 87, 90, 210, 251
  reserve, 79–81, 85, 88, 90, 98, 104
  scheme, 215
  time, 62
  VPPs, 111, 113, 180, 183, 221, 222, 231, 232, 237
  VPPs optimal, 230, 231, 237, 247
Secondary frequency control (SFC), 132
Secondary frequency regulation, 154
Selling
  energy, 72, 85
  power, 72, 393
Sequential quadratic programming (SQP), 142

Shiftable loads, 63, 235
Shifting peak loads, 220
Sine charging, 303
Sinusoidal charging, 308
Small hydro plant (SHP), 2
Smart
 charging, 20
 grid, 213
 grids, 16, 137, 300, 301
 power distribution, 197
Smart distribution network (SDN), 41
Solar power, 210
Spinning reserve market, 39, 85, 90, 99, 215
Spot market, 182, 221
State of charge (SOC), 260
State of health (SOH), 259
Stochastic
 optimization, 41, 112, 216, 363
 scheduling, 215
Store energy, 38, 96
Superconductor magnetic energy storage (SMES), 140, 194
Supplying
 loads, 200
 power, 273
Support vector machine (SVM), 222, 325, 339, 351
Surplus
 power, 234
 profits, 215

T
Tabu search (TS), 142
Technical virtual power plant (TVPP), 8–10, 18, 41, 67–70, 72, 74, 81, 84, 136, 140, 141, 257, 274, 362
 application, 19
 energy scheduling, 95
 exchanged power, 96
 optimal, 98
 participation, 90, 97
 reserve scheduling, 97, 98
Thermal units, 116, 132, 380–382, 385, 389, 391, 392, 394
Total harmonic distortion (THD), 303, 304
Trading markets, 216
Transacted power, 275
Transmission system operator (TSO), 9, 131

U
Uncertainty
 affecting VPP control, 219
 alleviation, 267
 budget, 75, 103, 213
 concerns, 223
 condition, 204
 cost in VPPs, 212
 in
  electricity price, 230–232
  market price in VPPs, 237
  market prices, 100, 231
  power systems, 210
  RESs generation, 230
  VPP, 211
  VPP scheduling, 218
  wind power, 85
  wind power generation, 231
 limits, 196, 209
 load, 199
 market, 237
 mitigation, 209
 model, 219
 modeling, 75, 206, 210, 219
 modeling approaches, 210, 219
 profit, 231
 renewable energy, 197
 type, 214
 VPPs, 197
 wind, 194, 217, 218
 wind power, 218
Uncontrollable generating units, 38
Units
 aggregated, 5
 battery, 136
 controllable, 38
 conventional, 38, 40, 196
 energy, 38, 315, 377
 energy production, 123, 126
 industrial, 378
 operation, 237
 participates, 40
 power, 40, 203
 renewable, 42, 43
 RESs, 194
 virtual power plant, 378, 383
 wind, 41, 116, 389
Unserved energy cost, 212

Subject index

U

Upstream
  grids, 236, 247
  network, 164, 195, 200, 380, 382, 385, 389, 391, 392
Utilizing
  DERs characteristics, 274
  interruptible loads, 65

V

Value of lost load (VOLL), 65
Vehicle-to-grid (V2G), 301, 302
Virtual Fuel Cell Power Plant (VFCPP), 16
Virtual power plant (VPP), 1, 2, 5, 27, 31, 32, 37, 38, 59, 111, 113, 114, 133, 163, 179, 193, 194, 209–211, 229, 257, 343, 360, 361, 377
  energy, 216, 300, 379
  energy management, 308, 314, 315, 318, 362
  energy resources, 113
  formation, 195
  frequency regulation, 133, 157
  operation, 14, 142, 144, 158, 163, 218, 220, 222, 232, 274, 280, 285
  operational cost, 200
  optimal
    bidding strategy, 111
    dispatch, 230
    operation strategy, 276, 278
    scheduling, 230, 231, 237, 247
  participation, 103, 280, 289
  planning, 218
  power generation, 362
  power output, 134
  production, 52
  profit, 127
  profit maximization, 112
  profits, 217
  scheduling, 111, 113, 180, 183, 221, 222, 231, 232, 237
  uncertainty, 197
  units, 378, 383
Virtual synchronous generator (VSG), 136
Virtual synchronous machine (VSM), 12

W

Weighting coefficient method, 383, 387, 392–394
Wholesale
  auction VPPs, 10
  electricity market, 40, 68, 116, 141, 165, 168
  energy market, 67, 72, 182
  market, 38, 60, 67, 72, 79, 80, 85, 164, 165, 168, 171, 229
Wind
  areas, 39
  data, 326
  energy, 4, 200, 326, 337
  farm, 39, 41, 49, 116, 119, 120, 194, 378
  farm production, 41
  flow, 218
  generation, 138, 193, 218
  generation uncertainty, 220
  generators, 39, 133, 138, 218
  parks, 20
  power, 19, 28, 29, 101, 136, 210, 212, 215, 362, 378
    forecasting uncertainty, 218
    forecasts, 326
    generation, 103
    output, 211, 216
    plants, 136, 147–149, 215
    producer, 41
    uncertainty, 218
    uncertainty budget, 103
  speeds, 136, 325, 326, 330, 339
  uncertainty, 194, 217, 218
  units, 41, 116, 389
  velocity measurements, 336
Wind turbine generator (WTG), 29
Wind turbine (WT), 2, 60, 66, 112, 115, 116, 133, 136, 147, 148, 163, 164, 180, 181, 183, 184, 195, 201, 202, 209, 211, 229, 235, 273, 279, 289, 362, 364, 366
  manufacturers, 326
  power devices, 39
  rotor, 133

Printed in the United States
by Baker & Taylor Publisher Services